T0091467

Tensor Algebra and Analysis for Engineers

With Applications to Differential Geometry of Curves and Surfaces

Contemporary Mathematics and Its Applications: Monographs, Expositions and Lecture Notes

Print ISSN: 2591-7668
Online ISSN: 2591-7676

This series aims to inspire new curriculum and integrate current research into texts. Its aims and main scope are to publish:

- Cutting-edge Research Monographs
- Mathematical Plums
- Innovative Textbooks for capstone (special topics) undergraduate and graduate level courses
- Surveys on recent emergence of new topics in pure and applied mathematics
- Advanced undergraduate and graduate level textbooks that may initiate new directions and new courses within mathematics and applied mathematics curriculum
- Books emerging from important conferences and special occasions
- Lecture Notes on advanced topics

Monographs and textbooks on topics of interdisciplinary or cross-disciplinary interest are particularly suitable for the series.

Published

Vol. 5 *Tensor Algebra and Analysis for Engineers: With Applications to Differential Geometry of Curves and Surfaces*
by Paolo Vannucci

Vol. 4 *Frontiers in Entropy Across the Disciplines: Panorama of Entropy: Theory, Computation, and Applications*
edited by Willi Freeden & M Zuhair Nashed

Vol. 3 *Introduction to Algebraic Coding Theory*
by Tzuong-Tsieng Moh

More information on this series can also be found at
https://www.worldscientific.com/series/cmameln

Contemporary Mathematics and Its Applications
Monographs, Expositions and Lecture Notes
Vol. **5**

Tensor Algebra and Analysis for Engineers

With Applications to Differential Geometry of Curves and Surfaces

Paolo Vannucci
Université de Versailles et Saint-Quentin-en-Yvelines, France

NEW JERSEY • LONDON • SINGAPORE • BEIJING • SHANGHAI • HONG KONG • TAIPEI • CHENNAI • TOKYO

Published by

World Scientific Publishing Co. Pte. Ltd.
5 Toh Tuck Link, Singapore 596224
USA office: 27 Warren Street, Suite 401-402, Hackensack, NJ 07601
UK office: 57 Shelton Street, Covent Garden, London WC2H 9HE

Library of Congress Cataloging-in-Publication Data
Names: Vannucci, Paolo, author.
Title: Tensor algebra and analysis for engineers : with applications to differential geometry of curves and surfaces / author, Paolo Vannucci, Université de Versailles Saint-Quentin-en-Yvelines, France.
Description: Singapore ; Hackensack, NJ : World Scientific Publishing Co. Pte. Ltd., [2023] | Series: Contemporary mathematics and its applications. Monographs, expositions and lecture notes, 2591-7668 ; vol. 5 | Includes bibliographical references and index.
Identifiers: LCCN 2022037837 | ISBN 9789811264801 (hardcover) | ISBN 9789811264818 (ebook for institutions) | ISBN 9789811264825 (ebook for individuals)
Subjects: LCSH: Engineering mathematics. | Tensor algebra. | Surfaces--Mathematical models.
Classification: LCC TA347.L5 V36 2023 | DDC 620.001/51--dc23/eng/20221011
LC record available at https://lccn.loc.gov/2022037837

British Library Cataloguing-in-Publication Data
A catalogue record for this book is available from the British Library.

For any available supplementary material, please visit
https://www.worldscientific.com/worldscibooks/10.1142/13099#t=suppl

Desk Editors: Soundararajan Raghuraman/Lai Fun Kwong

Typeset by Stallion Press
Email: enquiries@stallionpress.com

Preface

This textbook, addressed to graduate students and young researchers in mechanics, has been developed from the class notes of different courses in continuum mechanics that I have been delivering for several years as part of the Master's program in MMM: Mathematical Methods for Mechanics at the University Paris-Saclay.

Far from being exhaustive, as any primer text, the intention of this book is to introduce students in mechanics and engineering to the mathematical language and tools that are necessary for a modern approach to continuum theoretical and applied mechanics. The presentation of the matter is hence tailored for this scientific community, and necessarily, it is different, in terms of language and objectives, from that normally proposed to students of other disciplines, such as physics, especially general relativity, or pure mathematics.

What has motivated me to write this textbook is the idea of collecting in a single, introductory book a set of results and tools useful for studies in mechanics and presenting them in a modern, succinct way. Almost all the results and theorems are proved, and the reader is guided along a *tour* that starts from vectors and ends at the differential geometry of surfaces, passing through the algebra of tensors of second and fourth orders, the differential geometry of curves, the tensor analysis for fields and deformations, and the use of curvilinear coordinates.

Some topics are specially treated, such as rotations, the algebra of fourth-order tensors, which is fundamental for the mechanics of modern materials, or the properties of differential operators. Some other topics are intentionally omitted because they are less important to continuum mechanics or too advanced for an introductory text. Though in some modern texts, tensors are directly presented in the most general setting of curvilinear coordinates, I preferred here to choose a more traditional approach, introducing first the tensors in Cartesian coordinates, normally used for classical problems. Then, an entire chapter is devoted to the passage to curvilinear coordinates and to the formalism of co- and contravariant components.

The tensor theory and results are specially applied to introduce some subjects concerning differential geometry of curves and surfaces. Also, in this case, the presentation is mainly intended for applications to continuum mechanics and, in particular, in view of courses on slender beams or thin shells. All the presentation of the topics of differential geometry is extensively based on tensor algebra and analysis.

More than a hundred exercises are proposed to the reader, many of them completing the theoretical part through new results and proofs. All the exercises are entirely developed and solved at the end of the book in order to provide the reader with thorough support for his learning.

In Chapter 1, vectors and points are introduced and also, with a small anticipation of some results of the second chapter, applied vectors are visited. Chapter 2 is completely devoted to the algebra of second-rank tensors and the succeeding Chapter 3 to that of fourth-rank tensors. Intentionally, these are the only two types of tensors introduced in the book: They are the most important tensors in mechanics, and they allow us to represent deformation, stress, and the constitutive laws. I preferred not to introduce tensors in an absolutely general way but to go directly to the most important tensors for applications in mechanics; for the same reason, the algebra of other tensors, namely of third-rank tensors, is not presented in this primer text.

The analysis of tensors is done using first-differential geometry of curves, in Chapter 4, for differentiation and integration with respect to only one variable, then introducing the differential operators for fields and deformations, in Chapter 5.

Then, a generalization of second-rank tensor algebra and analysis in the sense of the use of curvilinear coordinates is presented in Chapter 6, where the notion of metric tensor, co- and contravariant components, and Christoffel's symbols are introduced.

Finally, Chapter 7 is entirely devoted to an introduction to the differential geometry of surfaces. Classical topics such as the first and second fundamental forms of a surface, the different types of curvatures, the Gauss–Weingarten equations, or the concepts of minimal surfaces, geodesics, and the Gauss–Codazzi conditions are presented, with all these topics being of great interest in mechanics.

I tried to write a coherent, almost self-contained manual of mathematical tools for graduate students in mechanics with the hope of helping young students progress in their studies. The exposition is as simple as possible, sober, and sometimes minimalist. I intentionally avoided burdening the language and the text with nonessential details and considerations, but I have always tried to grasp the essence of a result and its usefulness.

It is my most sincere hope that the reader who dares to persevere through the pages of this book will find a benefit to his studies in continuous mechanics. This is, eventually, the goal of this primer text.

About the Author

Paolo Vannucci is a full professor of mechanics at the University of Versailles & Saint Quentin (UVSQ), part of the University of Paris-Saclay. He is the author of one book, a co-author of five books, and the author or a co-author of more than 100 papers published in international journals and presented in conference proceedings. He is a member of the editorial board of five international journals and a reviewer for the journal *Mathematical Reviews*. He is currently the director of the master's program in Mathematical Methods in Mechanics at the University of Paris-Saclay. As a professor, he teaches several courses, including continuum mechanics, analytical mechanics, the theory of structures, differential geometry, fluid–structure interactions, and anisotropic mechanics. His research concerns different topics in mechanics and applied mathematics: numerical methods for nonlinear mechanics, description of anisotropic phenomena by means of tensor invariants, optimization of structures by meta-heuristics (genetic algorithms and PSO), optimal design of anisotropic structures, effects of ultimate loads (wind storms and explosions) on monumental structures, etc.

Acknowledgments

I am indebted to many persons for the topics of this book. Professor E. G. G. Virga, University of Pavia, introduced me to tensor algebra during my PhD at the University of Pisa many years ago.

Then, I have had the privilege of collaborating with Professor G. Verchery, at the University of Burgundy, who introduced me to the representation methods based upon tensor invariants. This has been very useful for developing some results in the algebra of fourth-rank tensors.

I also wish to thank Professor P. M. Mariano, University of Florence, who always pushed me to go forward and to consider problems of modern mechanics; many of the discussions I had with him have been very important to me.

I have also had many interesting and useful discussions with Professor J. Lerbet, University of Evry, and with Doctor C. Fourcade, of Renault S.A.; I wish to thank them sincerely.

I am also grateful to Prof. A. Frediani, who has been for me an example of honesty in teaching and science, and I cannot forget Prof. P. Villaggio, my PhD director, who passed by some years ago: his teaching and personality leaved an indelebile trace in all my life of scientist.

Finally, I wish to thank my wife, Carla, my daughter, Bianca, and my son, Alessandro. Without them, nothing would have been possible; because of them many things happen.

Contents

Preface v

About the Author ix

Acknowledgments xi

List of Symbols xvii

1. Points and Vectors **1**

1.1 Points and vectors 1
1.2 Scalar product, distance, orthogonality 3
1.3 Basis of $\mathcal{V}$, expression of the scalar product 5
1.4 Applied vectors 7
1.5 Exercises 12

2. Second-Rank Tensors **15**

2.1 Second-rank tensors 15
2.2 Dyads, tensor components 16
2.3 Tensor product 17
2.4 Transpose, symmetric and skew tensors 18
2.5 Trace, scalar product of tensors 19
2.6 Spherical and deviatoric parts 21
2.7 Determinant, inverse of a tensor 22
2.8 Eigenvalues and eigenvectors of a tensor 27
2.9 Skew tensors and cross product 30
2.10 Orientation of a basis 35

2.11 Rotations . . . 36
2.12 Reflexions . . . 50
2.13 Polar decomposition . . . 51
2.14 Exercises . . . 54

3. Fourth-Rank Tensors **57**
3.1 Fourth-rank tensors . . . 57
3.2 Dyads, tensor components . . . 58
3.3 Conjugation product, transpose, symmetries . . . 59
3.4 Trace and scalar product of fourth-rank tensors . . . 62
3.5 Projectors and identities . . . 63
3.6 Orthogonal conjugator . . . 66
3.7 Rotations and symmetries . . . 67
3.8 The Kelvin formalism . . . 69
3.9 The polar formalism for plane tensors . . . 71
3.10 Exercises . . . 72

4. Tensor Analysis: Curves **75**
4.1 Curves of points, vectors and tensors . . . 75
4.2 Differentiation of curves . . . 76
4.3 Integral of a curve of vectors and length of a curve . . . 80
4.4 The Frenet–Serret basis . . . 84
4.5 Curvature of a curve . . . 86
4.6 The Frenet–Serret formula . . . 88
4.7 The torsion of a curve . . . 90
4.8 Osculating sphere and circle . . . 91
4.9 Evolute, involute and envelopes of plane curves . . . 93
4.10 The theorem of Bonnet . . . 95
4.11 Canonic equations of a curve . . . 96
4.12 Exercises . . . 98

5. Tensor Analysis: Fields **103**
5.1 Scalar, vector and tensor fields . . . 103
5.2 Differentiation of fields, differential operators . . . 103
5.3 Properties of the differential operators . . . 105
5.4 Theorems on fields . . . 111
5.5 Differential operators in Cartesian coordinates . . . 115
5.6 Differential operators in cylindrical coordinates . . . 115
5.7 Differential operators in spherical coordinates . . . 120
5.8 Exercises . . . 123

6. Curvilinear Coordinates **125**

6.1 Introduction 125
6.2 Curvilinear coordinates, metric tensor 125
6.3 Co- and contravariant components 128
6.4 Spatial derivatives of fields in curvilinear coordinates . . . 135
6.5 Exercises 139

7. Surfaces in $\mathcal{E}$ **141**

7.1 Surfaces in $\mathcal{E}$, coordinate lines and tangent planes 141
7.2 Surfaces of revolution 143
7.3 Ruled surfaces 146
7.4 First fundamental form of a surface 146
7.5 Second fundamental form of a surface 149
7.6 Curvatures of a surface 151
7.7 The theorem of Rodrigues 154
7.8 Classification of the points of a surface 156
7.9 Developable surfaces 158
7.10 Points of a surface of revolution 159
7.11 Lines of curvature, conjugated directions, asymptotic directions 161
7.12 Dupin's conical curves 162
7.13 The Gauss–Weingarten equations 163
7.14 The *theorema egregium* 165
7.15 Minimal surfaces 166
7.16 Geodesics 169
7.17 The Gauss–Codazzi compatibility conditions 174
7.18 Exercises 179

Suggested Texts 183

Solutions to the Exercises 185

Index 207

List of Symbols

$:=$: definition symbol

$|$: such that

$\exists!$: exists and is unique

$\mathbb{R}$: set of real numbers

$\mathcal{E}$: ordinary 3D Euclidean space

$\mathcal{V}$: vector space of translations, associated with $\mathcal{E}$

$Lin(\mathcal{V})$: vector space of second-rank tensors

$\mathbb{L}\mathrm{in}(\mathcal{V})$: vector space of fourth-rank tensors

x, y, z etc.: scalars (elements of $\mathbb{R}$)

p, q, r etc.: points (elements of $\mathcal{E}$)

$\mathbf{u}, \mathbf{v}, \mathbf{w}$ etc.: vectors (elements of $\mathcal{V}$)

$\mathbf{u} = p - q$ etc.: vector difference of two points of $\mathcal{E}$

$\mathbf{L}, \mathbf{M}, \mathbf{N}$ etc.: second-rank tensors (elements of $Lin(\mathcal{V})$)

$\mathbb{L}, \mathbb{M}, \mathbb{N}$ etc.: fourth-rank tensors (elements of $\mathbb{L}\mathrm{in}(\mathcal{V})$)

$\mathbf{u} \cdot \mathbf{v}$ etc.: scalar product of vectors

$\mathbf{L} \cdot \mathbf{M}$ etc.: scalar product of second-rank tensors

$\mathbf{u} \times \mathbf{v}$ etc.: cross product of vectors

$\mathbf{u} \otimes \mathbf{v}$ etc.: dyad (second-rank tensor) of vectors

$u = |\mathbf{u}|$: norm of a vector

$L = |\mathbf{L}|$: norm of a tensor

$|p - q|, |\mathbf{u} - \mathbf{v}|, |\mathbf{L} - \mathbf{M}|$: distance of points, vectors or tensors

$\det \mathbf{L}$: determinant of a second-rank tensor $\mathbf{L}$

$\mathbf{L}^\top$: transpose of a second-rank tensor $\mathbf{L}$

$\mathbf{L}^{-1}$: inverse of a second-rank tensor $\mathbf{L}$

$\mathbf{L}^{-\top} = (\mathbf{L}^\top)^{-1} = (\mathbf{L}^{-1})^\top$

$Sym(\mathcal{V})$: subspace of $Lin(\mathcal{V})$ of symmetric second-rank tensors

$Skw(\mathcal{V})$: subspace of $Lin(\mathcal{V})$ of skew second-rank tensors

$Sph(\mathcal{V})$: subspace of $Lin(\mathcal{V})$ of spherical second-rank tensors

$Dev(\mathcal{V})$: subspace of $Lin(\mathcal{V})$ of deviatoric second-rank tensors

$Orth(\mathcal{V})$: subspace of $Lin(\mathcal{V})$ of orthogonal second-rank tensors

$Orth(\mathcal{V})^+$: subspace of $Lin(\mathcal{V})$ of rotation second-rank tensors

$\mathcal{S}$: unit sphere of $\mathcal{V}$: $\mathcal{S} = \{\mathbf{u} \in \mathcal{V} | \, |\mathbf{u}| = 1\}$

δ_{ij}: Kronecker's delta

$\Re$: real part of a complex quantity

$\Im$: imaginary part of a complex quantity

$\mathbb{L} = \mathbf{A} \boxtimes \mathbf{B}$: conjugation product of two second-rank tensors

$\mathbb{S}^{sph}$: spherical projector

$\mathbb{D}^{dev}$: deviatoric projector

$\mathbb{T}^{trp}$: transpose projector

$\mathbb{S}^{sym}$: symmetry projector

$\mathbb{W}^{skw}$: antisymmetry projector

$\mathbb{I}$: identity of $\mathbb{L}\mathrm{in}(\mathcal{V})$

$\mathbb{I}^s$: restriction of $\mathbb{I}$ to symmetric tensors of $Lin(\mathcal{V})$

Chapter 1

Points and Vectors

1.1 Points and vectors

We consider in the following a *point space* $\mathcal{E}$ whose elements are points p. In classical mechanics, $\mathcal{E}$ is identified with the *Euclidean three-dimensional space*, wherein events are intended to be set. On $\mathcal{E}$, we admit the existence of an operation, the *difference* of any couple of its elements:

$$q - p, \;\; p, q \in \mathcal{E}.$$

We associate with $\mathcal{E}$ a *vector space* $\mathcal{V}$ whose dimension is $\dim\mathcal{V} = 3$ and whose elements are vectors $\mathbf{v}$ representing *translations over* $\mathcal{E}$:

$$\forall p, q \in \mathcal{E}, \;\; \exists! \; \mathbf{v} \in \mathcal{V} | \; q - p = \mathbf{v}.$$

Any element $\mathbf{v} \in \mathcal{V}$ is hence a transformation over $\mathcal{E}$ that can be written using the previous definition as

$$\forall \mathbf{v} \in \mathcal{V}, \;\; \mathbf{v} : \mathcal{E} \rightarrow \mathcal{E} | \; q = \mathbf{v}(p) \;\; \rightarrow \;\; q = p + \mathbf{v}.$$

We remark that the result of the application of the translation $\mathbf{v}$ depends upon the argument p:

$$q = p + \mathbf{v} \neq p_1 + \mathbf{v} = q_1,$$

whose geometric meaning is depicted in Fig. 1.1. Unlike the difference, the sum of two points is not defined and is meaningless.

We define the *sum of two vectors* $\mathbf{u}$ and $\mathbf{v}$ as the vector $\mathbf{w}$ such that

$$(\mathbf{u} + \mathbf{v})(p) = \mathbf{u}(\mathbf{v}(p)) = \mathbf{w}(p).$$

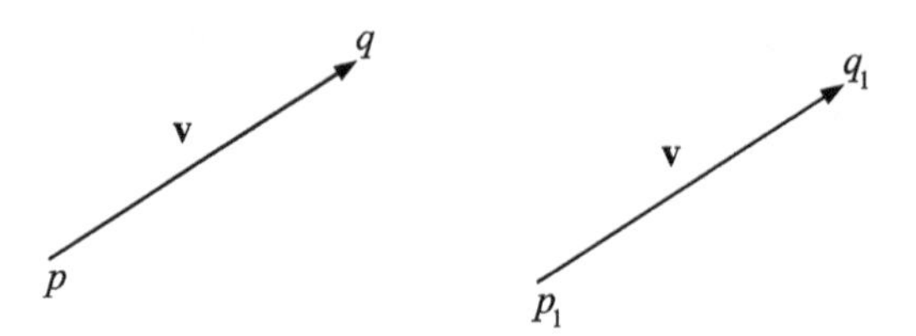

Figure 1.1: Same translation applied to two different points.

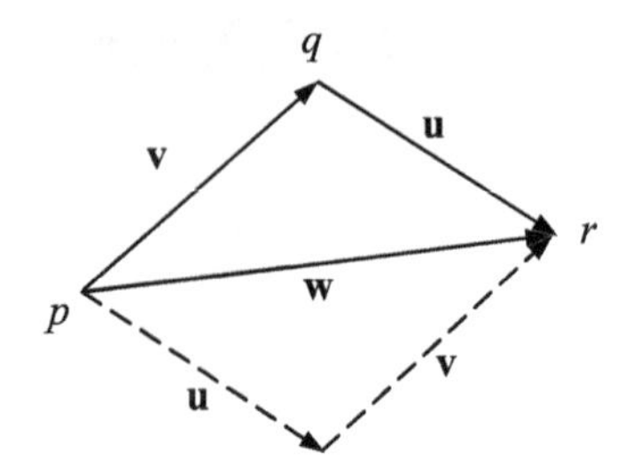

Figure 1.2: Sum of two vectors: the parallelogram rule.

This means that if

$$q = \mathbf{v}(p) = p + \mathbf{v},$$

then

$$r = \mathbf{u}(q) = q + \mathbf{u} = \mathbf{w}(p),$$

see Fig. 1.2, which shows that the above definition actually coincides with the *parallelogram rule* and that

$$\mathbf{u} + \mathbf{v} = \mathbf{v} + \mathbf{u},$$

as is obvious, for the sum over a vector space commutes. It is evident that the sum of more than two vectors can be defined iteratively, summing up a vector at a time to the sum of the previous vectors.

The *null vector* $\mathbf{o}$ is defined as the difference of any two coincident points:

$$\mathbf{o} := p - p \quad \forall p \in \mathcal{E};$$

$\mathbf{o}$ is unique and the only vector such that

$$\mathbf{v} + \mathbf{o} = \mathbf{v} \quad \forall \mathbf{v} \in \mathcal{V}.$$

In fact,

$$\forall p \in \mathcal{E}, \quad \mathbf{v}+\mathbf{o}=\mathbf{v}+p-p \ \rightarrow \ p+\mathbf{v}+\mathbf{o}=p+\mathbf{v} \quad \Longleftrightarrow \quad \mathbf{v}+\mathbf{o}=\mathbf{v}.$$

A *linear combination* of n vectors $\mathbf{v}_i$ is defined as the vector[1]

$$\mathbf{w} := k_i\mathbf{v}_i, \quad k_i \in \mathbb{R}, \quad i=1,\ldots,n.$$

The $n+1$ vectors $\mathbf{w}$, $\mathbf{v}_i$, $i=1,\ldots,n$, are said to be *linearly independent* if there does not exist a set of n scalars k_i such that the above equation is satisfied and are said to be *linearly dependent* in the opposite case.

1.2 Scalar product, distance, orthogonality

A *scalar product* on a vector space is a positive definite, symmetric, bilinear form. A *form* ω is a function

$$\omega : \mathcal{V} \times \mathcal{V} \rightarrow \mathbb{R},$$

i.e. ω operates on a couple of vectors to give a real number, a *scalar*. We indicate the scalar product of two vectors $\mathbf{u}$ and $\mathbf{v}$ as[2]

$$\omega(\mathbf{u},\mathbf{v}) = \mathbf{u}\cdot\mathbf{v}.$$

The properties of *bilinearity* prescribe that, $\forall \mathbf{u},\mathbf{v} \in \mathcal{V}$ and $\forall \alpha,\beta \in \mathbb{R}$,

$$\mathbf{u}\cdot(\alpha\mathbf{v}+\beta\mathbf{w}) = \alpha\mathbf{u}\cdot\mathbf{v}+\beta\mathbf{u}\cdot\mathbf{w},$$
$$(\alpha\mathbf{u}+\beta\mathbf{v})\cdot\mathbf{w} = \alpha\mathbf{u}\cdot\mathbf{w}+\beta\mathbf{v}\cdot\mathbf{w},$$

while *symmetry* implies that

$$\mathbf{u}\cdot\mathbf{v} = \mathbf{v}\cdot\mathbf{u} \quad \forall \mathbf{u},\mathbf{v} \in \mathcal{V}.$$

Finally, the *positive definiteness* means that

$$\mathbf{v}\cdot\mathbf{v} > 0 \ \forall \mathbf{v} \in \mathcal{V}, \quad \mathbf{v}\cdot\mathbf{v} = 0 \quad \Longleftrightarrow \quad \mathbf{v} = \mathbf{o}.$$

Any two vectors $\mathbf{u},\mathbf{v} \in \mathcal{V}$ are said to be *orthogonal* $\Longleftrightarrow$

$$\mathbf{u}\cdot\mathbf{v} = 0.$$

---

[1]We adopt here and in the following the *Einstein notation* for summations: All the times when an index is repeated in a monomial, then the summation with respect to that index, called the *dummy index*, is understood, e.g. $k_i\mathbf{v}_i = \sum_i k_i\mathbf{v}_i$. We then say that the index i is *saturated*. If a repeated index is underlined, then it is not a dummy index, i.e. there is no summation.

[2]The scalar product $\omega(\mathbf{u},\mathbf{v})$ is also indicated as $< \mathbf{u},\mathbf{v} >$.

Thanks to the properties of the scalar product, we can define the *Euclidean norm* of a vector $\mathbf{v}$ as the nonnegative scalar, denoted equivalently by v or $|\mathbf{v}|$:

$$v = |\mathbf{v}| := \sqrt{\mathbf{v} \cdot \mathbf{v}}.$$

Theorem 1. *The norm of a vector has the following properties:* $\forall \mathbf{u}, \mathbf{v} \in \mathcal{V}, k \in \mathbb{R}$,

$$\begin{aligned} &|\mathbf{u} \cdot \mathbf{v}| \leq u\ v \quad \text{(Schwarz's inequality)};\\ &|\mathbf{u} + \mathbf{v}| \leq u + v \quad \text{(Minkowski's triangular inequality)};\\ &|k\mathbf{v}| = |k|v. \end{aligned}$$

Proof. *Schwarz's inequality*: It is sufficient to prove that

$$(\mathbf{u} \cdot \mathbf{v})^2 \leq \mathbf{u} \cdot \mathbf{u}\ \mathbf{v} \cdot \mathbf{v}.$$

Let $x = \mathbf{v} \cdot \mathbf{v}$ and $y = -\mathbf{u} \cdot \mathbf{v}$. Then, by the positive definiteness of the scalar product, we get

$$(x\mathbf{u} + y\mathbf{v}) \cdot (x\mathbf{u} + y\mathbf{v}) \geq 0,$$

which implies that

$$\begin{aligned} x^2\mathbf{u} \cdot \mathbf{u} + 2xy\mathbf{u} \cdot \mathbf{v} + y^2\mathbf{v} \cdot \mathbf{v} = (\mathbf{v} \cdot \mathbf{v})^2\mathbf{u} \cdot \mathbf{u} - 2\mathbf{v} \cdot \mathbf{v}(\mathbf{u} \cdot \mathbf{v})^2 \\ + \mathbf{v} \cdot \mathbf{v}(\mathbf{u} \cdot \mathbf{v})^2 \geq 0; \end{aligned}$$

supposing $\mathbf{v} \neq \mathbf{o}$ (otherwise, the proof is trivial), we get the thesis on dividing by $\mathbf{v} \cdot \mathbf{v}$.

Minkowski's inequality: Because the two members of the inequality to be proved are nonnegative, it is sufficient to prove that

$$(\mathbf{u} + \mathbf{v}) \cdot (\mathbf{u} + \mathbf{v}) \leq (u + v)^2 = u^2 + 2uv + v^2.$$

This can be proved easily:

$$\begin{aligned} (\mathbf{u} + \mathbf{v}) \cdot (\mathbf{u} + \mathbf{v}) &= \mathbf{u} \cdot \mathbf{u} + 2\mathbf{u} \cdot \mathbf{v} + \mathbf{v} \cdot \mathbf{v} = u^2 + 2\mathbf{u} \cdot \mathbf{v} + v^2 \\ &\leq u^2 + 2|\mathbf{u} \cdot \mathbf{v}| + v^2 \leq u^2 + 2uv + v^2, \end{aligned}$$

in which the last operation follows from Schwarz's inequality.

The proof of the third property is immediate, it is sufficient to use the same definition of norm. □

We define the *distance between any two points* p and $q \in \mathcal{E}$ the scalar

$$d(p, q) := |p - q| = |q - p|.$$

Similarly, the *distance between any two vectors* $\mathbf{u}$ and $\mathbf{v} \in \mathcal{V}$ is defined as

$$d(\mathbf{u},\mathbf{v}) := |\mathbf{u}-\mathbf{v}| = |\mathbf{v}-\mathbf{u}|.$$

Two points or two vectors are *coincident* if and only if their distance is null.

The *unit sphere* $\mathcal{S}$ *of* $\mathcal{V}$ is defined as the set of all the vectors whose norm is one:

$$\mathcal{S} := \{\mathbf{v} \in \mathcal{V} |\ v = 1\}.$$

1.3 Basis of $\mathcal{V}$, expression of the scalar product

There is a general way to define a *basis* for a vector space of any kind. We limit the introduction of the concept of basis to the case of $\mathcal{V}$ only, which is of interest in classical mechanics. Generally, a *basis* $\mathcal{B}$ of $\mathcal{V}$ is any set of three linearly independent vectors $\mathbf{e}_i, i = 1, 2, 3$, of $\mathcal{V}$:

$$\mathcal{B} = \{\mathbf{e}_1, \mathbf{e}_2, \mathbf{e}_3\}.$$

The introduction of a basis for $\mathcal{V}$ is useful for representing vectors. In fact, once a basis $\mathcal{B}$ is fixed, any vector $\mathbf{v} \in \mathcal{V}$ can be represented as a linear combination of the vectors of the basis, where the coefficients v_i of the linear combination are the *Cartesian components* of $\mathbf{v}$:

$$\mathbf{v} = v_i\mathbf{e}_i = v_1\mathbf{e}_1 + v_2\mathbf{e}_2 + v_3\mathbf{e}_3.$$

Though the choice of the elements of a basis is completely arbitrary, the only condition being their linear independency, we use in the following only *orthonormal bases*, which are bases composed of mutually orthogonal vectors of $\mathcal{S}$, i.e. satisfying

$$\mathbf{e}_i \cdot \mathbf{e}_j = \delta_{ij},$$

where the symbol δ_{ij} is the so-called *Kronecker's delta*:

$$\delta_{ij} = \begin{cases} 1 & \text{if } i = j, \\ 0 & \text{if } i \neq j. \end{cases}$$

The use of orthonormal bases has great advantages; namely, it allows us to give a very simple rule for the calculation of the scalar product:

$$\mathbf{u} \cdot \mathbf{v} = u_i\mathbf{e}_i \cdot v_j\mathbf{e}_j = u_iv_j\delta_{ij} = u_iv_i = u_1v_1 + u_2v_2 + u_3v_3.$$

In particular, it is

$$\mathbf{v} \cdot \mathbf{e}_i = v_k \mathbf{e}_k \cdot \mathbf{e}_i = v_k \delta_{ik} = v_i, \quad i = 1, 2, 3.$$

So, the Cartesian components of a vector are the projection of the vector on the three vectors of the basis $\mathcal{B}$; such quantities are the *director cosines* of $\mathbf{v}$ in the basis $\mathcal{B}$. In fact, if θ is the angle formed by two vectors $\mathbf{u}$ and $\mathbf{v}$, then

$$\mathbf{u} \cdot \mathbf{v} = u\ v\ \cos\theta.$$

This relation is used to define the angle between two vectors,

$$\theta = \arccos \frac{\mathbf{u} \cdot \mathbf{v}}{u\ v},$$

which can be proved easily: Given two vectors $\mathbf{u}$ and $\mathbf{v}$, we look for $c \in \mathbb{R}$ such that the vector $\mathbf{u} - c\mathbf{v}$ is orthogonal to $\mathbf{v}$:

$$(\mathbf{u} - c\mathbf{v}) \cdot \mathbf{v} = 0 \iff c = \frac{\mathbf{u} \cdot \mathbf{v}}{\mathbf{v} \cdot \mathbf{v}} = \frac{\mathbf{u} \cdot \mathbf{v}}{v^2}.$$

Now, if $\mathbf{u}$ is inclined at θ on $\mathbf{v}$, its projection u_v on the direction of $\mathbf{v}$ is

$$u_v = u\ \cos\theta,$$

and, by construction (see Fig. 1.3), it is also

$$u_v = c\ v.$$

So,

$$c = \frac{u}{v} \cos\theta \ \rightarrow\ \frac{u}{v} \cos\theta = \frac{\mathbf{u} \cdot \mathbf{v}}{v^2} \ \Rightarrow\ \cos\theta = \frac{\mathbf{u} \cdot \mathbf{v}}{u\ v}.$$

We remark that while the scalar product, being an intrinsic operation, does not change with a change of basis, the components v_i of a vector are

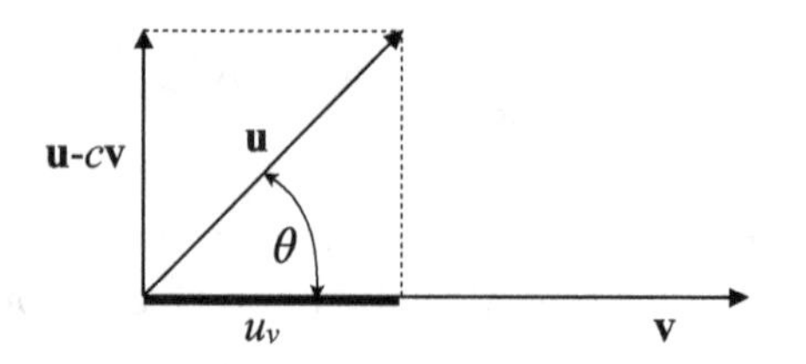

Figure 1.3: Angle between two vectors.

not intrinsic quantities, but they are basis-dependent: A change of the basis makes the components change. The way this change is done is introduced in Section 2.11.

A *frame* $\mathcal{R}$ for $\mathcal{E}$ is composed of a point $o \in \mathcal{E}$, the *origin*, and a basis $\mathcal{B}$ of $\mathcal{V}$:

$$\mathcal{R} := \{o, \mathcal{B}\} = \{o; \mathbf{e}_1, \mathbf{e}_2, \mathbf{e}_3\}.$$

The use of a frame for $\mathcal{E}$ is useful for determining the position of a point p, which can be done through its *Cartesian coordinates* x_i, defined as the components in $\mathcal{B}$ of the vector $p - o$:

$$x_i := (p - o) \cdot \mathbf{e}_i, \quad i = 1, 2, 3.$$

Of course, the coordinates x_i of a point $p \in \mathcal{E}$ depend upon the choice of o and $\mathcal{B}$.

1.4 Applied vectors

We introduce now a set of definitions, concepts, and results that are widely used in physics, especially in mechanics. For that, we need to anticipate some results that are introduced in the next chapter, namely that of cross product, in Section 2.9, and of complementary projector, in Exercise 2, Chapter 2. This slight deviation from the good rule of consistent progression in stating the results is justified by the fact that, actually, the matter presented hereafter is still that of vectors. The reader can, of course, come back to the topics of this section once they have studied Chapter 2.

We call *applied vector* $\mathbf{v}^p$ a vector $\mathbf{v}$ associated to a point $p \in \mathcal{E}$. In physics, the concept of applied vector[3] is often employed, for example, to represent forces[4]. We define the *resultant* of a system of n applied vectors $\mathbf{v}_i^p$ as the vector

$$\mathbf{R} := \sum_{i=1}^{n} \mathbf{v}_i^p.$$

We define the *moment of an applied vector* $\mathbf{v}^p$ about a point o, called the *center of the moment*, the vector

$$\mathbf{M}_o := (p - o) \times \mathbf{v}^p,$$

[3] In the literature, applied vectors are also called *bound vectors*.

[4] The fact that in classical mechanics forces can be represented by vectors is actually a fundamental postulate of physics. Forces are vectors that cannot be considered belonging to the translation space $\mathcal{V}$; nevertheless, the definitions and results found earlier are also valid for vectors $\notin \mathcal{V}$.

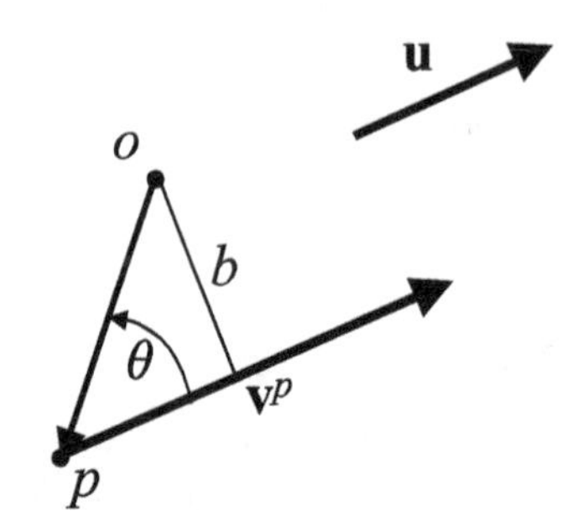

Figure 1.4: Moment arm of an applied vector.

and the *resultant moment of a system of n applied vectors* $\mathbf{v}_i^p$ about a point o the vector

$$\mathbf{M}_o^r := \sum_{i=1}^{n} (p_i - o) \times \mathbf{v}_i^p.$$

We remark that $\mathbf{R}, \mathbf{M}_o$, and $\mathbf{M}_o^r$ are *not* applied vectors.

If $\mathbf{u} \in \mathcal{S} |\ \mathbf{u} \times \mathbf{v}^p = \mathbf{o}$, then

$$b := |(\mathbf{I} - \mathbf{u} \otimes \mathbf{u})(p - o)| = |p - o| \sin\theta$$

is called the *moment arm* of $\mathbf{v}^p$ with respect to the center o. It measures the distance of o from the *line of action*, i.e. the line passing through p and parallel to $\mathbf{v}^p$, cf. Fig. 1.4.

Theorem 2 (Transport of moment). *if $\mathbf{M}_{o_1}$ is the moment of an applied vector $\mathbf{v}^p$ about a center o_1, the moment $\mathbf{M}_{o_2}$ of $\mathbf{v}^p$ about another center o_2 is*

$$\mathbf{M}_{o_2} = \mathbf{M}_{o_1} + (o_1 - o_2) \times \mathbf{v}^p.$$

Proof. Referring to Fig. 1.5,

$$\begin{aligned}
\mathbf{M}_{o_2} &= (p - o_2) \times \mathbf{v}^p \\
&= (p - o_1 + o_1 - o_2) \times \mathbf{v}^p = (p - o_1) \times \mathbf{v}^p + (o_1 - o_2) \times \mathbf{v}^p \\
&= \mathbf{M}_{o_1} + (o_1 - o_2) \times \mathbf{v}^p.
\end{aligned}$$

□

A consequence of this theorem is that $\mathbf{M}_{o_1} = \mathbf{M}_{o_2} \iff \mathbf{v}^p \times (o_1 - o_2)$, i.e. if $\mathbf{v}^p$ and $o_1 - o_2$ are parallel. It follows from this that the moment of an applied vector does not change when calculated about the points of a straight line parallel to the vector itself or, more importantly, if $\mathbf{v}^p$ is translated along its line of action.

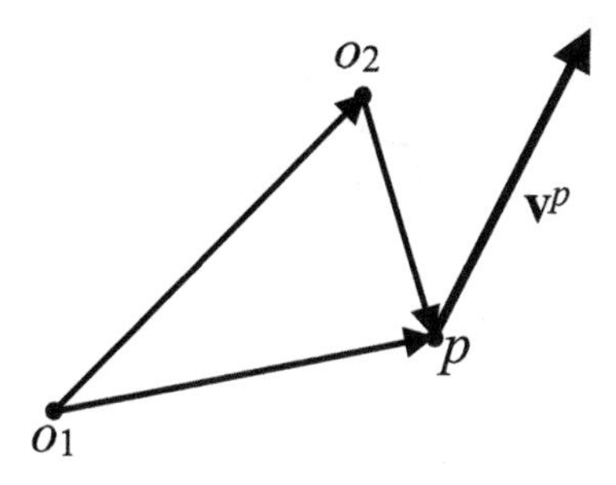

Figure 1.5: Scheme for the transposition of the moment.

The above theorem can be extended to the resultant moment of a system of applied vectors to give (the proof is quite similar)

$$\mathbf{M}^r_{o_2} = \mathbf{M}^r_{o_1} + (o_1 - o_2) \times \mathbf{R}. \tag{1.1}$$

Also, in this case, the resultant moment does not change when $\mathbf{R}$ and $o_1 - o_2$ are parallel vectors, but not exclusively, as another possibility is that $\mathbf{R} = \mathbf{o}$: For the systems of applied vectors with null resultant, the resultant moment is invariant with respect to the center of the moment.

An interesting relation can be found if the two members of the last equation are projected onto $\mathbf{R}$, which gives

$$\mathbf{M}^r_{o_1} \cdot \mathbf{R} = \mathbf{M}^r_{o_2} \cdot \mathbf{R} : \tag{1.2}$$

The projection of the resultant moment onto the direction of $\mathbf{R}$ does not depend upon the center of the moment.

A particularly important case of the system with a null resultant is that of a *couple*, which is composed of two opposite vectors $\mathbf{v}$ and $-\mathbf{v}$, which are applied to two points p and q:

$$\mathbf{v}^p = -\mathbf{v}^q.$$

Of course, by definition, $\mathbf{R} = \mathbf{o}$ for any couple and, as a consequence, the resultant moment $\mathbf{M}^r$ of a couple, called the *moment of the couple* and simply denoted by $\mathbf{M}$, is independent of the center of the moment (that is why the index denoting the center of the moment is omitted): Referring to Fig. 1.6,

$$\begin{aligned}\mathbf{M} &= (p - o) \times \mathbf{v}^p + (q - o) \times \mathbf{v}^q = (p - o) \times \mathbf{v} - (q - o) \times \mathbf{v}\\ &= ((p - o) - (q - o)) \times \mathbf{v} = (p - q) \times \mathbf{v}.\end{aligned}$$

If $\mathbf{u} \in \mathcal{S} |\ \mathbf{u} \times \mathbf{v} = \mathbf{o}$, then

$$b_c :\doteq |(\mathbf{I} - \mathbf{u} \otimes \mathbf{u})(p - q)| = |p - q| \sin\theta$$

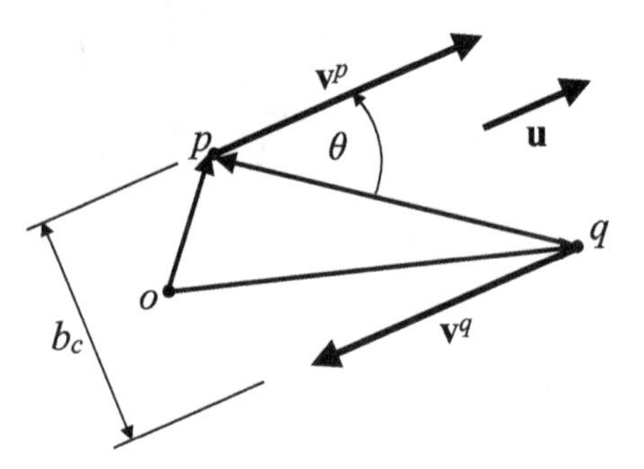

Figure 1.6: Scheme of a couple.

is the *couple arm*. We then have

$$M = |(p-q) \times \mathbf{v}| = |p-q| v \sin\theta = b_c v.$$

The *central axis* $\mathbf{A}$ of a system of n applied vectors with $\mathbf{R} \neq \mathbf{o}$ is the axis such that

$$\mathbf{M}_a^r \times \mathbf{R} = \mathbf{o} \quad \forall a \in \mathcal{A}.$$

Theorem 3 (Existence and uniqueness of the central axis). *The central axis of a system of n vectors exists and is unique.*

Proof. Existence: we need at least a point $a \in \mathcal{E} |\ \mathbf{M}_a^R = k\mathbf{R}, \quad k \in \mathbb{R} \Rightarrow \mathbf{M}_a^r \times \mathbf{R} = \mathbf{o}$. From Eq. (1.1), $\forall o \in \mathcal{E}$, we get

$$\mathbf{M}_a^r \times \mathbf{R} = \mathbf{M}_o^r \times \mathbf{R} + ((o-a) \times \mathbf{R}) \times \mathbf{R} = \mathbf{M}_o^r \times \mathbf{R} - \mathbf{R}^2(o-a) + (\mathbf{R} \cdot (o-a))\mathbf{R}.$$

Then, if we take for $o-a$ the vector

$$o - a = \frac{\mathbf{M}_o^r \times \mathbf{R}}{\mathbf{R}^2},$$

it is evident that we get

$$\mathbf{M}_a^r \times \mathbf{R} = \mathbf{o}.$$

Hence, the point

$$a = o - \frac{\mathbf{M}_o^r \times \mathbf{R}}{\mathbf{R}^2} \in \mathcal{A}.$$

So, because $\mathbf{M}^r$ does not change when calculated with respect to the points of an axis parallel to $\mathbf{R}$, $\mathcal{A}$ is the axis passing through a and parallel to $\mathbf{R}$ whose equation is

$$p = a + t\,\mathbf{R} = o - \frac{\mathbf{M}_o^r \times \mathbf{R}}{\mathbf{R}^2} + t\,\mathbf{R}, \quad t \in \mathbb{R}.$$

Uniqueness: Suppose another axis $\hat{\mathcal{A}} \neq \mathcal{A}$ exists, which is necessarily parallel to $\mathcal{A}$. If $q \in \hat{\mathcal{A}}$, again using Eq. (1.1), we get

$$\mathbf{M}_q^r \times \mathbf{R} = \mathbf{M}_a^r \times \mathbf{R} + ((a - q) \times \mathbf{R}) \times \mathbf{R}.$$

In this equation, the left-hand side and the first term on the right-hand side are null by the definition of central axis. Because $(a-q) \times \mathbf{R}$ is perpendicular to $\mathbf{R}$ and $\mathbf{R} \neq \mathbf{o}$ by hypothesis, the left-hand side is null if and only if $a = q \Rightarrow \hat{\mathcal{A}} = \mathcal{A}$. □

The central axis has another remarkable property.

Theorem 4 (Property of minimum of the central axis). *The points of the central axis minimize the resultant moment.*

Proof. When $\mathbf{M}^r$ is calculated about a point $a \in \mathcal{A}$, it is parallel to $\mathbf{R}$, which is not the case for any point $q \notin \mathcal{A}$. In this last case, hence, $\mathbf{M}^r$ also has a component orthogonal to $\mathbf{R}$. Then, by virtue of the invariance of the projection of $\mathbf{M}^r$ onto $\mathbf{R}$, Eq. (1.2), $\mathbf{M}^r$ gets its minimum value when calculated about the points of $\mathcal{A}$. □

Let us now consider the case of systems for which

$$\mathbf{M}_o^r \cdot \mathbf{R} = 0 \quad \forall o \in \mathcal{E}.$$

This is namely the case of systems of coplanar or parallel vectors (cf. Exercise 1.5). Because in this case, for the points $a \in \mathcal{A}$, it must be at the same time $\mathbf{M}_a^r \cdot \mathbf{R} = 0$ and $\mathbf{M}_a^r \times \mathbf{R} = \mathbf{o}$, so the only possibility is that

$$\mathbf{M}_a^r = \mathbf{o} \quad \forall a \in \mathcal{A},$$

i.e. in this case, $\mathcal{A}$ is the axis of points that make the resultant moment vanish.

Two systems of applied vectors are *equivalent* if they have the same resultant $\mathbf{R}$ and the same resultant moment $\mathbf{M}_o^r$ about any center $o \in \mathcal{E}$. The equivalence does not depend upon the center o. In fact, by Eq. (1.1), if two systems have the same $\mathbf{R}$ and the same $\mathbf{M}_{o_1}^r$, with o_1 a given point, then also $\mathbf{M}_{o_2}^r$ will be the same $\forall o_2 \in \mathcal{E}$.

Theorem 5 (Reduction of a system of applied vectors). *A system of applied vectors is always equivalent to the system composed of the resultant* $\mathbf{R}$ *applied at a point* o *and by a couple with moment* $\mathbf{M} = \mathbf{M}_o^r$, *with* o *any point of* $\mathcal{E}$.

Proof. By construction, $\mathbf{R}$ is the same for the two systems; moreover, for the equivalent system (resultant plus couple), it is

$$\mathbf{M} + (o - o) \times \mathbf{R} = \mathbf{M}.$$

So, if the couple has a moment $\mathbf{M} = \mathbf{M}_o^r$, the two systems are equivalent. □

In practice, this theorem affirms that it is always possible to reduce a system of n applied vectors to only an applied vector equal to $\mathbf{R}$ and to a couple or, if one of the two vectors composed of the couple is applied to the same point of $\mathbf{R}$, to two applied vectors. It is worth noting that the equivalence of the two systems is preserved if a vector is translated along its line of action because in such a case, $\mathbf{R}$ and $\mathbf{M}_o^r$ do not change.

Finally, a system of n applied vectors is said to be *equilibrated* if

$$\mathbf{R} = \mathbf{o}, \quad \mathbf{M}_o^r = \mathbf{o} \quad \forall o \in \mathcal{E}.$$

We note that because $\mathbf{R} = \mathbf{o}$, the center o can be any point of $\mathcal{E}$.

1.5 Exercises

1. Prove that the null vector is unique.
2. Prove that the null vector is orthogonal to any vector.
3. Prove that the norm of the null vector is zero.

4. Prove that
$$\mathbf{u} \cdot \mathbf{v} = 0 \iff |\mathbf{u} - \mathbf{v}| = |\mathbf{u} + \mathbf{v}| \;\; \forall \mathbf{u}, \mathbf{v} \in \mathcal{V}.$$
5. Prove the *linear forms representation theorem*: Let $\psi : \mathcal{V} \to \mathbb{R}$ be a linear function. Then, $\exists!\ \mathbf{u} \in \mathcal{V}$ such that
$$\psi(\mathbf{v}) = \mathbf{u} \cdot \mathbf{v} \;\; \forall \mathbf{v} \in \mathcal{V}.$$
6. Consider a point p and two noncollinear vectors $\mathbf{u}, \mathbf{v} \in \mathcal{S}$ at p. Show that a vector $\mathbf{w}$ is the bisector of the angle formed by $\mathbf{u}$ and $\mathbf{v}$ if and only if $\mathbf{w} \cdot \mathbf{u} = \mathbf{w} \cdot \mathbf{v}$.
7. Show that in the case of systems composed of coplanar or parallel applied vectors with $\mathbf{R} \neq \mathbf{o}, \mathbf{M}_o^r \cdot \mathbf{R} = 0 \; \forall o \in \mathcal{E}$.
8. Prove that any system of applied vectors with $\mathbf{R} = \mathbf{o}$ is equivalent to a couple.
9. Prove that a system of applied vectors all passing through a point p is equivalent to $\mathbf{R}$ applied to p.
10. Prove that if for a system of applied vectors $\mathbf{M}_o^r = \mathbf{o}$, then the system is equivalent to $\mathbf{R}$ applied to o. Then, show that if $o \in \mathcal{A}$, this is the case of coplanar or parallel vectors.
11. Prove that a system of applied vectors is equilibrated if and only if any equivalent system is equilibrated.
12. Prove that two applied vectors form an equilibrated system if and only if they are two opposite vectors applied to the same point.
13. Prove that a system of applied vectors is equilibrated if all the vectors pass through the same point and $\mathbf{R} = \mathbf{o}$.

Chapter 2

Second-Rank Tensors

2.1 Second-rank tensors

A *second-rank tensor* $\mathbf{L}$ is any linear application from $\mathcal{V}$ to $\mathcal{V}$:

$$\mathbf{L}:\mathcal{V}\rightarrow\mathcal{V}\mid\mathbf{L}(\alpha_i\mathbf{u}_i)=\alpha_i\mathbf{L}\mathbf{u}_i\ \forall\alpha_i\in\mathbb{R},\ \mathbf{u}_i\in\mathcal{V},\ i=1,\ldots,n.$$

Though here, $\mathcal{V}$ indicates the vector space of translations over $\mathcal{E}$, the definition of tensor[1] is more general, and in particular, $\mathcal{V}$ can be any vector space.

Defining the *sum of two tensors* as

$$(\mathbf{L}_1+\mathbf{L}_2)\mathbf{u}=\mathbf{L}_1\mathbf{u}+\mathbf{L}_2\mathbf{u}\ \ \forall\mathbf{u}\in\mathcal{V},\tag{2.1}$$

the *product of a scalar by a tensor* as

$$(\alpha\mathbf{L})\mathbf{u}=\alpha(\mathbf{L}\mathbf{u})\ \ \forall\alpha\in\mathbb{R},\mathbf{u}\in\mathcal{V},$$

and the *null tensor* $\mathbf{O}$ as the unique tensor such that

$$\mathbf{O}\mathbf{u}=\mathbf{o}\ \forall\mathbf{u}\in\mathcal{V},$$

then the set of all the tensors $\mathbf{L}$ that operate on $\mathcal{V}$ forms a vector space, denoted by $Lin(\mathcal{V})$. We define the *identity tensor* $\mathbf{I}$ as the unique tensor such that

$$\mathbf{I}\mathbf{u}=\mathbf{u}\ \ \forall\mathbf{u}\in\mathcal{V}.$$

[1]We consider, for the time being, only second-rank tensors that constitute a very important set of operators in classical and continuum mechanics. In the following, we also introduce fourth-rank tensors.

Different operations can be defined for the second-rank tensors. We consider all of them in the following sections.

2.2 Dyads, tensor components

For any couple of vectors $\mathbf{u}$ and $\mathbf{v}$, the *dyad*[2] $\mathbf{u} \otimes \mathbf{v}$ is a tensor defined by

$$(\mathbf{u} \otimes \mathbf{v})\mathbf{w} := \mathbf{v} \cdot \mathbf{w}\ \mathbf{u} \quad \forall \mathbf{w} \in \mathcal{V}.$$

The application defined above is actually a tensor because of the bilinearity of the scalar product. The introduction of dyads allows us to express any tensor as a linear combination of dyads. In fact, it can be proved that if $\mathcal{B} = \{\mathbf{e}_1, \mathbf{e}_2, \mathbf{e}_3\}$ is a basis of $\mathcal{V}$, then the set of nine dyads,

$$\mathcal{B}^2 = \{\mathbf{e}_i \otimes \mathbf{e}_j,\ i, j = 1, 2, 3\},$$

is a basis of $Lin(\mathcal{V})$ so that $\dim(Lin(\mathcal{V})) = 9$. This implies that any tensor, $\mathbf{L} \in Lin(\mathcal{V})$, can be expressed as

$$\mathbf{L} = L_{ij}\ \mathbf{e}_i \otimes \mathbf{e}_j, \quad i, j = 1, 2, 3,$$

where L_{ij}s are the nine *Cartesian components* of $\mathbf{L}$ with respect to $\mathcal{B}^2$. L_{ij}s can be calculated easily:

$$\mathbf{e}_i \cdot \mathbf{L}\mathbf{e}_j = \mathbf{e}_i \cdot L_{hk}\mathbf{e}_h \otimes \mathbf{e}_k\ \mathbf{e}_j = L_{hk}\mathbf{e}_i \cdot \mathbf{e}_h\ \mathbf{e}_k \cdot \mathbf{e}_j = L_{hk}\delta_{ih}\delta_{jk} = L_{ij}.$$

The above expression is sometimes called the *canonical decomposition of a tensor*. The components of a dyad can be computed as follows:

$$(\mathbf{u} \otimes \mathbf{v})_{ij} = \mathbf{e}_i \cdot (\mathbf{u} \otimes \mathbf{v})\ \mathbf{e}_j = \mathbf{u} \cdot \mathbf{e}_i\ \mathbf{v} \cdot \mathbf{e}_j = u_i\ v_j. \tag{2.2}$$

The components of a vector $\mathbf{v}$, resulting from the application of a tensor $\mathbf{L}$ on a vector $\mathbf{u}$, can now be calculated:

$$\mathbf{v} = \mathbf{L}\mathbf{u} = L_{ij}(\mathbf{e}_i \otimes \mathbf{e}_j)(u_k\mathbf{e}_k) = L_{ij}u_k\delta_{jk}\mathbf{e}_i = L_{ij}u_j\mathbf{e}_i \ \rightarrow\ v_i = L_{ij}u_j. \tag{2.3}$$

Depending upon two indices, any second-rank tensor $\mathbf{L}$ can be represented by a matrix whose entries are the Cartesian components of $\mathbf{L}$ in the

---

[2] In some texts, the dyad is also called the *tensor product*; we prefer to use the term dyad because the term tensor product can be ambiguous, as it is used to denote the product of two tensors, see Section 2.3.

basis $\mathcal{B}$:

$$\mathbf{L} = \begin{bmatrix} L_{11} & L_{12} & L_{13} \\ L_{21} & L_{22} & L_{23} \\ L_{31} & L_{32} & L_{33} \end{bmatrix}.$$

Because any $\mathbf{u} \in \mathcal{V}$, depending upon only one index, can be represented by a column vector, Eq. (2.3) represents actually the classical operation of the multiplication of a 3×3 matrix by a 3×1 vector.

2.3 Tensor product

The *tensor product* of $\mathbf{L}_1$ and $\mathbf{L}_2 \in Lin(\mathcal{V})$ is defined by

$$(\mathbf{L}_1\mathbf{L}_2)\mathbf{v} := \mathbf{L}_1(\mathbf{L}_2\mathbf{v}) \ \ \forall \mathbf{v} \in \mathcal{V}.$$

By linearity and Eq. (2.1), $\forall \mathbf{L}, \mathbf{L}_1, \mathbf{L}_2 \in Lin(\mathcal{V}), \mathbf{u} \in \mathcal{V}$, we get

$$\begin{aligned}
&[\mathbf{L}(\mathbf{L}_1 + \mathbf{L}_2)]\mathbf{v} = \mathbf{L}[(\mathbf{L}_1 + \mathbf{L}_2)\mathbf{v}] = \mathbf{L}(\mathbf{L}_1\mathbf{v} + \mathbf{L}_2\mathbf{v}) \\
&\quad = \mathbf{L}\mathbf{L}_1\mathbf{v} + \mathbf{L}\mathbf{L}_2\mathbf{v} = (\mathbf{L}\mathbf{L}_1 + \mathbf{L}\mathbf{L}_2)\mathbf{v} \ \rightarrow \ \mathbf{L}(\mathbf{L}_1 + \mathbf{L}_2) \\
&\quad = \mathbf{L}\mathbf{L}_1 + \mathbf{L}\mathbf{L}_2.
\end{aligned}$$

We remark that the tensor product is not symmetric:

$$\mathbf{L}_1\mathbf{L}_2 \neq \mathbf{L}_2\mathbf{L}_1;$$

however, by the same definition of the identity tensor and of tensor product,

$$\mathbf{IL} = \mathbf{LI} = \mathbf{L} \ \forall \mathbf{L} \in Lin(\mathcal{V}).$$

The Cartesian components of a tensor $\mathbf{L} = \mathbf{AB}$ can be calculated using Eq. (2.3):

$$\begin{aligned}
L_{ij} &= \mathbf{e}_i \cdot (\mathbf{AB})\mathbf{e}_j = \mathbf{e}_i \cdot \mathbf{A}(\mathbf{B}\mathbf{e}_j) = \mathbf{e}_i \cdot \mathbf{A}(B_{hk}(\mathbf{e}_j)_k \ \mathbf{e}_h) = B_{hk}\delta_{jk}\mathbf{e}_i \cdot \mathbf{A}\mathbf{e}_h \\
&= B_{hk}\delta_{jk}\mathbf{e}_i \cdot (A_{pq}(\mathbf{e}_h)_q \ \mathbf{e}_p) = A_{pq}B_{hk}\delta_{jk}\delta_{qh}\delta_{ip} = A_{ih}B_{hj}.
\end{aligned}$$

The above result simply corresponds to the row–column multiplication of two matrices. Using that, the following two identities can be readily shown:

$$\begin{aligned}
&(\mathbf{a} \otimes \mathbf{b})(\mathbf{c} \otimes \mathbf{d}) = \mathbf{b} \cdot \mathbf{c}(\mathbf{a} \otimes \mathbf{d}) \ \ \forall \mathbf{a}, \mathbf{b}, \mathbf{c}, \mathbf{d} \in \mathcal{V}, \\
&\mathbf{A}(\mathbf{a} \otimes \mathbf{b}) = (\mathbf{A}\mathbf{a}) \otimes \mathbf{b} \ \ \forall \mathbf{a}, \mathbf{b} \in \mathcal{V}, \ \mathbf{A} \in Lin(\mathcal{V}).
\end{aligned} \tag{2.4}$$

Finally, the symbol $\mathbf{L}^2$ is normally used to denote, in short, the product $\mathbf{LL}$, $\forall \mathbf{L} \in Lin(\mathcal{V})$.

2.4 Transpose, symmetric and skew tensors

For any tensor $\mathbf{L} \in Lin(\mathcal{V})$, there exists just one tensor $\mathbf{L}^\top$, called the *transpose* of $\mathbf{L}$, such that

$$\mathbf{u} \cdot \mathbf{L}\mathbf{v} = \mathbf{v} \cdot \mathbf{L}^\top \mathbf{u} \quad \forall \mathbf{u}, \mathbf{v} \in \mathcal{V}. \tag{2.5}$$

The transpose of the transpose of $\mathbf{L}$ is $\mathbf{L}$:

$$\mathbf{u} \cdot \mathbf{L}\mathbf{v} = \mathbf{v} \cdot \mathbf{L}^\top \mathbf{u} = \mathbf{u} \cdot (\mathbf{L}^\top)^\top \mathbf{v} \quad \Rightarrow \quad (\mathbf{L}^\top)^\top = \mathbf{L}.$$

The Cartesian components of $\mathbf{L}^\top$ are obtained by swapping the indices of the components of $\mathbf{L}$:

$$L_{ij}^\top = \mathbf{e}_i \cdot \mathbf{L}^\top \mathbf{e}_j = \mathbf{e}_j \cdot (\mathbf{L}^\top)^\top \mathbf{e}_i = \mathbf{e}_j \cdot \mathbf{L}\mathbf{e}_i = L_{ji}.$$

It is immediate to show that

$$(\mathbf{A} + \mathbf{B})^\top = \mathbf{A}^\top + \mathbf{B}^\top \quad \forall \mathbf{A}, \mathbf{B} \in Lin(\mathcal{V}),$$

while

$$\mathbf{u} \cdot (\mathbf{AB})\mathbf{v} = \mathbf{B}\mathbf{v} \cdot \mathbf{A}^\top \mathbf{u} = \mathbf{v} \cdot \mathbf{B}^\top \mathbf{A}^\top \mathbf{u} \quad \Rightarrow \quad (\mathbf{AB})^\top = \mathbf{B}^\top \mathbf{A}^\top.$$

Moreover,

$$\mathbf{u} \cdot (\mathbf{a} \otimes \mathbf{b})\mathbf{v} = \mathbf{a} \cdot \mathbf{u}\ \mathbf{b} \cdot \mathbf{v} = \mathbf{v} \cdot (\mathbf{b} \otimes \mathbf{a})\mathbf{u} \quad \Rightarrow \quad (\mathbf{a} \otimes \mathbf{b})^\top = \mathbf{b} \otimes \mathbf{a}. \tag{2.6}$$

A tensor $\mathbf{L}$ is *symmetric* $\Longleftrightarrow$

$$\mathbf{L} = \mathbf{L}^\top.$$

In such a case, because $L_{ij} = L_{ij}^\top$, we have

$$L_{ij} = L_{ji}.$$

A symmetric tensor is hence represented, in a given basis, by a symmetric matrix and has only six independent Cartesian components. Applying Eq. (2.5) to $\mathbf{I}$, it is immediately recognized that the identity tensor is symmetric: $\mathbf{I} = \mathbf{I}^\top$.

A tensor $\mathbf{L}$ is *antisymmetric* or *skew* $\Longleftrightarrow$

$$\mathbf{L} = -\mathbf{L}^\top.$$

In this case, because $L_{ij} = -L^\top_{ij}$, we have (no summation on the index i, see footnote 1, Chapter 1)

$$L_{ij} = -L_{ji} \;\Rightarrow\; L_{\underline{ii}} = 0\ \forall i = 1, 2, 3.$$

A skew tensor is hence represented, in a given basis, by an antisymmetric matrix whose components on the diagonal are identically null in any basis; finally, a skew tensor only depends upon three independent Cartesian components.

If we denote by $Sym(\mathcal{V})$ the set of all the symmetric tensors and by $Skw(\mathcal{V})$ that of all the skew tensors, then it is evident that $\forall \alpha, \beta, \lambda, \mu \in \mathbb{R}$,

$$\begin{gathered}
Sym(\mathcal{V}) \cap Skw(\mathcal{V}) = \mathbf{O},\\
\alpha\mathbf{A} + \beta\mathbf{B} \in Sym(\mathcal{V})\ \forall \mathbf{A}, \mathbf{B} \in Sym(\mathcal{V}),\\
\lambda\mathbf{L} + \mu\mathbf{M} \in Skw(\mathcal{V})\ \forall \mathbf{L}, \mathbf{M} \in Skw(\mathcal{V}),
\end{gathered}$$

so $Sym(\mathcal{V})$ and $Skw(\mathcal{V})$ are vector subspaces of $Lin(\mathcal{V})$ with $\dim(Sym(\mathcal{V})) = 6$, while $\dim(Skw(\mathcal{V})) = 3$.

Any tensor $\mathbf{L}$ can be decomposed into the sum of a symmetric, $\mathbf{L}^s$, and an antisymmetric, $\mathbf{L}^a$, tensor:

$$\mathbf{L} = \mathbf{L}^s + \mathbf{L}^a,$$

with

$$\mathbf{L}^s = \frac{\mathbf{L} + \mathbf{L}^\top}{2} \quad \in Sym(\mathcal{V})$$

and

$$\mathbf{L}^a = \frac{\mathbf{L} - \mathbf{L}^\top}{2} \quad \in Skw(\mathcal{V})$$

so that, finally,

$$Lin(\mathcal{V}) = Sym(\mathcal{V}) \oplus Skw(\mathcal{V}).$$

2.5 Trace, scalar product of tensors

There exists one and only one linear form

$$\mathrm{tr} : Lin(\mathcal{V}) \to \mathbb{R},$$

called the *trace*, such that

$$\mathrm{tr}(\mathbf{a} \otimes \mathbf{b}) = \mathbf{a} \cdot \mathbf{b} \quad \forall \mathbf{a}, \mathbf{b} \in \mathcal{V}.$$

For the same definition that has been given without making use of any basis of $\mathcal{V}$, the trace of a tensor is a *tensor invariant*, i.e. a quantity extracted from a tensor that does not depend upon the basis.

Linearity implies that

$$\text{tr}(\alpha\mathbf{A}+\beta\mathbf{B})=\alpha\text{tr}\mathbf{A}+\beta\text{tr}\mathbf{B}\ \ \forall\alpha,\beta\in\mathbb{R},\ \ \mathbf{A},\mathbf{B}\in Lin(\mathcal{V}).$$

It is just linearity to give the rule for calculating the trace of a tensor $\mathbf{L}$:

$$\text{tr}\mathbf{L}=\text{tr}(L_{ij}\mathbf{e}_i\otimes\mathbf{e}_j)=L_{ij}\text{tr}(\mathbf{e}_i\otimes\mathbf{e}_j)=L_{ij}\ \mathbf{e}_i\cdot\mathbf{e}_j=L_{ij}\delta_{ij}=L_{ii}. \quad (2.7)$$

A tensor is hence an operator whose sum of the components on the diagonal,

$$\text{tr}\mathbf{L}=L_{11}+L_{22}+L_{33},$$

is constant, regardless of the basis.

Following the same procedure above, it is readily seen that

$$\text{tr}\mathbf{L}^\top=\text{tr}\mathbf{L},$$

which implies, by linearity, that

$$\text{tr}\mathbf{L}=0\ \ \forall\mathbf{L}\in Skw(\mathcal{V}). \quad (2.8)$$

The *scalar product* of tensors $\mathbf{A}$ and $\mathbf{B}$ is a positive definite, symmetric bilinear form defined by

$$\mathbf{A}\cdot\mathbf{B}=\text{tr}(\mathbf{A}^\top\mathbf{B}).$$

This definition implies that, $\forall\mathbf{L},\mathbf{M},\mathbf{N}\in Lin(\mathcal{V}),\ \alpha,\beta\in\mathbb{R}$,

$$\begin{aligned}
&\mathbf{L}\cdot(\alpha\mathbf{M}+\beta\mathbf{N})=\alpha\mathbf{L}\cdot\mathbf{M}+\beta\mathbf{L}\cdot\mathbf{N},\\
&(\alpha\mathbf{L}+\beta\mathbf{M})\cdot\mathbf{N}=\alpha\mathbf{L}\cdot\mathbf{N}+\beta\mathbf{M}\cdot\mathbf{N},\\
&\mathbf{L}\cdot\mathbf{M}=\mathbf{M}\cdot\mathbf{L},\\
&\mathbf{L}\cdot\mathbf{L}>0\ \ \forall\mathbf{L}\in Lin(\mathcal{V}),\ \ \mathbf{L}\cdot\mathbf{L}=0\iff\mathbf{L}=\mathbf{O}.
\end{aligned}$$

These properties give the rule for computing the scalar product of two tensors $\mathbf{A}$ and $\mathbf{B}$:

$$\begin{aligned}
\mathbf{A}\cdot\mathbf{B}&=A_{ij}(\mathbf{e}_i\otimes\mathbf{e}_j)\cdot B_{hk}(\mathbf{e}_h\otimes\mathbf{e}_k)=A_{ij}B_{hk}(\mathbf{e}_i\otimes\mathbf{e}_j)\cdot(\mathbf{e}_h\otimes\mathbf{e}_k)\\
&=A_{ij}B_{hk}\ \text{tr}[(\mathbf{e}_i\otimes\mathbf{e}_j)^\top(\mathbf{e}_h\otimes\mathbf{e}_k)]=A_{ij}B_{hk}\ \text{tr}[(\mathbf{e}_j\otimes\mathbf{e}_i)(\mathbf{e}_h\otimes\mathbf{e}_k)]\\
&=A_{ij}B_{hk}\ \text{tr}[\mathbf{e}_i\cdot\mathbf{e}_h(\mathbf{e}_j\otimes\mathbf{e}_k)]=A_{ij}B_{hk}\ \mathbf{e}_i\cdot\mathbf{e}_h\ \mathbf{e}_j\cdot\mathbf{e}_k\\
&=A_{ij}B_{hk}\delta_{ih}\delta_{jk}=A_{ij}B_{ij}.
\end{aligned}$$

As in the case of vectors, the scalar product of two tensors is equal to the sum of the products of the corresponding components. In a similar manner, or using Eq. $(2.4)_1$, it is easily shown that, $\forall \mathbf{a}, \mathbf{b}, \mathbf{c}, \mathbf{d} \in \mathcal{V}$,

$$(\mathbf{a} \otimes \mathbf{b}) \cdot (\mathbf{c} \otimes \mathbf{d}) = \mathbf{a} \cdot \mathbf{c} \ \mathbf{b} \cdot \mathbf{d} = a_i b_j c_i d_j,$$

while by the same definition of the tensor scalar product,

$$\text{tr}\mathbf{L} = \mathbf{I} \cdot \mathbf{L} \quad \forall \mathbf{L} \in Lin(\mathcal{V}).$$

Similar to vectors, we define the *Euclidean norm* of a tensor $\mathbf{L}$ the nonnegative scalar, denoted either by L or $|\mathbf{L}|$:

$$L = |\mathbf{L}| = \sqrt{\mathbf{L} \cdot \mathbf{L}} = \sqrt{\text{tr}(\mathbf{L}^\top \mathbf{L})} = \sqrt{L_{ij} L_{ij}}$$

and the *distance* $d(\mathbf{L}, \mathbf{M})$ of two tensors $\mathbf{L}$ and $\mathbf{M}$ the norm of the tensor difference:

$$d(\mathbf{L}, \mathbf{M}) := |\mathbf{L} - \mathbf{M}| = |\mathbf{M} - \mathbf{L}|.$$

2.6 Spherical and deviatoric parts

Let $\mathbf{L} \in Sym(\mathcal{V})$; the *spherical part* of $\mathbf{L}$ is defined by

$$\mathbf{L}^{sph} := \frac{1}{3} \text{tr}\mathbf{L} \ \mathbf{I}$$

and the *deviatoric part* by

$$\mathbf{L}^{dev} := \mathbf{L} - \mathbf{L}^{sph}$$

so that

$$\mathbf{L} = \mathbf{L}^{sph} + \mathbf{L}^{dev}.$$

We remark that

$$\text{tr}\mathbf{L}^{sph} = \frac{1}{3} \text{tr}\mathbf{L} \ \text{tr}\mathbf{I} = \text{tr}\mathbf{L} \ \Rightarrow \ \text{tr}\mathbf{L}^{dev} = 0,$$

i.e. the deviatoric part is a traceless tensor. Let $\mathbf{A}, \mathbf{B} \in Lin(\mathcal{V})$, then

$$\mathbf{A}^{sph} \cdot \mathbf{B}^{dev} = \frac{1}{3} \text{tr}\mathbf{A} \ \mathbf{I} \cdot \mathbf{B}^{dev} = \frac{1}{3} \text{tr}\mathbf{A} \ \text{tr}\mathbf{B}^{dev} = 0, \tag{2.9}$$

i.e. any spherical tensor is orthogonal to any deviatoric tensor.

The sets

$$Sph(\mathcal{V}) := \left\{ \mathbf{A}^{sph} \in Lin(\mathcal{V}) |\ \mathbf{A}^{sph} = \frac{1}{3}\text{tr}\mathbf{A}\mathbf{I}\ \forall \mathbf{A} \in Lin(\mathcal{V}) \right\},$$
$$Dev(\mathcal{V}) := \left\{ \mathbf{A}^{dev} \in Lin(\mathcal{V}) |\ \mathbf{A}^{dev} = \mathbf{A} - \mathbf{A}^{sph}\ \forall \mathbf{A} \in Lin(\mathcal{V}) \right\}$$

form two subspaces of $Lin(\mathcal{V})$; the proof is left to the reader. For what is proved above, $Sph(\mathcal{V})$ and $Dev(\mathcal{V})$ are two *mutually orthogonal subspaces* of $Lin(\mathcal{V})$.

2.7 Determinant, inverse of a tensor

The reader is probably familiar with the concept of determinant of a matrix. We show here that the determinant of a second-rank tensor can be defined intrinsically and that it corresponds with the determinant of the matrix that represents it in any basis of $\mathcal{V}$. For this purpose, we first need to introduce a mapping:

$$\omega : \mathcal{V} \times \mathcal{V} \times \mathcal{V} \to \mathbb{R}$$

is a *skew trilinear form* if $\omega(\mathbf{u}, \mathbf{v}, \cdot), \omega(\mathbf{u}, \cdot, \mathbf{v})$, and $\omega(\cdot, \mathbf{u}, \mathbf{v})$ are linear forms on $\mathcal{V}$ and if

$$\omega(\mathbf{u}, \mathbf{v}, \mathbf{w}) = -\omega(\mathbf{v}, \mathbf{u}, \mathbf{w}) = -\omega(\mathbf{u}, \mathbf{w}, \mathbf{v}) = -\omega(\mathbf{w}, \mathbf{v}, \mathbf{u})\ \forall \mathbf{u}, \mathbf{v}, \mathbf{w} \in \mathcal{V}. \tag{2.10}$$

Using this definition, we can state the following.

Theorem 6. *Three vectors are linearly independent if and only if every skew trilinear form of them is not null.*

Proof. In fact, let $\mathbf{u} = \alpha\mathbf{v} + \beta\mathbf{w}$, then for any skew trilinear form ω,

$$\omega(\mathbf{u}, \mathbf{v}, \mathbf{w}) = \omega(\alpha\mathbf{v} + \beta\mathbf{w}, \mathbf{v}, \mathbf{w}) = \alpha\omega(\mathbf{v}, \mathbf{v}, \mathbf{w}) + \beta\omega(\mathbf{w}, \mathbf{v}, \mathbf{w}) = 0$$

because of Eq. (2.10) applied to the permutation of the positions of the two $\mathbf{u}$ and the two $\mathbf{w}$.

□

It is evident that the set of all the skew trilinear forms is a vector space and that we denote by Ω, whose null element is the *null form* ω_0,

$$\omega_0(\mathbf{u},\mathbf{v},\mathbf{w}) = 0 \ \forall \mathbf{u},\mathbf{v},\mathbf{w} \in \mathcal{V}.$$

For a given $\omega(\mathbf{u},\mathbf{v},\mathbf{w}) \in \Omega$, any $\mathbf{L} \in Lin(\mathcal{V})$ induces another form $\omega_L(\mathbf{u},\mathbf{v},\mathbf{w}) \in \Omega$, defined as

$$\omega_L(\mathbf{u},\mathbf{v},\mathbf{w}) = \omega(\mathbf{L}\mathbf{u},\mathbf{L}\mathbf{v},\mathbf{L}\mathbf{w}) \ \forall \mathbf{u},\mathbf{v},\mathbf{w} \in \mathcal{V}.$$

A key point[3] for the following developments is that $\dim \Omega = 1$.

This means that $\forall \omega_1, \omega_2 \neq \omega_0 \in \Omega, \exists \lambda \in \mathbb{R}$ such that

$$\omega_2(\mathbf{u},\mathbf{v},\mathbf{w}) = \lambda \omega_1(\mathbf{u},\mathbf{v},\mathbf{w}) \ \forall \mathbf{u},\mathbf{v},\mathbf{w} \in \mathcal{V}.$$

So, $\forall \mathbf{L} \in Lin(\mathcal{V})$, there must exist $\lambda_L \in \mathbb{R}$ such that

$$\omega(\mathbf{L}\mathbf{u},\mathbf{L}\mathbf{v},\mathbf{L}\mathbf{w}) = \omega_L(\mathbf{u},\mathbf{v},\mathbf{w}) = \lambda_L \ \omega(\mathbf{u},\mathbf{v},\mathbf{w}) \ \forall \mathbf{u},\mathbf{v},\mathbf{w} \in \mathcal{V}. \tag{2.11}$$

The scalar[4] λ_L is the *determinant of* $\mathbf{L}$, and in the following, it will be denoted as $\det \mathbf{L}$. The determinant of a tensor $\mathbf{L}$ is an intrinsic quantity of $\mathbf{L}$, i.e. it does not depend upon the particular form ω, nor on the basis of $\mathcal{V}$. In fact, we have never introduced, so far, a basis for defining $\det \mathbf{L}$, hence it cannot depend upon the choice of a basis for $\mathcal{V}$, i.e. $\det \mathbf{L}$ is *tensor-invariant.*

Then, if ω^a and $\omega^b \in \Omega$, because $\dim \Omega = 1$, there exists $k \in \mathbb{R},\ k \neq 0$, such that

$$\begin{aligned}
\omega^b(\mathbf{u},\mathbf{v},\mathbf{w}) &= k\ \omega^a(\mathbf{u},\mathbf{v},\mathbf{w}) \ \forall \mathbf{u},\mathbf{v},\mathbf{w} \in \mathcal{V} \Rightarrow \\
\omega^b(\mathbf{L}\mathbf{u},\mathbf{L}\mathbf{v},\mathbf{L}\mathbf{w}) &= k\ \omega^a(\mathbf{L}\mathbf{u},\mathbf{L}\mathbf{v},\mathbf{L}\mathbf{w}) \rightarrow \\
\omega^b_L(\mathbf{u},\mathbf{v},\mathbf{w}) &= k\ \omega^a_L(\mathbf{u},\mathbf{v},\mathbf{w}).
\end{aligned}$$

Moreover, by Eq. (2.11), we get

$$\begin{aligned}
\omega^a(\mathbf{L}\mathbf{u},\mathbf{L}\mathbf{v},\mathbf{L}\mathbf{w}) &= \omega^a_L(\mathbf{u},\mathbf{v},\mathbf{w}) = \lambda^a_L \omega^a(\mathbf{u},\mathbf{v},\mathbf{w}), \\
\omega^b(\mathbf{L}\mathbf{u},\mathbf{L}\mathbf{v},\mathbf{L}\mathbf{w}) &= \omega^b_L(\mathbf{u},\mathbf{v},\mathbf{w}) = \lambda^b_L \omega^b(\mathbf{u},\mathbf{v},\mathbf{w})
\end{aligned}$$

[3] The proof of this statement is rather involved and outside of our scope; the interested reader is referred to the classical textbook by Halmos on linear algebra, Section 31 (see the bibliography). The theory of the determinants is developed in Section 53.

[4] More precisely, $\det \mathbf{L}$ is the function that associates a scalar with each tensor (Halmos, Section 53). We can, however, for the sake of practice, identify $\det \mathbf{L}$ with the scalar associated with $\mathbf{L}$, without consequences for our purposes.

so that

$$\lambda_L^b k\ \omega^a(\mathbf{u},\mathbf{v},\mathbf{w}) = \lambda_L^b \omega^b(\mathbf{u},\mathbf{v},\mathbf{w}) = \omega_L^b(\mathbf{u},\mathbf{v},\mathbf{w})$$
$$= k\ \omega_L^a(\mathbf{u},\mathbf{v},\mathbf{w}) = \lambda_L^a k\ \omega^a(\mathbf{u},\mathbf{v},\mathbf{w}) \iff \lambda_L^a = \lambda_L^b,$$

which proves that $\det \mathbf{L}$ does not depend upon the skew trilinear form but only upon $\mathbf{L}$.

The definition given for $\det \mathbf{L}$ allows us to prove some important properties. First of all,

$$\det \mathbf{O} = 0;$$

in fact, $\forall \omega \in \Omega$,

$$\det \mathbf{O}\ \omega(\mathbf{u},\mathbf{v},\mathbf{w}) = \omega(\mathbf{O}\mathbf{u},\mathbf{O}\mathbf{v},\mathbf{O}\mathbf{w}) = \omega(\mathbf{o},\mathbf{o},\mathbf{o}) = 0\ \forall \mathbf{u},\mathbf{v},\mathbf{w} \in \mathcal{V}$$

because ω operates on three identical, i.e. linearly dependent, vectors. Moreover, if $\mathbf{L} = \mathbf{I}$, then

$$\det \mathbf{I}\ \omega(\mathbf{u},\mathbf{v},\mathbf{w}) = \omega(\mathbf{I}\mathbf{u},\mathbf{I}\mathbf{v},\mathbf{I}\mathbf{w}) = \omega(\mathbf{u},\mathbf{v},\mathbf{w})$$

if and only if

$$\det \mathbf{I} = 1. \tag{2.12}$$

A third property is that $\forall \mathbf{a},\mathbf{b} \in \mathcal{V}$,

$$\det(\mathbf{a} \otimes \mathbf{b}) = 0. \tag{2.13}$$

In fact, if $\mathbf{L} = \mathbf{a} \otimes \mathbf{b}$, then

$$\det \mathbf{L}\ \omega(\mathbf{u},\mathbf{v},\mathbf{w}) = \omega(\mathbf{L}\mathbf{u},\mathbf{L}\mathbf{v},\mathbf{L}\mathbf{w}) = \omega((\mathbf{b}\cdot\mathbf{u})\mathbf{a},(\mathbf{b}\cdot\mathbf{v})\mathbf{a},(\mathbf{b}\cdot\mathbf{w})\mathbf{a}) = 0$$

because the three vectors on which $\omega \in \Omega$ operates are linearly dependent; with $\mathbf{u},\mathbf{v}$, and $\mathbf{w}$ being arbitrary, this implies Eq. (2.13).

An important result is the following.

Theorem 7 (Theorem of Binet). $\forall \mathbf{A},\mathbf{B} \in Lin(\mathcal{V})$,

$$\det(\mathbf{A}\mathbf{B}) = \det \mathbf{A} \det \mathbf{B}. \tag{2.14}$$

Proof. $\forall \omega \in \Omega$ and $\forall \mathbf{u},\mathbf{v},\mathbf{w} \in \mathcal{V}$,

$$\lambda_{AB}\omega(\mathbf{u},\mathbf{v},\mathbf{w}) = \omega(\mathbf{AB}\mathbf{u},\mathbf{AB}\mathbf{v},\mathbf{AB}\mathbf{w}) = \omega(\mathbf{A}(\mathbf{B}\mathbf{u}),\mathbf{A}(\mathbf{B}\mathbf{v}),\mathbf{A}(\mathbf{B}\mathbf{w})) =$$
$$\lambda_A\omega(\mathbf{B}\mathbf{u},\mathbf{B}\mathbf{v},\mathbf{B}\mathbf{w}) = \lambda_A\lambda_B\omega(\mathbf{u},\mathbf{v},\mathbf{w}) \iff \lambda_{AB} = \lambda_A\lambda_B,$$

which proves the theorem. □

A tensor $\mathbf{L}$ is called *singular* if $\det\mathbf{L}=0$, otherwise it is *non-singular*.

Considering Eq. (2.11), with some effort but without major difficulties, one can see that, if in a basis $\mathcal{B}$ of $\mathcal{V}$, we have $\mathbf{L}=L_{ij}\mathbf{e}_i\otimes\mathbf{e}_j$, then

$$\det\mathbf{L}=\sum_{\pi\in\mathcal{P}_3}\epsilon_{\pi(1),\pi(2),\pi(3)}L_{1,\pi(1)}L_{2,\pi(2)}L_{3,\pi(3)},$$

where $\mathcal{P}_3$ is the set of all the permutations π of $\{1,2,3\}$ and the $\epsilon_{i,j,k}$s are the components of *Ricci's alternator*[5]:

$$\epsilon_{i,j,k}:=\begin{cases}1 & \text{if } \{i,j,k\} \text{ is an even permutation of } \{1,2,3\},\\ 0 & \text{if } \{i,j,k\} \text{ is not a permutation of } \{1,2,3\},\\ -1 & \text{if } \{i,j,k\} \text{ is an odd permutation of } \{1,2,3\}.\end{cases}$$

The above rule for $\det\mathbf{L}$ coincides with that for calculating the determinant of the matrix whose entries are L_{ij}s. This shows that, once a basis $\mathcal{B}$ for $\mathcal{V}$ is chosen, $\det\mathbf{L}$ coincides with the determinant of the matrix representing it in $\mathcal{B}$ and, finally, that

$$\begin{aligned}\det\mathbf{L}=&\,L_{11}L_{22}L_{33}+L_{12}L_{23}L_{31}+L_{13}L_{32}L_{21}\\&-L_{11}L_{23}L_{32}-L_{22}L_{13}L_{31}-L_{33}L_{12}L_{21}.\end{aligned}\tag{2.15}$$

This result shows immediately that $\forall\mathbf{L}\in Lin(\mathcal{V})$, and regardless of $\mathcal{B}$, we have

$$\det\mathbf{L}^\top=\det\mathbf{L}.\tag{2.16}$$

Using Eq. (2.15), it is not difficult to show that, $\forall\alpha\in\mathbb{R}$,

$$\det(\mathbf{I}+\alpha\mathbf{L})=1+\alpha I_1+\alpha^2I_2+\alpha^3I_3,\tag{2.17}$$

where I_1,I_2, and I_3 are the three *principal invariants* of $\mathbf{L}$:

$$I_1=\mathrm{tr}\mathbf{L},\quad I_2=\frac{\mathrm{tr}^2\mathbf{L}-\mathrm{tr}\mathbf{L}^2}{2},\quad I_3=\det\mathbf{L}.\tag{2.18}$$

---

[5]We recall that a *permutation* of an ordered set of n objects is *even* if it can be obtained as the product of an even number of *transpositions*, i.e. exchange of places, of any couple of its elements and it is *odd* if the number of transpositions is odd. For the set $\{1,2,3\}$, the even permutations are $\{1,2,3\},\{3,1,2\},\{2,3,1\}$, while the odd ones are $\{2,1,3\},\{1,3,2\},\{3,2,1\}$; any triplet having at least a repeated number is not a permutation.

A tensor $\mathbf{L} \in Lin(\mathcal{V})$ is said to be *invertible* if there is a tensor $\mathbf{L}^{-1} \in Lin(\mathcal{V})$, called the *inverse* of $\mathbf{L}$, such that

$$\mathbf{L}\mathbf{L}^{-1} = \mathbf{L}^{-1}\mathbf{L} = \mathbf{I}. \tag{2.19}$$

If $\mathbf{L}$ is invertible, then $\mathbf{L}^{-1}$ is unique. By the above definition, if $\mathbf{L}$ is invertible, then

$$\mathbf{u}_1 = \mathbf{L}\mathbf{u} \Rightarrow \mathbf{u} = \mathbf{L}^{-1}\mathbf{u}_1.$$

Theorem 8. *Any invertible tensor maps triples of linearly independent vectors into triples of still linearly independent vectors.*

Proof. Let $\mathbf{L}$ be an invertible tensor and $\mathbf{u}_1 = \mathbf{L}\mathbf{u}, \mathbf{v}_1 = \mathbf{L}\mathbf{v}, \mathbf{w}_1 = \mathbf{L}\mathbf{w}$, where $\mathbf{u}, \mathbf{v}$, and $\mathbf{w}$ are three linearly independent vectors. Let us suppose that there exist $h, k \in \mathbb{R}$ such that

$$\mathbf{u}_1 = h\mathbf{v}_1 + k\mathbf{w}_1.$$

Then, because $\mathbf{L}$ is invertible,

$$\mathbf{L}^{-1}\mathbf{u}_1 = \mathbf{L}^{-1}(h\mathbf{v}_1 + k\mathbf{w}_1) = h\mathbf{L}^{-1}\mathbf{v}_1 + k\mathbf{L}^{-1}\mathbf{w}_1 = h\mathbf{v} + k\mathbf{w},$$

which goes against the hypothesis. Consequently, $\mathbf{u}_1, \mathbf{v}_1$, and $\mathbf{w}_1$ are linearly independent. □

This result, along with the definition of a determinant, Eq. (2.11), and Theorem 6, proves the following.

Theorem 9 (Invertibility theorem). $\mathbf{L} \in Lin(\mathcal{V})$ *is invertible* $\iff$ $\det \mathbf{L} \neq 0$.

Using the theorem of Binet, Theorem 7, along with Eqs. (2.12) and (2.19), we get

$$\det \mathbf{L}^{-1} = \frac{1}{\det \mathbf{L}}.$$

Equation (2.19) applied to $\mathbf{L}^{-1}$, along with the uniqueness of the inverse, gives immediately that

$$(\mathbf{L}^{-1})^{-1} = \mathbf{L},$$

while

$$\mathbf{B}^{-1}\mathbf{A}^{-1} = \mathbf{B}^{-1}\mathbf{A}^{-1}\mathbf{A}\mathbf{B}(\mathbf{A}\mathbf{B})^{-1} = (\mathbf{A}\mathbf{B})^{-1}.$$

The operations of transpose and inversion commute:

$$\mathbf{L}^\top(\mathbf{L}^\top)^{-1} = \mathbf{I} = \mathbf{L}^{-1}\mathbf{L} = \mathbf{I}^\top = (\mathbf{L}^{-1}\mathbf{L})^\top = \mathbf{L}^\top(\mathbf{L}^{-1})^\top \Rightarrow$$

$$(\mathbf{L}^{-1})^\top = (\mathbf{L}^\top)^{-1} := \mathbf{L}^{-\top}.$$

2.8 Eigenvalues and eigenvectors of a tensor

If there exists a $\lambda \in \mathbb{R}$ and a $\mathbf{v} \in \mathcal{V}$, except the null vector, such that

$$\mathbf{L}\mathbf{v} = \lambda\mathbf{v}, \tag{2.20}$$

then λ is an *eigenvalue* and $\mathbf{v}$ an *eigenvector*, relative to λ, of $\mathbf{L}$. It is immediate to observe that, thanks to linearity, any eigenvector $\mathbf{v}$ of $\mathbf{L}$ is determined to within a multiplier, i.e. that $k\mathbf{v}$ is an eigenvector of $\mathbf{L}$ too $\forall k \in \mathbb{R}$. Often, the multiplier k is fixed in such a way that $|\mathbf{v}| = 1$.

To determine the eigenvalues and eigenvectors of a tensor, we rewrite Eq. (2.20) as

$$(\mathbf{L} - \lambda\mathbf{I})\mathbf{v} = \mathbf{o}. \tag{2.21}$$

The condition for this homogeneous system having a non-null solution is

$$\det(\mathbf{L} - \lambda\mathbf{I}) = 0;$$

this is the so-called *characteristic* or *Laplace's equation*. In the case of a second-rank tensor over $\mathcal{V}$, the Laplace's equation is an algebraic equation of degree three with real coefficients. The roots of the Laplace's equation are the eigenvalues of $\mathbf{L}$; because the components of $\mathbf{L}$, and hence the coefficients of the characteristic equation, are all real, then the eigenvalues of $\mathbf{L}$ are all real or one real and two complex conjugate.

For any eigenvalue λ_i, $i = 1, 2, 3$, of $\mathbf{L}$, the corresponding eigenvectors $\mathbf{v}_i$ can be found by solving Eq. (2.21), once we set $\lambda = \lambda_i$.

The *proper space* of $\mathbf{L}$ relative to λ is the subspace of $Lin(\mathcal{V})$ composed of all the vectors that satisfy Eq. (2.21). The *multiplicity* of λ is the dimension of its proper space, while the *spectrum* of $\mathbf{L}$ is the set composed of all of its eigenvalues, each one with its multiplicity.

$\mathbf{L}^\top$ has the same eigenvalues of $\mathbf{L}$ because the Laplace's equation is the same in both the cases:

$$\det(\mathbf{L}^\top - \lambda\mathbf{I}) = \det(\mathbf{L}^\top - \lambda\mathbf{I}^\top) = \det(\mathbf{L} - \lambda\mathbf{I})^\top = \det(\mathbf{L} - \lambda\mathbf{I}).$$

However, this is not the case for the eigenvectors that are generally different, as a numerical example can show.

Developing the Laplace's equation, it is easy to show that it can be written as

$$\det(\mathbf{L} - \lambda\mathbf{I}) = -\lambda^3 + I_1\lambda^2 - I_2\lambda + I_3 = 0,$$

which is merely an application of Eq. (2.17). If we denote $\mathbf{L}^3 = \mathbf{LLL}$, using Eq. (2.18), one can prove the following.

Theorem 10 (Cayley–Hamilton theorem). $\forall \mathbf{L} \in Lin(\mathcal{V})$,

$$\mathbf{L}^3 - I_1\mathbf{L}^2 + I_2\mathbf{L} - I_3\mathbf{I} = \mathbf{O}.$$

A *quadratic form* defined by $\mathbf{L}$ is any form $\omega : \mathcal{V} \times \mathcal{V} \to \mathbb{R}$ of the type

$$\omega = \mathbf{v} \cdot \mathbf{L}\mathbf{v};$$

if $\omega > 0\ \forall \mathbf{v} \in \mathcal{V},\ \omega = 0 \iff \mathbf{v} = \mathbf{o}$, then ω and $\mathbf{L}$ are said to be *positive definite*. The eigenvalues of a positive definite tensor are positive. In fact, if λ is an eigenvalue of $\mathbf{L}$, which is positive definite, and $\mathbf{v}$ its eigenvector, then

$$\mathbf{v} \cdot \mathbf{L}\mathbf{v} = \mathbf{v} \cdot \lambda\mathbf{v} = \lambda\mathbf{v}^2 > 0 \iff \lambda > 0.$$

Let $\mathbf{v}_1$ and $\mathbf{v}_2$ be two eigenvectors of a symmetric tensor $\mathbf{L}$ relative to the eigenvalues λ_1 and λ_2, respectively, with $\lambda_1 \neq \lambda_2$. Then,

$$\lambda_1\mathbf{v}_1 \cdot \mathbf{v}_2 = \mathbf{L}\mathbf{v}_1 \cdot \mathbf{v}_2 = \mathbf{L}\mathbf{v}_2 \cdot \mathbf{v}_1 = \lambda_2\mathbf{v}_2 \cdot \mathbf{v}_1 \iff \mathbf{v}_1 \cdot \mathbf{v}_2 = 0.$$

Actually, symmetric tensors have a particular importance, specified by the following.

Theorem 11 (Spectral theorem). *The eigenvectors of a symmetric tensor form a basis of $\mathcal{V}$.*

This theorem[6] is of paramount importance in linear algebra: It proves that the eigenvalues of a symmetric tensor $\mathbf{L}$ are real valued and, remembering the definition of eigenvalues and eigenvectors, Eq. (2.20), that there exists a basis $\mathcal{B}_N = \{\mathbf{u}_1, \mathbf{u}_2, \mathbf{u}_3\}$ of $\mathcal{V}$ composed of eigenvectors of $\mathbf{L}$, i.e. by vectors that are mutually orthogonal and that remain mutually orthogonal once transformed by $\mathbf{L}$. Such a basis is called the *normal basis*.

If $\lambda_i, i = 1, 2, 3$, are the eigenvalues of $\mathbf{L}$, then the components of $\mathbf{L}$ in $\mathcal{B}_N$ are

$$L_{ij} = \mathbf{u}_i \cdot \mathbf{L}\mathbf{u}_j = \mathbf{u}_i \cdot \lambda_j\mathbf{u}_j = \lambda_j\delta_{ij},$$

so finally in $\mathcal{B}_N$, we have

$$\mathbf{L} = \lambda_i\mathbf{e}_i \otimes \mathbf{e}_i,$$

[6]The proof of the spectral theorem is omitted here; the interested reader can find a proof of it in the classical text by Halmos, p. 155, see the suggested texts.

i.e. $\mathbf{L}$ is diagonal and is completely represented by its eigenvalues. In addition, it is easy to check that

$$I_1 = \lambda_1 + \lambda_2 + \lambda_3, \ I_2 = \lambda_1\lambda_2 + \lambda_2\lambda_3 + \lambda_3\lambda_1, \ I_3 = \lambda_1\lambda_2\lambda_3.$$

A tensor with a unique eigenvalue λ of multiplicity three is said to be *spherical*; in such a case, any basis of $\mathcal{V}$ is $\mathcal{B}_N$ and

$$\mathbf{L} = \lambda\mathbf{I}.$$

Eigenvalues and eigenvectors also have another important property: Let us consider the quadratic form $\omega := \mathbf{v} \cdot \mathbf{L}\mathbf{v}, \ \forall \mathbf{v} \in \mathcal{S}$, defined by a symmetric tensor $\mathbf{L}$. We look for the directions $\mathbf{v} \in \mathcal{S}$, whereupon ω is stationary. Then, we have to solve the constrained problem

$$\nabla_{\mathbf{v}}(\mathbf{v} \cdot \mathbf{L}\mathbf{v}) = \mathbf{o}, \quad \mathbf{v} \in \mathcal{S}.$$

Using Lagrange's multiplier technique, we solve the equivalent problem

$$\nabla_{(\mathbf{v},\lambda)}(\mathbf{v} \cdot \mathbf{L}\mathbf{v} - \lambda(\mathbf{v}^2 - 1)) = 0,$$

which restitutes the equation

$$\mathbf{L}\mathbf{v} = \lambda\mathbf{v}$$

and the constraint $|\mathbf{v}| = 1$. The above equation is exactly the one defining the eigenvalue problem of $\mathbf{L}$: The stationary values (i.e. the maximum and minimum) of ω hence correspond to two eigenvalues of $\mathbf{L}$ and the directions $\mathbf{v}$, whereupon the stationarity coincides with the respective eigenvectors.

Two tensors $\mathbf{A}$ and $\mathbf{B}$ are said to be *coaxial* if they have the same normal basis $\mathcal{B}_N$, i.e. if they share the same eigenvectors. Let $\mathbf{u}$ be an eigenvector of $\mathbf{A}$, relative to the eigenvalue λ_A, and of $\mathbf{B}$, relative to λ_B. Then,

$$\mathbf{ABu} = \mathbf{A}\lambda_B\mathbf{u} = \lambda_B\mathbf{Au} = \lambda_A\lambda_B\mathbf{u} = \lambda_A\mathbf{Bu} = \mathbf{B}\lambda_A\mathbf{u} = \mathbf{BAu},$$

which shows, on the one hand, that $\mathbf{Bu}$ is also an eigenvector of $\mathbf{A}$, relative to the same eigenvalue λ_A; in the same way, of course, $\mathbf{Au}$ is an eigenvector of $\mathbf{B}$ relative to λ_B. In other words, this shows that $\mathbf{B}$ leaves unchanged any proper space of $\mathbf{A}$ and vice versa. On the other hand, we see that, at least for what concerns the eigenvectors, two tensors commute if and only if they are coaxial. Because any vector can be written as a linear combination of the vectors of $\mathcal{B}_N$, and for the linearity of tensors, we have finally proved the following.

Theorem 12 (Commutation theorem). *Two tensors commute if and only if they are coaxial.*

2.9 Skew tensors and cross product

Because $\dim(\mathcal{V}) = \dim(Skw(\mathcal{V})) = 3$, an isomorphism can be established between $\mathcal{V}$ and $Skw(\mathcal{V})$, i.e. between vectors and skew tensors. We establish hence a way to associate in a unique way a vector to any skew tensor and inversely. For this purpose, we first introduce the following.

Theorem 13. *The spectrum of any tensor* $\mathbf{W} \in Skw(\mathcal{V})$ *is* $\{0\}$ *and the dimension of its proper space is 1.*

Proof. This theorem states that zero is the only real eigenvalue of any skew tensor and that its multiplicity is 1. In fact, let $\mathbf{w}$ be an eigenvector of $\mathbf{W}$ relative to the eigenvector λ. Then,

$$\begin{aligned}\lambda^2\mathbf{w}^2 &= \mathbf{W}\mathbf{w}\cdot\mathbf{W}\mathbf{w} = \mathbf{w}\cdot\mathbf{W}^\top\mathbf{W}\mathbf{w} = -\mathbf{w}\cdot\mathbf{W}\mathbf{W}\mathbf{w}\\ &= -\mathbf{w}\cdot\mathbf{W}(\lambda\mathbf{w}) = -\lambda\mathbf{w}\cdot\mathbf{W}\mathbf{w} = -\lambda^2\mathbf{w}^2 \iff \lambda = 0.\end{aligned}$$

Then, if $\mathbf{W} \neq \mathbf{O}$, its rank is necessarily 2 because $\det\mathbf{W} = 0\ \forall\mathbf{W} \in Skw(\mathcal{V})$; hence, the equation

$$\mathbf{W}\mathbf{w} = \mathbf{o} \tag{2.22}$$

has ∞^1 solutions, i.e. the multiplicity of λ is 1, which proves the theorem. □

The last equation also shows the way the isomorphism is constructed: In fact, using Eq. (2.22), it is easy to check that if $\mathbf{w} = (a, b, c)$, then

$$\mathbf{w} = (a, b, c) \iff \mathbf{W} = \begin{bmatrix} 0 & -c & b \\ c & 0 & -a \\ -b & a & 0 \end{bmatrix}. \tag{2.23}$$

The proper space of $\mathbf{W}$ is called the *axis of* $\mathbf{W}$, and it is indicated by $\mathcal{A}(\mathbf{W})$:

$$\mathcal{A}(\mathbf{W}) := \{\mathbf{u} \in \mathcal{V} |\ \mathbf{W}\mathbf{u} = \mathbf{o}\}.$$

The consequence of what is shown above is that $\dim \mathcal{A}(\mathbf{W}) = 1$. With regard to Eq. (2.23), one can easily check that the equation

$$\mathbf{u} \cdot \mathbf{u} = \frac{1}{2}\mathbf{W} \cdot \mathbf{W} \tag{2.24}$$

is satisfied only by $\mathbf{w}$ and by its opposite $-\mathbf{w}$. Because both these vectors belong to $\mathcal{A}(\mathbf{W})$, choosing one of them corresponds to choosing an orientation for $\mathcal{E}$, see the next section. We always make our choice according to Eq. (2.23), which fixes once and for all the isomorphism between $\mathcal{V}$ and $Skw(\mathcal{V})$ that corresponds to any vector $\mathbf{w}$ with one and only one *axial tensor* $\mathbf{W}$ and vice versa, any skew tensor $\mathbf{W}$ with a unique *axial vector* $\mathbf{w}$.

It is worth noting that the above isomorphism between the vector spaces $\mathcal{V}$ and $Skw(\mathcal{V})$ implies that to any linear combination of vectors $\mathbf{a}$ and $\mathbf{b}$ corresponds an equal linear combination of the corresponding axial tensors $\mathbf{W}_a$ and $\mathbf{W}_b$ and vice versa, i.e. $\forall a, b \in \mathbb{R}$,

$$\mathbf{w} = \alpha\mathbf{a} + \beta\mathbf{b} \iff \mathbf{W} = \alpha\mathbf{W}_a + \beta\mathbf{W}_b, \tag{2.25}$$

where $\mathbf{W}$ is the axial tensor of $\mathbf{w}$. Such a property is immediately checked using Eq. (2.23).

It is useful, for further development, to calculate the powers of $\mathbf{W}$:

$$\mathbf{W}^2 = \mathbf{W}\mathbf{W} = -\mathbf{W}^\top(-\mathbf{W}^\top) = (\mathbf{W}\mathbf{W})^\top = (\mathbf{W}^2)^\top, \tag{2.26}$$

i.e. $\mathbf{W}^2$ is symmetric. Moreover, if we take $\mathbf{w} \in \mathcal{S}$, which is always possible because eigenvectors are determined to within an arbitrary multiplier,

$$\begin{aligned}\mathbf{W}^2\mathbf{u} &= \mathbf{W}\mathbf{W}\mathbf{u} = \mathbf{w} \times (\mathbf{w} \times \mathbf{u}) = \mathbf{w} \cdot \mathbf{u}\mathbf{w} - \mathbf{w} \cdot \mathbf{w}\mathbf{u} \\ &= -(\mathbf{I} - \mathbf{w} \otimes \mathbf{w})\mathbf{u} \ \Rightarrow\ \mathbf{W}^2 = -(\mathbf{I} - \mathbf{w} \otimes \mathbf{w});\end{aligned} \tag{2.27}$$

we remark that $\mathbf{W}^2\mathbf{u}$ gives the opposite of the projection of any vector $\mathbf{u} \in \mathcal{V}$ onto the direction orthogonal to $\mathbf{w}$, see Exercise 2.14.

Applying recursively the previous results,

$$\begin{aligned}\mathbf{W}^3 &= \mathbf{W}\mathbf{W}^2 = -\mathbf{W}(\mathbf{I} - \mathbf{w} \otimes \mathbf{w}) = -\mathbf{W} + (\mathbf{W}\mathbf{w}) \otimes \mathbf{w} = -\mathbf{W}, \\ \mathbf{W}^4 &= \mathbf{W}\mathbf{W}^3 = -\mathbf{W}^2, \\ \mathbf{W}^5 &= \mathbf{W}\mathbf{W}^4 = -\mathbf{W}^3,\end{aligned} \tag{2.28}$$

etc.

An important property of any couple axial tensor $\mathbf{W}$ – axial vector $\mathbf{w} \in \mathcal{S}$ is

$$\mathbf{W}\mathbf{W} = -\frac{1}{2}|\mathbf{W}|^2(\mathbf{I} - \mathbf{w} \otimes \mathbf{w}), \tag{2.29}$$

while Eq. (2.24) can be generalized to any two axial couples $\mathbf{w}_1, \mathbf{W}_1$ and $\mathbf{w}_2, \mathbf{W}_2$:

$$\mathbf{w}_1 \cdot \mathbf{w}_2 = \frac{1}{2}\mathbf{W}_1 \cdot \mathbf{W}_2.$$

The proof of these two last properties is rather easy and left to the reader.

We define the *cross product* of two vectors $\mathbf{a}$ and $\mathbf{b}$ the vector

$$\mathbf{a} \times \mathbf{b} = \mathbf{W}_a\mathbf{b},$$

where $\mathbf{W}_a$ is the axial tensor of $\mathbf{a}$. If $\mathbf{a} = (a_1, a_2, a_3)$ and $\mathbf{b} = (b_1, b_2, b_3)$, then by Eq. (2.23), we get

$$\mathbf{a} \times \mathbf{b} = (a_2b_3 - a_3b_2, a_3b_1 - a_1b_3, a_1b_2 - a_2b_1).$$

It is immediate to check that such a result can also be obtained using Ricci's alternator,

$$\mathbf{a} \times \mathbf{b} = \epsilon_{ijk}a_jb_k\mathbf{e}_i, \tag{2.30}$$

or even by computing the symbolic determinant,

$$\mathbf{a} \times \mathbf{b} = \det \begin{bmatrix} \mathbf{e}_1 & \mathbf{e}_2 & \mathbf{e}_3 \\ a_1 & a_2 & a_3 \\ b_1 & b_2 & b_3 \end{bmatrix}.$$

The cross product is bilinear: $\forall \mathbf{a}, \mathbf{b}, \mathbf{u} \in \mathcal{V},\ \alpha, \beta \in \mathbb{R}$,

$$(\alpha\mathbf{a} + \beta\mathbf{b}) \times \mathbf{u} = \alpha\mathbf{a} \times \mathbf{u} + \beta\mathbf{b} \times \mathbf{u},$$
$$\mathbf{u} \times (\alpha\mathbf{a} + \beta\mathbf{b}) = \alpha\mathbf{u} \times \mathbf{a} + \beta\mathbf{u} \times \mathbf{b}.$$

In fact, the first equation above is a consequence of Eq. (2.25), while the second one is a simple application to axial tensors of the same definition of tensor.

Three important results concerning the cross product are stated by the following theorems.

Theorem 14 (Condition of parallelism). *Two vectors* $\mathbf{a}$ *and* $\mathbf{b}$ *are parallel, i.e.* $\mathbf{b} = k\mathbf{a}$, $k \in \mathbb{R} \iff$

$$\mathbf{a} \times \mathbf{b} = \mathbf{o}.$$

Proof. This property is actually a consequence of the fact that any eigenvalue of a tensor is determined to within a multiplier:

$$\mathbf{a} \times \mathbf{b} = \mathbf{W}_a\mathbf{b} = \mathbf{o} \iff \mathbf{b} = k\mathbf{a}, \ k \in \mathbb{R},$$

for Theorem 13. □

Theorem 15 (Orthogonality property).

$$\mathbf{a} \times \mathbf{b} \cdot \mathbf{a} = \mathbf{a} \times \mathbf{b} \cdot \mathbf{b} = 0. \tag{2.31}$$

Proof.

$$\mathbf{a} \times \mathbf{b} \cdot \mathbf{a} = \mathbf{W}_a\mathbf{b} \cdot \mathbf{a} = \mathbf{b} \cdot \mathbf{W}_a^\top\mathbf{a} = -\mathbf{b} \cdot \mathbf{W}_a\mathbf{a} = -\mathbf{b} \cdot \mathbf{o} = 0,$$
$$\mathbf{a}\times\mathbf{b} \cdot \mathbf{b} = \mathbf{W}_a\mathbf{b} \cdot \mathbf{b} = \mathbf{b} \cdot \mathbf{W}_a^\top\mathbf{b} = -\mathbf{b} \cdot \mathbf{W}_a\mathbf{b} \iff \mathbf{a} \times \mathbf{b} \cdot \mathbf{b} = 0.$$

□

Theorem 16. $\mathbf{a} \times \mathbf{b}$ *is the axial vector of the tensor* $(\mathbf{b} \otimes \mathbf{a} - \mathbf{a} \otimes \mathbf{b})$.

Proof. First of all, by Eq. (2.6), we see that

$$(\mathbf{b} \otimes \mathbf{a} - \mathbf{a} \otimes \mathbf{b}) \in Skew(\mathcal{V}).$$

Then,

$$(\mathbf{b} \otimes \mathbf{a} - \mathbf{a} \otimes \mathbf{b})(\mathbf{a} \times \mathbf{b}) = \mathbf{a} \cdot \mathbf{a} \times \mathbf{b}\ \mathbf{b} - \mathbf{b} \cdot \mathbf{a} \times \mathbf{b}\ \mathbf{a} = 0$$

for Theorem 15. □

Theorem 16 allows us to show another important result about cross product.

Theorem 17 (Antisymmetry of the cross product). *The cross product is antisymmetric*:

$$\mathbf{a} \times \mathbf{b} = -\mathbf{b} \times \mathbf{a}\ \forall \mathbf{a}, \mathbf{b} \in \mathcal{V}. \tag{2.32}$$

Proof. Let $\mathbf{W}_1 = (\mathbf{b} \otimes \mathbf{a} - \mathbf{a} \otimes \mathbf{b})$ be the axial tensor of $\mathbf{a} \times \mathbf{b}$ and $\mathbf{W}_2 = (-\mathbf{a} \otimes \mathbf{b} + \mathbf{b} \otimes \mathbf{a})$ that of $-\mathbf{b} \times \mathbf{a}$. Evidently, $\mathbf{W}_1 = \mathbf{W}_2$, which implies Eq. (2.32) for the isomorphism between $\mathcal{V}$ and $Lin(\mathcal{V})$. □

This property and, again, Theorem 16 lets us derive the formula for the *double cross product*:

$$\begin{aligned}\mathbf{u}\times(\mathbf{v}\times\mathbf{w}) &= -(\mathbf{v}\times\mathbf{w})\times\mathbf{u}\\ &= -(\mathbf{w}\otimes\mathbf{v}-\mathbf{v}\otimes\mathbf{w})\mathbf{u}=\mathbf{u}\cdot\mathbf{w}\;\mathbf{v}-\mathbf{u}\cdot\mathbf{v}\;\mathbf{w}.\end{aligned} \tag{2.33}$$

Another interesting result concerns the *mixed product*:

$$\mathbf{u}\times\mathbf{v}\cdot\mathbf{w}=\mathbf{W}_u\mathbf{v}\cdot\mathbf{w}=-\mathbf{v}\cdot\mathbf{W}_u\mathbf{w}=-\mathbf{v}\cdot\mathbf{u}\times\mathbf{w}=\mathbf{w}\times\mathbf{u}\cdot\mathbf{v}, \tag{2.34}$$

and similarly,

$$\mathbf{u}\times\mathbf{v}\cdot\mathbf{w}=\mathbf{v}\times\mathbf{w}\cdot\mathbf{u}.$$

Using this last result, we can obtain a formula for the norm of a cross product; if $\mathbf{a}=a\;\mathbf{e}_a$ and $\mathbf{b}=b\;\mathbf{e}_b$, with $\mathbf{e}_a,\mathbf{e}_b\in\mathcal{S}$, are two vectors forming the angle θ, then

$$\begin{aligned}&(\mathbf{a}\times\mathbf{b})\cdot(\mathbf{a}\times\mathbf{b})=\mathbf{a}\times\mathbf{b}\cdot(\mathbf{a}\times\mathbf{b})=(\mathbf{a}\times\mathbf{b})\times\mathbf{a}\cdot\mathbf{b}\\ &\quad=-\mathbf{a}\times(\mathbf{a}\times\mathbf{b})\cdot\mathbf{b}=(-\mathbf{a}\cdot\mathbf{b}\;\mathbf{a}+\mathbf{a}^2\;\mathbf{b})\cdot\mathbf{b}=\mathbf{b}\cdot(\mathbf{a}^2\mathbf{I}-\mathbf{a}\otimes\mathbf{a})\mathbf{b}\\ &\quad=a^2\;\mathbf{b}\cdot(\mathbf{I}-\mathbf{e}_a\otimes\mathbf{e}_a)\mathbf{b}=a^2b^2\;\mathbf{e}_b\cdot(\mathbf{I}-\mathbf{e}_a\otimes\mathbf{e}_a)\mathbf{e}_b\\ &\quad=a^2b^2(1-\cos^2\theta)=a^2b^2\sin^2\theta\rightarrow|\mathbf{a}\times\mathbf{b}|=ab\sin\theta.\end{aligned} \tag{2.35}$$

So, the norm of a cross product can be interpreted, geometrically, as the area of the parallelogram spanned by the two vectors. As a consequence, the absolute value of the mixed product (2.34) measures the volume of the prism delimited by three non-coplanar vectors, cf. Fig. 2.1.

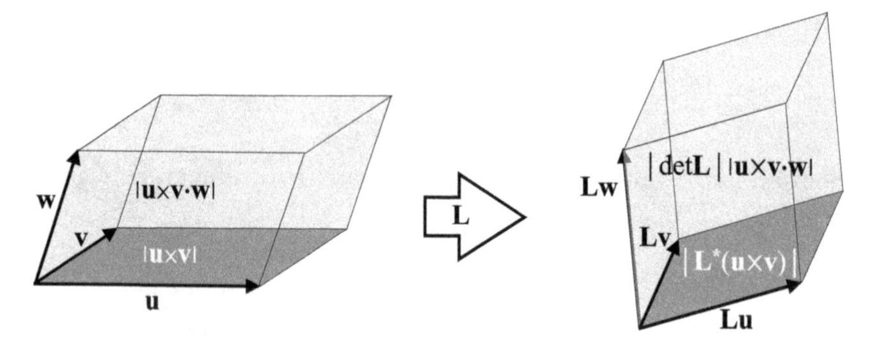

Figure 2.1: Geometrical meaning of the cross and mixed products before (left) and after (right) the application of a tensor $\mathbf{L}$ on the vectors $\mathbf{u},\mathbf{v},\mathbf{w}$.

Because the cross product is antisymmetric and the scalar one is symmetric, it is easy to check that the form

$$\beta(\mathbf{u}, \mathbf{v}, \mathbf{w}) = \mathbf{u} \times \mathbf{v} \cdot \mathbf{w}$$

is a skew trilinear form. Then, by Eq. (2.11), we get

$$\mathbf{Lu} \times \mathbf{Lv} \cdot \mathbf{Lw} = \det \mathbf{L}\ \mathbf{u} \times \mathbf{v} \cdot \mathbf{w}. \tag{2.36}$$

Following the interpretation given above for the absolute value of the mixed product, we can conclude that $|\det \mathbf{L}|$ can be interpreted as a coefficient of volume expansion[7] cf. again Fig. 2.1. A geometrical interpretation can then be given to the case of a non-invertible tensor, i.e. of $\det \mathbf{L} = 0$: It crushes a prism into a flat region (the three original vectors become coplanar, i.e. linearly dependent).

The *adjugate* of $\mathbf{L}$ is the tensor

$$\mathbf{L}^* := (\det \mathbf{L})\mathbf{L}^{-\top}.$$

From Eq. (2.36), we get hence

$$\det \mathbf{L}\ \mathbf{u} \times \mathbf{v} \cdot \mathbf{w} = \mathbf{Lu} \times \mathbf{Lv} \cdot \mathbf{Lw} = \mathbf{L}^\top(\mathbf{Lu} \times \mathbf{Lv}) \cdot \mathbf{w} \quad \forall \mathbf{w}$$

$$\Rightarrow \mathbf{Lu} \times \mathbf{Lv} = \mathbf{L}^*(\mathbf{u} \times \mathbf{v}).$$

2.10 Orientation of a basis

It is immediate to observe that a basis $\mathcal{B} = \{\mathbf{e}_1, \mathbf{e}_2, \mathbf{e}_3\}$ can be oriented in two opposite ways[8]: For example, once two unit mutually orthogonal vectors $\mathbf{e}_1$ and $\mathbf{e}_2$ are chosen, there are two opposite unit vectors perpendicular to both $\mathbf{e}_1$ and $\mathbf{e}_2$ that can be chosen to form $\mathcal{B}$.

We say that $\mathcal{B}$ is *positively oriented* or *right-handed* if

$$\mathbf{e}_1 \times \mathbf{e}_2 \cdot \mathbf{e}_3 = 1,$$

while $\mathcal{B}$ is *negatively oriented* or *left-handed* if

$$\mathbf{e}_1 \times \mathbf{e}_2 \cdot \mathbf{e}_3 = -1.$$

Schematically, a right-handed basis is represented in Fig. 2.2, where a left-handed basis is represented too with a dashed $\mathbf{e}_3$.

---

[7] This result is classical and fundamental for the analysis of deformation in continuum mechanics.

[8] It is evident that this is also true for one- and two-dimensional vector spaces.

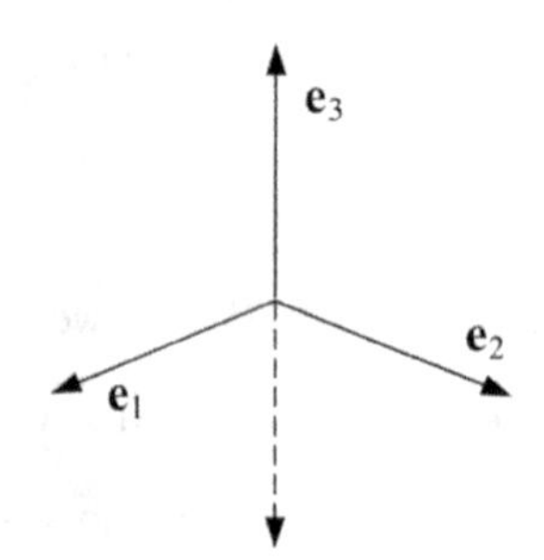

Figure 2.2: Right- and left-handed bases.

With a right-handed basis, by definition, the axial tensors of the three vectors of the basis are

$$\begin{aligned}\mathbf{W}_{\mathbf{e}_1} &= \mathbf{e}_3 \otimes \mathbf{e}_2 - \mathbf{e}_2 \otimes \mathbf{e}_3,\\ \mathbf{W}_{\mathbf{e}_2} &= \mathbf{e}_1 \otimes \mathbf{e}_3 - \mathbf{e}_3 \otimes \mathbf{e}_1,\\ \mathbf{W}_{\mathbf{e}_3} &= \mathbf{e}_2 \otimes \mathbf{e}_1 - \mathbf{e}_1 \otimes \mathbf{e}_2.\end{aligned}$$

2.11 Rotations

In the previous chapter, we have seen that the elements of $\mathcal{V}$ represent translations over $\mathcal{E}$. A *rotation*, i.e. a rigid rotation of the space, is an operation that transforms any two vectors $\mathbf{u}, \mathbf{v} \in \mathcal{V}$ into two other vectors $\hat{\mathbf{u}}, \hat{\mathbf{v}} \in \mathcal{V}$ in such a way that

$$u = \hat{u}, \quad v = \hat{v}, \quad \mathbf{u} \cdot \mathbf{v} = \hat{\mathbf{u}} \cdot \hat{\mathbf{v}}, \tag{2.37}$$

i.e. a rotation is a transformation that preserves norms and angles. Because a rotation is a transformation from $\mathcal{V}$ to $\mathcal{V}$, rotations are tensors, so we can write

$$\hat{\mathbf{v}} = \mathbf{R}\mathbf{v},$$

with $\mathbf{R}$ the *rotation tensor* or simply *rotation*.

Conditions (2.37) impose some restrictions on $\mathbf{R}$:

$$\hat{\mathbf{u}} \cdot \hat{\mathbf{v}} = \mathbf{R}\mathbf{u} \cdot \mathbf{R}\mathbf{v} = \mathbf{u} \cdot \mathbf{R}^\top \mathbf{R}\mathbf{v} = \mathbf{u} \cdot \mathbf{v} \iff \mathbf{R}^\top \mathbf{R} = \mathbf{I} = \mathbf{R}\mathbf{R}^\top.$$

A tensor that preserves the angles belongs to $Orth(\mathcal{V})$, the subspace of *orthogonal tensors*; we leave to the reader the proof that $Orth(\mathcal{V})$ is actually a subspace of $Lin(\mathcal{V})$. Replacing in the above equation $\mathbf{v}$ with $\mathbf{u}$ shows immediately that an orthogonal tensor also preserves the norms. By the uniqueness of the inverse, we see that

$$\mathbf{R} \in Orth(\mathcal{V}) \iff \mathbf{R}^{-1} = \mathbf{R}^\top.$$

The above condition is not sufficient to characterize a rotation; in fact, a rotation must transform a right-handed basis into another right-handed basis, i.e. it must *preserve the orientation of the space*. This means that it must be

$$\hat{\mathbf{e}}_1 \times \hat{\mathbf{e}}_2 \cdot \hat{\mathbf{e}}_3 = \mathbf{R}\mathbf{e}_1 \times \mathbf{R}\mathbf{e}_2 \cdot \mathbf{R}\mathbf{e}_3 = \mathbf{e}_1 \times \mathbf{e}_2 \cdot \mathbf{e}_3.$$

By Eq. (2.36), we get hence the condition[9]

$$\det \mathbf{R}(\mathbf{e}_1 \times \mathbf{e}_2 \cdot \mathbf{e}_3) = \mathbf{e}_1 \times \mathbf{e}_2 \cdot \mathbf{e}_3 \iff \det \mathbf{R} = 1.$$

The tensors of $Orth(\mathcal{V})$ that have a determinant equal to 1 form the subspace of *proper rotations* or simply *rotations*, indicated by $Orth(\mathcal{V})^+$ or also by $SO(3)$. Only tensors of $Orth(\mathcal{V})^+$ represent rigid rotations of $\mathcal{E}$[10].

Theorem 18. *Each tensor* $\mathbf{R} \in Orth(\mathcal{V})$ *has the eigenvalue* ± 1, *with* $+1$ *for rotations.*

Proof. Let $\mathbf{u}$ be an eigenvector of $\mathbf{R} \in Orth(\mathcal{V})$ corresponding to the eigenvalue λ. Because $\mathbf{R}$ preserves the norm, we have

$$\mathbf{R}\mathbf{u} \cdot \mathbf{R}\mathbf{u} = \lambda^2 \mathbf{u}^2 = \mathbf{u}^2 \ \rightarrow \ \lambda^2 = 1.$$

We must now prove that there exists at least one real eigenvector λ. To this end, we consider the characteristic equation

$$f(\lambda) = \lambda^3 + k_1\lambda^2 + k_2\lambda + k_3 = 0,$$

whose coefficients k_i are real-valued because $\mathbf{R}$ has real-valued components. It is immediate to recognize that

$$\lim_{\lambda \to \pm\infty} f(\lambda) = \pm\infty.$$

---

[9]From the condition $\mathbf{R}^\top\mathbf{R} = \mathbf{I}$ and through Eq. (2.16) and the theorem of Binet, we recognize immediately that $\det \mathbf{R} = \pm 1 \ \ \forall \mathbf{R} \in Orth(\mathcal{V})$.

[10]A tensor $\mathbf{S} \in Orth(\mathcal{V})$ such that $\det \mathbf{S} = -1$ represents a transformation that changes the orientation of the space, like mirror symmetries do, see Section 2.12.

So, because $f(\lambda)$ is a real-valued continuous function, actually a polynomial of λ, there exists at least one $\lambda_1 \in \mathbb{R}$ such that

$$f(\lambda_1) = 0.$$

In addition, we already know that $\forall \mathbf{R} \in Orth(\mathcal{V}), \det \mathbf{R} = \pm 1$ and that, if $\lambda_i, i = 1, 2, 3$ are the eigenvalues of $\mathbf{R}$, then $\det \mathbf{R} = \lambda_1 \lambda_2 \lambda_3$. Hence, the following two are the possible cases:

(i) $\lambda_1 \in \mathbb{R}$ and $\lambda_2, \lambda_3 \in \mathbb{C}$, with $\lambda_3 = \overline{\lambda}_2$, the complex conjugate of λ_2;
(ii) $\lambda_i \in \mathbb{R}\ \forall i = 1, 2, 3$.

Let us consider the case of $\mathbf{R} \in Orth(\mathcal{V})^+$, i.e. a (proper) rotation $\rightarrow \det \mathbf{R} = 1$. Then, in the first case above,

$$\det \mathbf{R} = \lambda_1 \lambda_2 \overline{\lambda}_2 = \lambda_1 [\Re^2(\lambda_2) + \Im^2(\lambda_2)].$$

But

$$\Re^2(\lambda_2) + \Im^2(\lambda_2) = 1$$

because it is the square of the modulus of the complex eigenvalue λ_2. So, in this case,

$$\det \mathbf{R} = 1 \iff \lambda_1 = 1.$$

In the second case, $\lambda_i \in \mathbb{R}\ \forall i = 1, 2, 3$, either $\lambda_1 > 0, \lambda_2, \lambda_3 < 0$ or all of them are positive. Because the modulus of each eigenvalue must be equal to 1, either $\lambda_1 = 1$ or $\lambda_i = 1\ \forall i = 1, 2, 3$ (in this case, $\mathbf{R} = \mathbf{I}$).

Following the same steps, one can easily show that $\forall \mathbf{S} \in Orth(\mathcal{V})$ with $\det \mathbf{S} = -1$, there exists at least one real eigenvalue $\lambda_1 = -1$. □

Generally, a rotation tensor rotates the basis $\mathcal{B} = \{\mathbf{e}_1, \mathbf{e}_2, \mathbf{e}_3\}$ into the basis $\hat{\mathcal{B}} = \{\hat{\mathbf{e}}_1, \hat{\mathbf{e}}_2, \hat{\mathbf{e}}_3\}$:

$$\mathbf{R}\mathbf{e}_i = \hat{\mathbf{e}}_i \quad \forall i = 1, 2, 3 \ \Rightarrow \ R_{ij} = \mathbf{e}_i \cdot \mathbf{R}\mathbf{e}_j = \mathbf{e}_i \cdot \hat{\mathbf{e}}_j. \tag{2.38}$$

This result actually means that the jth column of $\mathbf{R}$ is composed of the components in the basis $\mathcal{B}$ of the vector $\hat{\mathbf{e}}_j$ of $\hat{\mathcal{B}}$. Because the two bases are orthonormal, such components are the director cosines of the axes of $\hat{\mathcal{B}}$ with respect to $\mathcal{B}$.

Geometrically, any rotation is characterized by an *axis of rotation* $\mathbf{w}$, $|\mathbf{w}| = 1$, and by an *amplitude* φ, i.e. the angle through which the space is rotated about $\mathbf{w}$. By definition, $\mathbf{w}$ is the (only) vector that is left unchanged by $\mathbf{R}$, i.e.

$$\mathbf{R}\mathbf{w} = \mathbf{w},$$

or, in other words, it is the eigenvector corresponding to the eigenvalue $+1$.

The question is then: How can a rotation tensor $\mathbf{R}$ be expressed by means of its geometrical parameters, $\mathbf{w}$ and φ? To this end, we have a fundamental theorem.

Theorem 19 (Euler's rotation representation theorem). $\forall \mathbf{R} \in Orth(\mathcal{V})^+$,

$$\mathbf{R} = \mathbf{I} + \sin\varphi \mathbf{W} + (1 - \cos\varphi)\mathbf{W}^2, \tag{2.39}$$

with φ *the rotation's amplitude and* $\mathbf{W}$ *the axial tensor of the rotation axis* $\mathbf{w}$.

Proof. We observe preliminarily that

$$\mathbf{R}\mathbf{w} = \mathbf{I}\mathbf{w} + \sin\varphi \mathbf{W}\mathbf{w} + (1 - \cos\varphi)\mathbf{W}\mathbf{W}\mathbf{w} = \mathbf{I}\mathbf{w} = \mathbf{w}, \tag{2.40}$$

i.e. that Eq. (2.39) actually defines a transformation that leaves unchanged the axis $\mathbf{w}$, like a rotation about $\mathbf{w}$ must do, and that $+1$ is an eigenvalue of $\mathbf{R}$.

We need now to prove that Eq. (2.39) actually represents a rotation tensor, i.e. we must prove that

$$\mathbf{R}\mathbf{R}^\top = \mathbf{I}, \quad \det \mathbf{R} = 1.$$

Through Eq. (2.28), we get

$$\begin{aligned}
\mathbf{R}\mathbf{R}^\top &= (\mathbf{I} + \sin\varphi \mathbf{W} + (1 - \cos\varphi)\mathbf{W}^2)(\mathbf{I} + \sin\varphi \mathbf{W} + (1 - \cos\varphi)\mathbf{W}^2)^\top \\
&= (\mathbf{I} + \sin\varphi \mathbf{W} + (1 - \cos\varphi)\mathbf{W}^2)(\mathbf{I} - \sin\varphi \mathbf{W} + (1 - \cos\varphi)\mathbf{W}^2) \\
&= \mathbf{I} + 2(1 - \cos\varphi)\mathbf{W}^2 - \sin^2\varphi \mathbf{W}^2 + (1 - \cos\varphi)^2\mathbf{W}^4 \\
&= \mathbf{I} + 2(1 - \cos\varphi)\mathbf{W}^2 - \sin^2\varphi \mathbf{W}^2 - (1 - \cos\varphi)^2\mathbf{W}^2 = \mathbf{I}.
\end{aligned}$$

Then, through Eq. (2.27), we obtain

$$\begin{aligned}
\mathbf{R} &= \mathbf{I} + \sin\varphi \mathbf{W} + (1 - \cos\varphi)\mathbf{W}^2 \\
&= \mathbf{I} + \sin\varphi \mathbf{W} - (1 - \cos\varphi)(\mathbf{I} - \mathbf{w} \otimes \mathbf{w}) \\
&= \cos\varphi \mathbf{I} + \sin\varphi \mathbf{W} + (1 - \cos\varphi)\mathbf{w} \otimes \mathbf{w}.
\end{aligned} \tag{2.41}$$

To go on, we need to express $\mathbf{W}$ and $\mathbf{w}\otimes\mathbf{w}$; if $\mathbf{w}=(w_1,w_2,w_3)$, then by Eq. (2.23), we have

$$\mathbf{W}=\begin{bmatrix}0 & -w_3 & w_2\\ w_3 & 0 & -w_1\\ -w_2 & w_1 & 0\end{bmatrix},$$

and by Eq. (2.2),

$$\mathbf{w}\otimes\mathbf{w}=\begin{bmatrix}w_1^2 & w_1w_2 & w_1w_3\\ w_1w_2 & w_2^2 & w_2w_3\\ w_1w_3 & w_2w_3 & w_3^2\end{bmatrix},$$

which on injecting into Eq. (2.41) gives

$$\mathbf{R}=\begin{bmatrix}\cos\varphi+(1-\cos\varphi)w_1^2 & -w_3\sin\varphi+w_1w_2(1-\cos\varphi) & w_2\sin\varphi+w_1w_3(1-\cos\varphi)\\ w_3\sin\varphi+w_1w_2(1-\cos\varphi) & \cos\varphi+(1-\cos\varphi)w_2^2 & -w_1\sin\varphi+w_2w_3(1-\cos\varphi)\\ -w_2\sin\varphi+w_1w_3(1-\cos\varphi) & w_1\sin\varphi+w_2w_3(1-\cos\varphi) & \cos\varphi+(1-\cos\varphi)w_3^2\end{bmatrix}. \tag{2.42}$$

This formula gives $\mathbf{R}$ as a function exclusively of $\mathbf{w}$ and φ, the geometrical elements of the rotation. Then,

$$\det\mathbf{R}=(w^2+(1-w^2)\cos\varphi)(\cos^2\varphi+w^2\sin^2\varphi),$$

and because $w=1,\det\mathbf{R}=1$, which proves that Eq. (2.39) actually represents a rotation.

We eventually need to prove that Eq. (2.39) represents the rotation about $\mathbf{w}$ of amplitude φ. To this end, we choose an orthonormal basis $\mathcal{B}=\{\mathbf{e}_1,\mathbf{e}_2,\mathbf{e}_3\}$ of $\mathcal{V}$ such that $\mathbf{w}=\mathbf{e}_3$, i.e. we analyze the particular case of a rotation of amplitude φ about $\mathbf{e}_3$. This is always possible thanks to the arbitrariness of the basis of $\mathcal{V}$. In such a case, Eq. (2.38) gives

$$\mathbf{R}=\begin{bmatrix}\cos\varphi & -\sin\varphi & 0\\ \sin\varphi & \cos\varphi & 0\\ 0 & 0 & 1\end{bmatrix}. \tag{2.43}$$

Moreover,

$$\mathbf{W}=\begin{bmatrix}0&-1&0\\1&0&0\\0&0&0\end{bmatrix},\quad \mathbf{w}\otimes\mathbf{w}=\begin{bmatrix}0&0&0\\0&0&0\\0&0&1\end{bmatrix},$$

$$\mathbf{W}^2=-(\mathbf{I}-\mathbf{w}\otimes\mathbf{w})=\begin{bmatrix}-1&0&0\\0&-1&0\\0&0&0\end{bmatrix}.$$

Hence,

$$\begin{aligned}\mathbf{I}+\sin\varphi\mathbf{W}+(1-\cos\varphi)\mathbf{W}^2&=\begin{bmatrix}1&0&0\\0&1&0\\0&0&1\end{bmatrix}+\sin\varphi\begin{bmatrix}0&-1&0\\1&0&0\\0&0&0\end{bmatrix}\\+(1-\cos\varphi)\begin{bmatrix}-1&0&0\\0&-1&0\\0&0&0\end{bmatrix}&=\begin{bmatrix}\cos\varphi&-\sin\varphi&0\\\sin\varphi&\cos\varphi&0\\0&0&1\end{bmatrix}=\mathbf{R}.\end{aligned}\tag{2.44}$$

□

Equation (2.39) gives another result: To obtain the inverse of $\mathbf{R}$, it is sufficient to change the sign of φ. In fact, because $\mathbf{W}\in Skw(\mathcal{V})$ and through Eq. (2.26),

$$\begin{aligned}\mathbf{R}^{-1}=\mathbf{R}^\top&=(\mathbf{I}+\sin\varphi\mathbf{W}+(1-\cos\varphi)\mathbf{W}^2)^\top\\&=\mathbf{I}+\sin\varphi\mathbf{W}^\top+(1-\cos\varphi)(\mathbf{W}^2)^\top\\&=\mathbf{I}-\sin\varphi\mathbf{W}+(1-\cos\varphi)\mathbf{W}^2\\&=\mathbf{I}+\sin(-\varphi)\mathbf{W}+(1-\cos(-\varphi))\mathbf{W}^2.\end{aligned}$$

The knowledge of the inverse of a rotation also allows us to perform the operation of *change of basis*, i.e. to determine the components of a vector or of a tensor in a basis $\hat{\mathcal{B}}=\{\hat{\mathbf{e}}_1,\hat{\mathbf{e}}_2,\hat{\mathbf{e}}_3\}$ rotated with respect to an original basis $\mathcal{B}=\{\mathbf{e}_1,\mathbf{e}_2,\mathbf{e}_3\}$ by a rotation $\mathbf{R}$ (in the following equations, the symbol $\hat{}$ indicates a quantity specified in the basis $\hat{\mathcal{B}}$). Considering that

$$\mathbf{e}_i=\mathbf{R}^{-1}\hat{\mathbf{e}}_i=\mathbf{R}^\top\hat{\mathbf{e}}_i=R^\top_{hk}(\hat{\mathbf{e}}_h\otimes\hat{\mathbf{e}}_k)\hat{\mathbf{e}}_i=R^\top_{hk}\delta_{ki}\hat{\mathbf{e}}_h,$$

we get, for a vector $\mathbf{u}$,

$$\mathbf{u}=u_i\mathbf{e}_i=R^\top_{ki}u_i\hat{\mathbf{e}}_k,$$

i.e.

$$\hat{u}_k = R^\top_{ki} u_i \ \rightarrow \ \hat{\mathbf{u}} = \mathbf{R}^\top \mathbf{u}.$$

We remark that, because $\mathbf{R}^\top = \mathbf{R}^{-1}$, the operation of change of basis is just the opposite of the rotation of the space (and actually, we have seen that it is sufficient to take the opposite of φ in Eq. (2.39) to get $\mathbf{R}^{-1}$).

For a second-rank tensor $\mathbf{L}$, we get

$$\mathbf{L} = L_{ij} \mathbf{e}_i \otimes \mathbf{e}_j = L_{ij} R^\top_{mi} \hat{\mathbf{e}}_m \otimes R^\top_{nj} \hat{\mathbf{e}}_n = R^\top_{mi} R^\top_{nj} L_{ij} \hat{\mathbf{e}}_m \otimes \hat{\mathbf{e}}_n,$$

i.e.

$$\hat{L}_{mn} = R^\top_{mi} R^\top_{nj} L_{ij} \ = R^\top_{mi} L_{ij} R_{jn} \ \rightarrow \ \hat{\mathbf{L}} = \mathbf{R}^\top \mathbf{L} \mathbf{R}.$$

We remark something that is typical of tensors: The components of a r-rank tensor in a rotated basis $\hat{\mathcal{B}}$ depend upon the rth powers of the director cosines of the axes of $\hat{\mathcal{B}}$, i.e. on the rth powers of the components R_{ij} of $\mathbf{R}$.

If a rotation tensor is known through its Cartesian components in a given basis $\mathcal{B}$, it is easy to calculate its geometrical elements: The rotation axis $\mathbf{w}$ is the eigenvector of $\mathbf{R}$ corresponding to the eigenvalue 1, so it is found by solving the equation

$$\mathbf{R}\mathbf{w} = \mathbf{w}$$

and then normalizing it, while the rotation amplitude φ can be found using (2.39) along with (2.27): Because the trace of a tensor is invariant, we get

$$\mathrm{tr}\mathbf{R} = 3 + (1 - \cos\varphi)\mathrm{tr}(-\mathbf{I} + \mathbf{w} \otimes \mathbf{w}) = 1 + 2\cos\varphi \ \rightarrow \ \varphi = \arccos\frac{\mathrm{tr}\mathbf{R} - 1}{2}.$$

It is interesting to consider the geometrical meaning of Eq. (2.39). For this purpose, we apply Eq. (2.39) to a vector $\mathbf{u}$, see Fig. 2.3:

$$\begin{aligned} \mathbf{R}\mathbf{u} &= (\mathbf{I} + \sin\varphi \mathbf{W} + (1 - \cos\varphi)\mathbf{W}^2)\mathbf{u} \\ &= \mathbf{u} + \sin\varphi \mathbf{w} \times \mathbf{u} + (1 - \cos\varphi)\mathbf{w} \times (\mathbf{w} \times \mathbf{u}). \end{aligned}$$

The rotated vector $\mathbf{R}\mathbf{u}$ is the sum of three vectors; in particular, $\sin\varphi\mathbf{W}\mathbf{u}$ is always orthogonal to $\mathbf{u}$, $\mathbf{w}$, and $(1 - \cos\varphi)\mathbf{W}^2\mathbf{u}$. If $\mathbf{u} \cdot \mathbf{w} = 0$, then $(1 - \cos\varphi)\mathbf{W}^2\mathbf{u}$ is also parallel to $\mathbf{u}$, see the sketch on the right in Fig. 2.3.

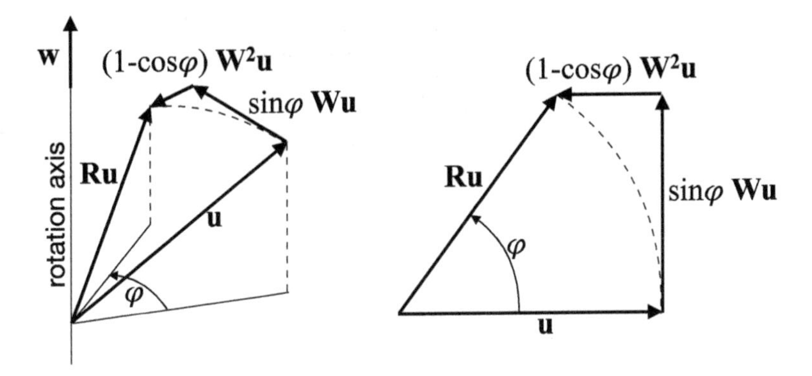

Figure 2.3: Rotation of a vector.

Let us consider now a composition of rotations. In particular, let us imagine that a vector $\mathbf{u}$ is rotated first by $\mathbf{R}_1$ around $\mathbf{w}_1$ through φ_1, then by $\mathbf{R}_2$ around $\mathbf{w}_2$ through φ_2. So, first, the vector $\mathbf{u}$ becomes the vector

$$\mathbf{u}_1 = \mathbf{R}_1\mathbf{u}.$$

Then, the vector $\mathbf{u}_1$ is rotated about $\mathbf{w}_2$ through φ_2 to become

$$\mathbf{u}_{12} = \mathbf{R}_2\mathbf{u}_1 = \mathbf{R}_2\mathbf{R}_1\mathbf{u}.$$

Let us now suppose that we change the order of the rotations: $\mathbf{R}_2$ first and then $\mathbf{R}_1$. The final result will be the vector

$$\mathbf{u}_{21} = \mathbf{R}_1\mathbf{R}_2\mathbf{u}. \tag{2.45}$$

Because the tensor product is not symmetric (i.e. it does not have the commutativity property), generally,[11]

$$\mathbf{u}_{12} \neq \mathbf{u}_{21}.$$

In other words, the order of the rotations matters: Changing the order of the rotations leads to a different final result. An example is shown in Fig. 2.4.

[11]We have seen, in Theorem 12, that two tensors commute $\Longleftrightarrow$ they are coaxial, i.e. if they have the same eigenvectors. Because the rotation axis is always a real eigenvector of a rotation tensor, if two tensors operate a rotation about different axes, they are not coaxial. Hence, the rotation tensors about different axes never commute.

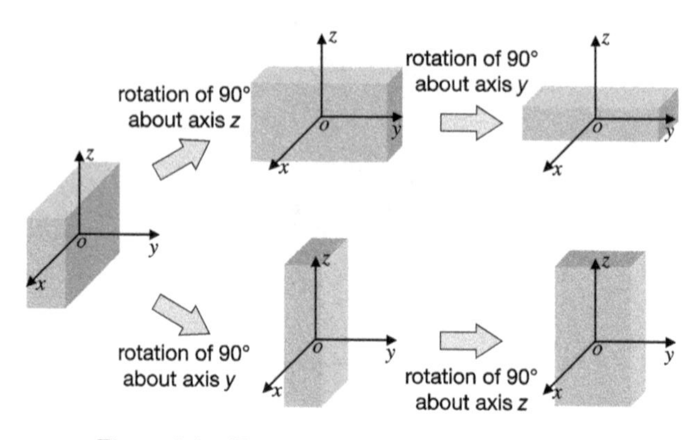

Figure 2.4: Non-commutativity of the rotations.

This is a fundamental difference between rotations and displacements that commute, see Fig. 1.2, because the composition of displacements is ruled by the sum of vectors:

$$\mathbf{w} = \mathbf{u} + \mathbf{v} = \mathbf{v} + \mathbf{u}. \tag{2.46}$$

This difference, which is a major point in physics, comes from the difference in the operators: vectors for the displacements and tensors for the rotations.

Any rotation can be specified by the knowledge of three parameters. This can be easily seen from Eq. (2.39): The parameters are the three components of $\mathbf{w}$ that are not independent because

$$w = |\mathbf{w}| = \sqrt{w_1^2 + w_2^2 + w_3^2} = 1$$

and by the amplitude angle φ. The choice of the parameters by which to express a rotation is not unique. Besides the use of the Cartesian components of $\mathbf{w}$ and φ, cf. Eq. (2.42), other choices are possible, let us see three of them:

(i) *Physical angles*: The rotation axis $\mathbf{w}$ is given through its spherical coordinates ψ, the *longitude*, $0 \leq \psi < 2\pi$, and θ, the *colatitude*, $0 \leq \theta \leq \pi$, see Fig. 2.5, the third parameter being the *rotation amplitude* φ. Then,

$$\mathbf{w} = (\sin\theta\cos\psi, \sin\theta\sin\psi, \cos\theta) \;\rightarrow\; \theta = \arccos w_3,\; \psi = \arctan\frac{w_2}{w_1},$$

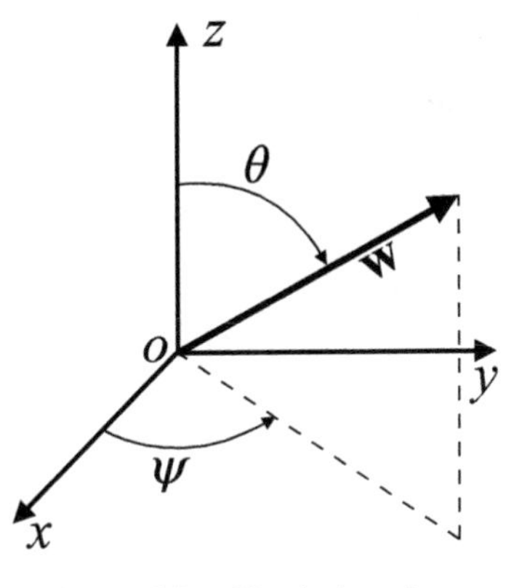

Figure 2.5: Physical angles.

and, Eq. (2.42),

$$\mathbf{R}=\begin{bmatrix} c\psi^2 s\theta^2+c\varphi(c\theta^2+s\psi^2 s\theta^2) & s\psi c\psi s\theta^2(1-c\varphi)-c\theta s\varphi & c\psi s\theta c\theta(1-c\varphi)+s\psi s\theta s\varphi \\ s\psi c\psi s\theta^2(1-c\varphi)+c\theta s\varphi & s\psi^2 s\theta^2+c\varphi(c\theta^2+c\psi^2 s\theta^2) & s\psi s\theta c\theta(1-c\varphi)-c\psi s\theta s\varphi \\ c\psi s\theta c\theta(1-c\varphi)-s\psi s\theta s\varphi & s\psi s\theta c\theta(1-c\varphi)+c\psi s\theta s\varphi & c\theta^2+c\varphi(c\psi^2 s\theta^2+s\psi^2 s\theta^2) \end{bmatrix},$$

where $c\psi=\cos\psi, s\psi=\sin\psi, c\theta=\cos\theta, s\theta=\sin\theta, c\varphi=\cos\varphi$, and $s\varphi=\sin\varphi$. We remark that all the components of $\mathbf{R}$ so expressed depend upon the first powers of the circular functions of φ. Hence, for what is said above, with this representation of the rotations, the components of a rotated r-rank tensor depend upon the rth power of the circular functions of φ, i.e. of the physical rotation, but not of ψ nor of θ.

(ii) *Euler's angles*: In this case, the three parameters are the amplitude of three particular rotations into which the rotation is decomposed. Such parameters are the angles ψ, the *precession*, θ, the *nutation*, and φ, the *proper rotation*, see Fig. 2.6. These three rotations are represented in Fig. 2.7. The first one, of amplitude ψ, is made about z to carry the axis x onto the *knots line* x_N, the line perpendicular to both the axes z and $\hat{z}$, and y onto $\overline{y}$; by Eq. (2.38), in the frame $\{x,y,z\}$, it is

$$\mathbf{R}_\psi=\begin{bmatrix}\cos\psi & -\sin\psi & 0\\ \sin\psi & \cos\psi & 0\\ 0 & 0 & 1\end{bmatrix}.$$

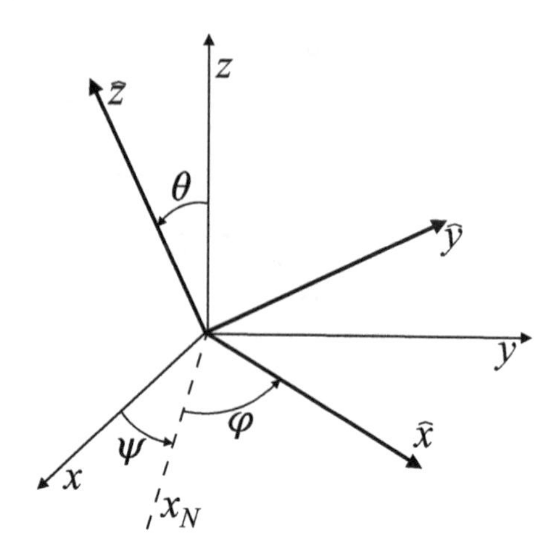

Figure 2.6: Euler's angles.

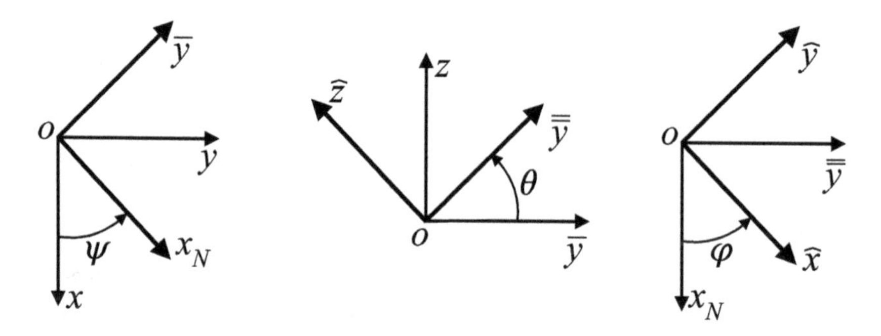

Figure 2.7: Euler's rotations, as seen from the respective axes of rotation.

The second one, of amplitude θ, is made about x_N to carry z onto $\hat{z}$; in the frame $\{x_N, \overline{y}, z\}$, it is

$$\mathbf{R}_\theta = \begin{bmatrix} 1 & 0 & 0 \\ 0 & \cos\theta & -\sin\theta \\ 0 & \sin\theta & \cos\theta \end{bmatrix},$$

while in the frame $\{x, y, z\}$,

$$\mathbf{R}_\theta^o = (\mathbf{R}_\psi^{-1})^\top \mathbf{R}_\theta \mathbf{R}_\psi^{-1} = \mathbf{R}_\psi \mathbf{R}_\theta \mathbf{R}_\psi^\top.$$

The last rotation, of amplitude φ, is made about $\hat{z}$ to carry x_N onto $\hat{x}$ and $\overline{\overline{y}}$ onto $\hat{y}$; in the frame $\{x_N, \overline{\overline{y}}, \hat{z}\}$, it is

$$\mathbf{R}_\varphi = \begin{bmatrix} \cos\varphi & -\sin\varphi & 0 \\ \sin\varphi & \cos\varphi & 0 \\ 0 & 0 & 1 \end{bmatrix},$$

while in $\{x, y, z\}$,

$$\mathbf{R}^o_\varphi = (\mathbf{R}_\psi^{-1})^\top (\mathbf{R}_\theta^{-1})^\top \mathbf{R}_\varphi \mathbf{R}_\theta^{-1} \mathbf{R}_\psi^{-1} = \mathbf{R}_\psi \mathbf{R}_\theta \mathbf{R}_\varphi \mathbf{R}_\theta^\top \mathbf{R}_\psi^\top.$$

Any vector $\mathbf{u}$ is transformed, by the global rotation, into the vector

$$\hat{\mathbf{u}} = \mathbf{R}\mathbf{u}.$$

But we can also write

$$\hat{\mathbf{u}} = \mathbf{R}^o_\varphi \overline{\overline{\mathbf{u}}},$$

where $\overline{\overline{\mathbf{u}}}$ is the vector transformed by the rotation $\mathbf{R}^o_\theta$,

$$\overline{\overline{\mathbf{u}}} = \mathbf{R}^o_\theta \overline{\mathbf{u}},$$

and $\overline{\mathbf{u}}$ is the vector transformed by the rotation $\mathbf{R}_\psi$:

$$\overline{\mathbf{u}} = \mathbf{R}_\psi \mathbf{u}.$$

Finally,

$$\hat{\mathbf{u}} = \mathbf{R}\mathbf{u} = \mathbf{R}^o_\varphi \mathbf{R}^o_\theta \mathbf{R}_\psi \mathbf{u} \;\rightarrow\; \mathbf{R} = \mathbf{R}^o_\varphi \mathbf{R}^o_\theta \mathbf{R}_\psi,$$

i.e. the global rotation tensor is obtained by composing, in the opposite order of execution of the rotations, the three tensors all expressed in the original basis. However,

$$\mathbf{R} = \mathbf{R}^o_\varphi \mathbf{R}^o_\theta \mathbf{R}_\psi = \mathbf{R}_\psi \mathbf{R}_\theta \mathbf{R}_\varphi \mathbf{R}_\theta^\top \mathbf{R}_\psi^\top \mathbf{R}_\psi \mathbf{R}_\theta \mathbf{R}_\psi^\top \mathbf{R}_\psi = \mathbf{R}_\psi \mathbf{R}_\theta \mathbf{R}_\varphi,$$

i.e. the global rotation tensor is also equal to the composition of the three rotations, in the order of execution, if the three rotations are expressed in their own particular bases. This result is general and not bounded to the Euler's rotations nor to three rotations.

Performing the tensor multiplications, we get

$$\mathbf{R} = \begin{bmatrix} \cos\psi\cos\varphi - \sin\psi\sin\varphi\cos\theta & -\cos\psi\sin\varphi - \sin\psi\cos\varphi\cos\theta & \sin\psi\sin\theta \\ \sin\psi\cos\varphi + \cos\psi\sin\varphi\cos\theta & -\sin\psi\sin\varphi + \cos\psi\cos\varphi\cos\theta & -\cos\psi\sin\theta \\ \sin\varphi\sin\theta & \cos\varphi\sin\theta & \cos\theta \end{bmatrix}.$$

The components of a vector $\mathbf{u}$ in the basis $\hat{\mathcal{B}}$ are then given by

$$\hat{\mathbf{u}} = \mathbf{R}^\top\mathbf{u} = \mathbf{R}_\varphi^\top\mathbf{R}_\theta^\top\mathbf{R}_\psi^\top\mathbf{u}$$

and those of a second-rank tensor by

$$\hat{\mathbf{L}} = \mathbf{R}^\top\mathbf{L}\mathbf{R} = \mathbf{R}_\varphi^\top\mathbf{R}_\theta^\top\mathbf{R}_\psi^\top\mathbf{L}\mathbf{R}_\psi\mathbf{R}_\theta\mathbf{R}_\varphi.$$

(iii) *Coordinate angles*: In this case, the rotation $\mathbf{R}$ is decomposed into three successive rotations α, β, and γ, respectively, about the axes x, y, and z of each rotation, i.e.

$$\mathbf{R} = \mathbf{R}_\alpha\mathbf{R}_\beta\mathbf{R}_\gamma$$

with

$$\mathbf{R}_\alpha = \begin{bmatrix} 1 & 0 & 0 \\ 0 & \cos\alpha & -\sin\alpha \\ 0 & \sin\alpha & \cos\alpha \end{bmatrix}, \quad \mathbf{R}_\beta = \begin{bmatrix} \cos\beta & 0 & -\sin\beta \\ 0 & 1 & 0 \\ \sin\beta & 0 & \cos\beta \end{bmatrix},$$

$$\mathbf{R}_\gamma = \begin{bmatrix} \cos\gamma & -\sin\gamma & 0 \\ \sin\gamma & \cos\gamma & 0 \\ 0 & 0 & 1 \end{bmatrix},$$

so finally,

$$\mathbf{R} = \begin{bmatrix} \cos\beta\cos\gamma & -\cos\beta\sin\gamma & -\sin\beta \\ \cos\alpha\sin\gamma - \sin\alpha\sin\beta\cos\gamma & \cos\alpha\cos\gamma + \sin\alpha\sin\beta\sin\gamma & -\sin\alpha\cos\beta \\ \sin\alpha\sin\gamma + \cos\alpha\sin\beta\cos\gamma & \sin\alpha\cos\gamma - \cos\alpha\sin\beta\sin\gamma & \cos\alpha\cos\beta \end{bmatrix}.$$

Let us now consider the case of *small rotations*, i.e. $|\varphi| \to 0$. In such a case,

$$\sin\varphi \simeq \varphi, \quad 1 - \cos\varphi \simeq 0,$$

and

$$\mathbf{R} \simeq \mathbf{I} + \varphi\mathbf{W},$$

i.e. in the small rotations approximation, any vector $\mathbf{u}$ is transformed into

$$\mathbf{R}\mathbf{u} \simeq (\mathbf{I} + \varphi\mathbf{W})\mathbf{u} = \mathbf{u} + \varphi\mathbf{w} \times \mathbf{u}, \tag{2.47}$$

i.e. by a skew tensor and not by a rotation tensor. The term $(1-\cos\varphi)\mathbf{W}^2\mathbf{u}$ has disappeared, as it is a higher-order infinitesimal quantity, and the term $\varphi\mathbf{w} \times \mathbf{u}$ is orthogonal to $\mathbf{u}$. Because $\varphi \to 0$, the arc is approximated by its tangent, the vector $\varphi\mathbf{w} \times \mathbf{u}$, see Fig. 2.8. Applying to Eq. (2.47) the procedure already seen for the composition of finite amplitude rotations, we get

$$\begin{aligned}
\mathbf{u}_1 &= \mathbf{R}_1\mathbf{u} = (\mathbf{I} + \varphi_1\mathbf{W}_1)\mathbf{u} = \mathbf{u} + \varphi_1\mathbf{w}_1 \times \mathbf{u},\\
\mathbf{u}_{21} &= \mathbf{R}_2\mathbf{u}_1 = (\mathbf{I} + \varphi_2\mathbf{W}_2)\mathbf{u}_1 = \mathbf{u}_1 + \varphi_2\mathbf{w}_2 \times \mathbf{u}_1\\
&= \mathbf{u} + \varphi_1\mathbf{w}_1 \times \mathbf{u} + \varphi_2\mathbf{w}_2 \times \mathbf{u}\\
&\quad + \varphi_1\varphi_2\mathbf{w}_2 \times (\mathbf{w}_1 \times \mathbf{u}).
\end{aligned}$$

If the order of the rotations is changed, the last term becomes $\varphi_1\varphi_2\mathbf{w}_1 \times (\mathbf{w}_2 \times \mathbf{u})$, which is, in general, different from $\varphi_1\varphi_2\mathbf{w}_2 \times (\mathbf{w}_1 \times \mathbf{u})$: To be

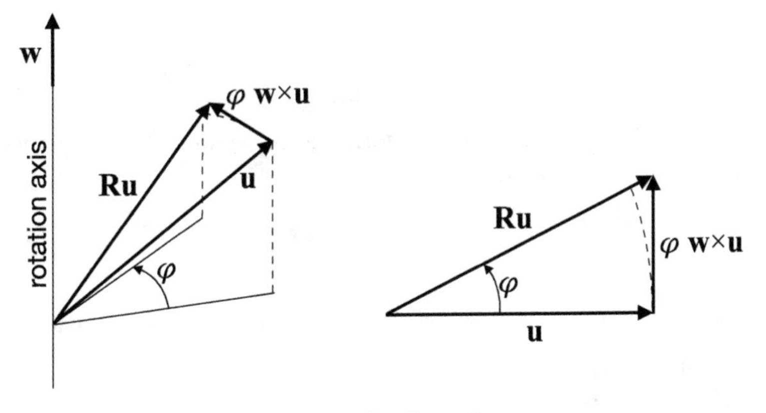

Figure 2.8: Small rotations.

precise, small rotations also do not commute.[12] However, for small rotations, $\varphi_1\varphi_2$ is negligible with respect to φ_1 and φ_2: In this approximation, small rotations commute. We remark that approximation (2.47) gives, for the displacements, a law that is quite similar to that of the velocities of the points of a rigid body:

$$\mathbf{v} = \mathbf{v}_0 + \boldsymbol{\omega} \times (p - o).$$

This is quite natural because

$$\omega = \frac{d\varphi}{dt},$$

i.e. a small amplitude rotation can be seen as a rotation made with finite angular velocity $\boldsymbol{\omega}$ in a small time interval dt.

2.12 Reflexions

Let us consider now tensors $\mathbf{S} \in Orth(\mathcal{V})$ that are not rotations, i.e. such that $\det \mathbf{S} = -1$. Let us call $\mathbf{S}$ an *improper rotation*. A particular improper rotation whose all eigenvalues are equal to -1 is the *inversion* or *reflexion tensor*:

$$\mathbf{S}_I = -\mathbf{I}.$$

The effect of $\mathbf{S}_I$ is to transform any basis $\mathcal{B}$ into the basis $-\mathcal{B}$, i.e. with all the basis vectors changed in orientation (or, equivalently, to change the sign of all the components of a vector). In other words, $\mathbf{S}_I$ changes the orientation of the space. This is also the effect of any other improper rotation $\mathbf{S}$ that can be decomposed into a proper rotation $\mathbf{R}$ followed by the reflexion $\mathbf{S}_I$[13]:

$$\mathbf{S} = \mathbf{S}_I\mathbf{R}. \tag{2.48}$$

Let $\mathbf{n} \in \mathcal{S}$, then

$$\mathbf{S}_R = \mathbf{I} - 2\mathbf{n} \otimes \mathbf{n} \tag{2.49}$$

is the tensor that operates the transformation of symmetry with respect to a plane orthogonal to $\mathbf{n}$. In fact,

$$\mathbf{S}_R\mathbf{n} = -\mathbf{n}, \quad \mathbf{S}_R\mathbf{m} = \mathbf{m} \quad \forall \mathbf{m} \in \mathcal{V} |\ \mathbf{m}\cdot\mathbf{n} = 0.$$

---

[12] This can happen for some vectors all the times when $\mathbf{w}_1 \cdot \mathbf{u} = \mathbf{w}_2 \cdot \mathbf{u}$, like for the case of a vector $\mathbf{u}$ orthogonal to both $\mathbf{w}_1$ and $\mathbf{w}_2$; however, this is no more than a curiosity, it has no importance in practice.

[13] The application of Binet's theorem shows immediately that $\det \mathbf{S} = -1$, while $\mathbf{S}_I\mathbf{R}(\mathbf{S}_I\mathbf{R})^\top = \mathbf{S}_I\mathbf{R}\mathbf{R}^\top\mathbf{S}_I^\top = -\mathbf{I}(-\mathbf{I})^\top = \mathbf{I}$: The decomposition in Eq. (2.48) actually gives an improper rotation.

$\mathbf{S}_R$ is an improper rotation; in fact, by Eq. (2.4),

$$\begin{aligned}(\mathbf{I}-2\mathbf{n}\otimes\mathbf{n})(\mathbf{I}-2\mathbf{n}\otimes\mathbf{n})^\top &= (\mathbf{I}-2\mathbf{n}\otimes\mathbf{n})(\mathbf{I}-2\mathbf{n}\otimes\mathbf{n})\\ &= \mathbf{I}-2\mathbf{n}\otimes\mathbf{n}-2\mathbf{n}\otimes\mathbf{n}+4(\mathbf{n}\otimes\mathbf{n})(\mathbf{n}\otimes\mathbf{n})=\mathbf{I},\end{aligned}$$

while by the same definition of trace and through Eqs. (2.13) and (2.17),

$$\begin{aligned}\det(\mathbf{I}-2\mathbf{n}\otimes\mathbf{n}) = 1-2\mathrm{tr}(\mathbf{n}\otimes\mathbf{n})+4\frac{\mathrm{tr}^2(\mathbf{n}\otimes\mathbf{n})-\mathrm{tr}(\mathbf{n}\otimes\mathbf{n})(\mathbf{n}\otimes\mathbf{n})}{2}\\ -8\det(\mathbf{n}\otimes\mathbf{n}) = -1.\end{aligned}$$

Let $\mathbf{S}=\mathbf{S}_I\mathbf{R}$ be an improper rotation, then

$$\begin{aligned}(\mathbf{S}\mathbf{u})\times(\mathbf{S}\mathbf{v}) &= (\mathbf{S}_I\mathbf{R}\mathbf{u})\times(\mathbf{S}_I\mathbf{R}\mathbf{v}) = \det(\mathbf{S}_I\mathbf{R})\left[(\mathbf{S}_I\mathbf{R})^{-1}\right]^\top(\mathbf{u}\times\mathbf{v})\\ &= \det\mathbf{S}_I\det\mathbf{R}(\mathbf{R}^{-1}\mathbf{S}_I^{-1})^\top(\mathbf{u}\times\mathbf{v})\\ &= -(-\mathbf{R}^{-1}\mathbf{I})^\top(\mathbf{u}\times\mathbf{v}) = \mathbf{R}(\mathbf{u}\times\mathbf{v}).\end{aligned}$$

The transformation by $\mathbf{S}$ of any vector $\mathbf{u}$ gives

$$\mathbf{S}\mathbf{u}=\mathbf{S}_I\mathbf{R}\mathbf{u}=-\mathbf{R}\mathbf{u},$$

i.e. it changes the orientation of the rotated vector; this is not the case when the same improper rotation transforms the vectors of a cross product: The rotated vector result of the cross product does not cause a change in orientation, i.e. the cross product is insensitive to a reflexion. That is why, to be precise, the result of a cross product is not a vector but a *pseudo-vector*: It behaves like a vector apart from the reflexions. For the same reason, a scalar result of a mixed product (scalar plus cross product of three vectors) is called a *pseudo-scalar* because in this case, the scalar result of the mixed product causes a change in sign under a reflexion, which can be checked easily.

2.13 Polar decomposition

Theorem 20 (Square root theorem). *Consider* $\mathbf{L}\in Sym(\mathcal{V})$ *and positive definite, then there exists a unique tensor* $\mathbf{U}\in Sym(\mathcal{V})$ *and positive definite such that*

$$\mathbf{L}=\mathbf{U}^2.$$

Proof. Existence: Consider $\mathbf{L}, \mathbf{U}, \mathbf{V} \in Sym(\mathcal{V})$ and positive definite and

$$\mathbf{L} = \omega_i \mathbf{e}_i \otimes \mathbf{e}_i$$

a spectral decomposition of $\mathbf{L}$, $\omega_i > 0\ \forall i$. Define $\mathbf{U}$ as

$$\mathbf{U} = \sqrt{\omega_i} \mathbf{e}_i \otimes \mathbf{e}_i;$$

then, by Eq. $(2.4)_1$, we get

$$\mathbf{U}^2 = \mathbf{L}.$$

Uniqueness: Suppose also that

$$\mathbf{V}^2 = \mathbf{L},$$

and let $\mathbf{e}$ be an eigenvector of $\mathbf{L}$ corresponding to the (positive) eigenvalue ω. Then, if $\lambda = \sqrt{\omega}$,

$$\mathbf{O} = (\mathbf{U}^2 - \lambda \mathbf{I})\mathbf{e} = (\mathbf{U} - \lambda \mathbf{I})(\mathbf{U} - \lambda \mathbf{I})\mathbf{e},$$

and once we set

$$\mathbf{v} = (\mathbf{U} - \lambda \mathbf{I})\mathbf{e},$$

we get

$$\mathbf{U}\mathbf{v} = -\lambda \mathbf{v} \ \Rightarrow\ \mathbf{v} = \mathbf{o} \ \Rightarrow\ \mathbf{U}\mathbf{e} = \lambda \mathbf{e}$$

because $\mathbf{U}$ is positive definite and $-\lambda$ cannot be an eigenvalue of $\mathbf{U}$ because $\lambda > 0$. In a similar way,

$$\mathbf{V}\mathbf{e} = \lambda \mathbf{e} \ \Rightarrow\ \mathbf{U}\mathbf{e} = \mathbf{V}\mathbf{e}$$

for every eigenvector $\mathbf{e}$ of $\mathbf{L}$. Because, based on spectral theorem, there exists a basis of eigenvectors of $\mathbf{L}$, $\mathbf{U} = \mathbf{V}$. □

We symbolically write that

$$\mathbf{U} = \sqrt{\mathbf{L}}.$$

For any $\mathbf{F} \in Lin(\mathcal{V})$, both $\mathbf{F}\mathbf{F}^\top$ and $\mathbf{F}^\top\mathbf{F}$ clearly $\in Sym(\mathcal{V})$. If in addition $\det \mathbf{F} > 0$, then

$$\mathbf{u} \cdot \mathbf{F}^\top \mathbf{F}\mathbf{u} = (\mathbf{F}\mathbf{u}) \cdot (\mathbf{F}\mathbf{u}) \geq 0,$$

with the zero value obtained $\iff$ $\mathbf{Fu} = \mathbf{o}$ and, because $\det \mathbf{F} > 0 \Rightarrow \mathbf{F}$ is invertible, $\iff$ $\mathbf{u} = \mathbf{o}$. As a consequence, $\mathbf{F}^\top \mathbf{F}$ is positive definite. In a similar way, it can be proved that $\mathbf{FF}^\top$ is also positive definite.

A particular tensor decomposition[14] is given by the following.

Theorem 21 (Polar decomposition theorem). $\forall \mathbf{F} \in Lin(\mathcal{V}) | \det \mathbf{F} > 0$ *exist and are uniquely determined by two positive definite tensors* $\mathbf{U}, \mathbf{V} \in Sym(\mathcal{V})$ *and a rotation* $\mathbf{R}$ *such that*

$$\mathbf{F} = \mathbf{RU} = \mathbf{VR}.$$

Proof. Uniqueness: Let $\mathbf{F} = \mathbf{RU}$ be a *right polar decomposition* of $\mathbf{F}$; because $\mathbf{R} \in Orth(\mathcal{V})^+$ and $\mathbf{U} \in Sym(\mathcal{V})$,

$$\mathbf{F}^\top \mathbf{F} = \mathbf{U}\mathbf{R}^\top \mathbf{R}\mathbf{U} = \mathbf{U}^2 \Rightarrow \mathbf{U} = \sqrt{\mathbf{F}^\top \mathbf{F}}.$$

By the square-root theorem, tensor $\mathbf{U}$ is unique, and because

$$\mathbf{R} = \mathbf{F}\mathbf{U}^{-1},$$

$\mathbf{R}$ is unique too.

Now, let $\mathbf{F} = \mathbf{VR}$ be a *left polar decomposition* of $\mathbf{F}$; by the same procedure, we get

$$\mathbf{FF}^\top = \mathbf{V}^2 \rightarrow \mathbf{V} = \sqrt{\mathbf{FF}^\top},$$

so $\mathbf{V}$ is unique, and also,

$$\mathbf{R} = \mathbf{V}^{-1}\mathbf{F}.$$

Existence: Let

$$\mathbf{U} = \sqrt{\mathbf{F}^\top \mathbf{F}},$$

so $\mathbf{U} \in Sym(\mathcal{V})$ and it is positive definite, and let

$$\mathbf{R} = \mathbf{F}\mathbf{U}^{-1}.$$

To prove that $\mathbf{F} = \mathbf{RU}$ is a right polar decomposition, we just have to show that $\mathbf{R} \in Orth(\mathcal{V})^+$. Since $\det \mathbf{F} > 0, \det \mathbf{U} > 0$ (the latter because

[14]This decomposition is fundamental to the theory of deformation of continuum bodies.

all the eigenvalues of $\mathbf{U}$ are strictly positive), by the theorem of Binet, also $\det \mathbf{R} > 0$. Then,

$$\mathbf{R}^\top \mathbf{R} = (\mathbf{F}\mathbf{U}^{-1})^\top (\mathbf{F}\mathbf{U}^{-1}) = \mathbf{U}^{-1}\mathbf{F}^\top \mathbf{F}\mathbf{U}^{-1} = \mathbf{U}^{-1}\mathbf{U}^2\mathbf{U}^{-1}$$
$$= \mathbf{I} \Rightarrow \mathbf{R} \in Orth(\mathcal{V})^+.$$

Now, let

$$\mathbf{V} = \mathbf{R}\mathbf{U}\mathbf{R}^\top,$$

then $\mathbf{V} \in Sym(\mathcal{V})$ and is positive definite, see Exercise 2.14, and

$$\mathbf{V}\mathbf{R} = \mathbf{R}\mathbf{U}\mathbf{R}^\top \mathbf{R} = \mathbf{R}\mathbf{U} = \mathbf{F},$$

which completes the proof. □

2.14 Exercises

1. Prove that

$$\mathbf{L}\mathbf{o} = \mathbf{o} \quad \forall \mathbf{L} \in Lin(\mathcal{V}).$$

2. Prove that, if a straight line r has the direction of $\mathbf{u} \in \mathcal{S}$, then the tensor giving the projection of a vector $\mathbf{v} \in \mathcal{V}$ on r is $\mathbf{u} \otimes \mathbf{u}$ (the *orthogonal projector*), while the one giving the projection on a direction orthogonal to r is $\mathbf{I} - \mathbf{u} \otimes \mathbf{u}$ (the *complementary projector*), see Fig. 2.9.
3. For any $\alpha \in \mathbb{R}, \mathbf{a}, \mathbf{b} \in \mathcal{V}$ and $\mathbf{A}, \mathbf{B} \in Lin(\mathcal{V})$, prove that

$$(\alpha \mathbf{A})^\top = \alpha \mathbf{A}^\top, \quad (\mathbf{A} + \mathbf{B})^\top = \mathbf{A}^\top + \mathbf{B}^\top, \quad (\mathbf{a} \otimes \mathbf{b})\mathbf{A} = \mathbf{a} \otimes (\mathbf{A}^\top \mathbf{b}).$$

4. Prove that

$$\mathbf{L} + \mathbf{O} = \mathbf{L} \quad \forall \mathbf{L} \in Lin(\mathcal{V}).$$

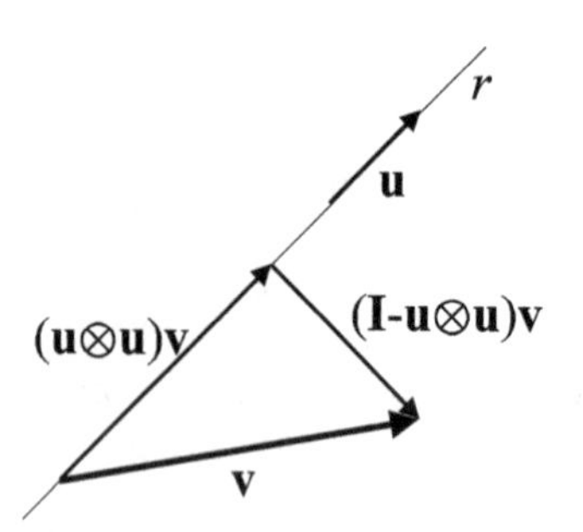

Figure 2.9: Projected vectors.

5. Prove that

$$\text{tr}\mathbf{I} = 3, \quad \text{tr}\mathbf{O} = 0.$$

6. Prove that, $\forall \mathbf{A}, \mathbf{B} \in Lin(\mathcal{V})$,

$$\text{tr}(\mathbf{AB}) = \text{tr}(\mathbf{BA}).$$

7. Prove that, $\forall \mathbf{L}, \mathbf{M}, \mathbf{N} \in Lin(\mathcal{V})$,

$$\mathbf{L}^\top \cdot \mathbf{M}^\top = \mathbf{L} \cdot \mathbf{M}, \quad \mathbf{LM} \cdot \mathbf{N} = \mathbf{L} \cdot \mathbf{NM}^\top = \mathbf{M} \cdot \mathbf{L}^\top \mathbf{N}.$$

8. Prove the assertions in Eq. (2.4).
9. Prove that any form defined by a tensor $\mathbf{L}$ can be written as a scalar product of tensors:

$$\mathbf{v} \cdot \mathbf{Lw} = \mathbf{L} \cdot \mathbf{v} \otimes \mathbf{w} \quad \forall \mathbf{v}, \mathbf{w} \in \mathcal{V}, \mathbf{L} \in Lin(\mathcal{V}).$$

10. Prove that $Sym(\mathcal{V})$ and $Skw(\mathcal{V})$ are orthogonal, i.e. prove that

$$\mathbf{A} \cdot \mathbf{B} = 0 \quad \forall \mathbf{A} \in Sym(\mathcal{V}), \ \mathbf{B} \in Skw(\mathcal{V}).$$

11. For any $\mathbf{L} \in Lin(\mathcal{V})$, prove that, if $\mathbf{A} \in Sym(\mathcal{V})$, then

$$\mathbf{A} \cdot \mathbf{L} = \mathbf{A} \cdot \mathbf{L}^s,$$

while if $\mathbf{B} \in Skw(\mathcal{V})$, then

$$\mathbf{B} \cdot \mathbf{L} = \mathbf{B} \cdot \mathbf{L}^a.$$

12. Let $\mathbf{A}, \mathbf{B}, \mathbf{C}, \mathbf{D} \in Lin(\mathcal{V})$; prove that

$$\mathbf{A} \cdot (\mathbf{BCD}) = (\mathbf{B}^\top \mathbf{A}) \cdot (\mathbf{CD}) = (\mathbf{AD}^\top) \cdot (\mathbf{BC}).$$

13. Prove that $\mathbf{L} \cdot \mathbf{W} = 0 \ \forall \mathbf{W} \in Skw(\mathcal{V}) \iff \mathbf{L} \in Sym(\mathcal{V})$.
14. Express by components the second principal invariant I_2 of a tensor $\mathbf{L}$.
15. Prove that if $\mathbf{a} = (a_1, a_2, a_3), \mathbf{b} = (b_1, b_2, b_3), \mathbf{c} = (c_1, c_2, c_3)$, then

$$\mathbf{a} \times \mathbf{b} \cdot \mathbf{c} = \det \begin{bmatrix} a_1 & a_2 & a_3 \\ b_1 & b_2 & b_3 \\ c_1 & c_2 & c_3 \end{bmatrix}.$$

16. Prove the uniqueness of the inverse tensor.
17. Show, using the Cartesian components, that all the dyads are singular.

18. Prove that if $\mathbf{L}$ is invertible and $\alpha \in \mathbb{R} - \{0\}$, then
$$(\alpha\mathbf{L})^{-1} = \alpha^{-1}\mathbf{L}^{-1}.$$
19. Prove that if $\mathbf{W}$ is the axial tensor of $\mathbf{w}$, then
$$\mathbf{WW} = -\frac{1}{2}|\mathbf{W}|^2(\mathbf{I} - \mathbf{w} \otimes \mathbf{w}).$$
20. Prove that for any two axial couples $\mathbf{w}_1, \mathbf{W}_1$ and $\mathbf{w}_2, \mathbf{W}_2$, we have
$$\mathbf{w}_1 \cdot \mathbf{w}_2 = \frac{1}{2}\mathbf{W}_1 \cdot \mathbf{W}_2.$$
21. Prove that $\forall \mathbf{u}, \mathbf{v} \in \mathcal{V}$, $\mathbf{u} \times \mathbf{v} = \mathbf{o} \iff \mathbf{u} \otimes \mathbf{v} \in Sym(\mathcal{V})$.
22. Let $\mathbf{L} \in Sym(\mathcal{V})$ and positive definite and $\mathbf{R} \in Orth(\mathcal{V})^+$, then prove that $\mathbf{RLR}^\top \in Sym(\mathcal{V})$ and that it is positive definite.
23. Prove that the spectrum of $\mathbf{L}^{sph}$ is composed of only
$$\lambda^{sph} = \frac{1}{3}\mathrm{tr}\mathbf{L}$$
and that any $\mathbf{u} \in \mathcal{S}$ is an eigenvector.
24. Prove that the eigenvalues λ^{dev} of $\mathbf{L}^{dev}$ are given by
$$\lambda^{dev} = \lambda - \lambda^{sph},$$
where λ is an eigenvalue of $\mathbf{L}$.

Chapter 3

Fourth-Rank Tensors

3.1 Fourth-rank tensors

A *fourth-rank tensor* $\mathbb{L}$ is any linear application from $Lin(\mathcal{V})$ to $Lin(\mathcal{V})$:

$$\mathbb{L} : Lin(\mathcal{V}) \to Lin(\mathcal{V}) \mid \mathbb{L}(\alpha_i \mathbf{A}_i) = \alpha_i \mathbb{L}\mathbf{A}_i \ \forall \alpha_i \in \mathbb{R}, \ \mathbf{A}_i \in Lin(\mathcal{V}),$$
$$i = 1, \ldots, n.$$

Defining the *sum of two fourth-rank tensors* as

$$(\mathbb{L}_1 + \mathbb{L}_2)\mathbf{A} = \mathbb{L}_1\mathbf{A} + \mathbb{L}_2\mathbf{A} \ \ \forall \mathbf{A} \in Lin(\mathcal{V}),$$

the *product of a scalar by a fourth-rank tensor* as

$$(\alpha\mathbb{L})\mathbf{A} = \alpha(\mathbb{L}\mathbf{A}) \ \ \forall \alpha \in \mathbb{R}, \mathbf{A} \in Lin(\mathcal{V}),$$

and the *null fourth-rank tensor* $\mathbb{O}$ as the unique tensor such that

$$\mathbb{O}\mathbf{A} = \mathbf{O} \ \forall \mathbf{A} \in Lin(\mathcal{V}),$$

then the set of all the tensors $\mathbb{L}$ that operate on $Lin(\mathcal{V})$ forms a vector space, denoted by $\mathbb{L}\mathrm{in}(\mathcal{V})$. We define the *fourth-rank identity tensor* $\mathbb{I}$ as a unique tensor such that

$$\mathbb{I}\mathbf{A} = \mathbf{A} \ \ \forall \mathbf{A} \in Lin(\mathcal{V}).$$

It is apparent that the algebra of fourth-rank tensors is similar to that of second-rank tensors, and in fact, several operations with fourth-rank tensors can be introduced in almost the same way, in some sense the operations shifting from $\mathcal{V}$ to $Lin(\mathcal{V})$. However, the algebra of fourth-rank tensors is richer than that of the second-rank ones, and some care must be taken.

In the following sections, we consider some of the operations that can be done with fourth-rank tensors.

3.2 Dyads, tensor components

For any couple of tensors $\mathbf{A}$ and $\mathbf{B} \in Lin(\mathcal{V})$, the (*tensor*) *dyad* $\mathbf{A} \otimes \mathbf{B}$ is the fourth-rank tensor defined by

$$(\mathbf{A} \otimes \mathbf{B})\mathbf{L} := \mathbf{B} \cdot \mathbf{L}\ \mathbf{A} \quad \forall \mathbf{L} \in Lin(\mathcal{V}).$$

The application defined above is actually a fourth-rank tensor because of the bilinearity of the scalar product of second-rank tensors. Applying this rule to the nine dyads of the basis $\mathcal{B}^2 = \{\mathbf{e}_i \otimes \mathbf{e}_j,\ i,j = 1,2,3\}$ of $Lin(\mathcal{V})$ leads to the introduction of 81 fourth-rank tensors,

$$\mathbf{e}_i \otimes \mathbf{e}_j \otimes \mathbf{e}_k \otimes \mathbf{e}_l := (\mathbf{e}_i \otimes \mathbf{e}_j) \otimes (\mathbf{e}_k \otimes \mathbf{e}_l),$$

that form a basis $\mathcal{B}^4 = \{\mathbf{e}_i \otimes \mathbf{e}_j \otimes \mathbf{e}_k \otimes \mathbf{e}_l,\ i,j = 1,2,3\}$ for $\mathbb{L}\mathrm{in}(\mathcal{V})$. We remark hence that $\dim(\mathbb{L}\mathrm{in}(\mathcal{V})) = 81$. A useful result is that

$$\begin{aligned}&(\mathbf{e}_i \otimes \mathbf{e}_j \otimes \mathbf{e}_k \otimes \mathbf{e}_l)(\mathbf{e}_p \otimes \mathbf{e}_q)\\ &\quad = (\mathbf{e}_k \otimes \mathbf{e}_l) \cdot (\mathbf{e}_p \otimes \mathbf{e}_q)(\mathbf{e}_i \otimes \mathbf{e}_j) = \delta_{kp}\delta_{lq}(\mathbf{e}_i \otimes \mathbf{e}_j). \qquad (3.1)\end{aligned}$$

Any fourth-rank tensor can be expressed as a linear combination (the *canonical decomposition*):

$$\mathbb{L} = L_{ijkl}\ \mathbf{e}_i \otimes \mathbf{e}_j \otimes \mathbf{e}_k \otimes \mathbf{e}_l, \quad i,j = 1,2,3,$$

where L_{ijkl}s are the 81 *Cartesian components* of $\mathbb{L}$ with respect to $\mathcal{B}^4$. L_{ijkl}s are defined by the operation

$$\begin{aligned}(\mathbf{e}_i \otimes \mathbf{e}_j) \cdot \mathbb{L}(\mathbf{e}_k \otimes \mathbf{e}_l) &= (\mathbf{e}_i \cdot \mathbf{e}_j) \cdot (L_{pqrs}\mathbf{e}_p \otimes \mathbf{e}_q \otimes \mathbf{e}_r \otimes \mathbf{e}_s)(\mathbf{e}_k \otimes \mathbf{e}_l)\\ &= (\mathbf{e}_i \otimes \mathbf{e}_j) \cdot (L_{pqrs}\delta_{rk}\delta_{sl}\mathbf{e}_p \otimes \mathbf{e}_q)\\ &= L_{pqrs}\delta_{rk}\delta_{sl}\delta_{ip}\delta_{jq} = L_{ijkl}.\end{aligned}$$

The components of a tensor dyad can be computed without any difficulty:

$$\begin{aligned}\mathbf{A} \otimes \mathbf{B} &= (A_{ij}\mathbf{e}_i \otimes \mathbf{e}_j) \otimes (B_{kl}\mathbf{e}_k \otimes \mathbf{e}_l) = A_{ij}B_{kl}\mathbf{e}_i \otimes \mathbf{e}_j \otimes \mathbf{e}_k \otimes \mathbf{e}_l\\ &\Rightarrow (\mathbf{A} \otimes \mathbf{B})_{ijkl} = A_{ij}B_{kl}\end{aligned}$$

so that, in particular,

$$((\mathbf{a} \otimes \mathbf{b}) \otimes (\mathbf{c} \otimes \mathbf{d}))_{ijkl} = a_i b_j c_k d_l.$$

Concerning the identity of $\mathbb{L}\mathrm{in}(\mathcal{V})$,

$$\begin{aligned}I_{ijkl} &= (\mathbf{e}_i \otimes \mathbf{e}_l) \cdot \mathbb{I}(\mathbf{e}_k \otimes \mathbf{e}_l) = (\mathbf{e}_i \otimes \mathbf{e}_j) \cdot (\mathbf{e}_k \otimes \mathbf{e}_l) = \mathbf{e}_i \cdot \mathbf{e}_k \mathbf{e}_j \cdot \mathbf{e}_l\\ &= \delta_{ik}\delta_{jl} \ \Rightarrow \mathbb{I} = \delta_{ik}\delta_{jl}(\mathbf{e}_i \otimes \mathbf{e}_l \otimes \mathbf{e}_k \otimes \mathbf{e}_l).\end{aligned}$$

The components of $\mathbf{A} \in Lin(\mathcal{V})$, resulting from the application of $\mathbb{L} \in \mathbb{L}\text{in}(\mathcal{V})$ on $\mathbf{B} \in Lin(\mathcal{V})$, can now be easily calculated:

$$\begin{aligned}\mathbf{A} = \mathbb{L}\mathbf{B} &= L_{ijkl}(\mathbf{e}_i \otimes \mathbf{e}_j \otimes \mathbf{e}_k \otimes \mathbf{e}_l)(B_{pq}\mathbf{e}_p \otimes \mathbf{e}_q) \\ &= L_{ijkl}B_{pq}\delta_{kp}\delta_{lq}(\mathbf{e}_i \otimes \mathbf{e}_j) \\ &= L_{ijkl}B_{kl}(\mathbf{e}_i \otimes \mathbf{e}_j) \ \Rightarrow\ A_{ij} = L_{ijkl}B_{kl}.\end{aligned} \tag{3.2}$$

Moreover,

$$\begin{aligned}\mathbb{L}(\mathbf{A} \otimes \mathbf{B})\mathbf{C} &= \mathbb{L}((\mathbf{A} \otimes \mathbf{B})\mathbf{C}) = \mathbb{L}(\mathbf{B} \cdot \mathbf{C}\mathbf{A}) = \mathbf{B} \cdot \mathbf{C}\, \mathbb{L}\mathbf{A} \\ &= ((\mathbb{L}\mathbf{A}) \otimes \mathbf{B})\mathbf{C} \ \Rightarrow \mathbb{L}(\mathbf{A} \otimes \mathbf{B}) = (\mathbb{L}\mathbf{A}) \otimes \mathbf{B}.\end{aligned}$$

Using this result and Eq. (3.1), we can determine the components of a product of fourth-rank tensors:

$$\begin{aligned}\mathbb{AB} &= A_{ijkl}(\mathbf{e}_i \otimes \mathbf{e}_j \otimes \mathbf{e}_k \otimes \mathbf{e}_l)B_{pqrs}(\mathbf{e}_p \otimes \mathbf{e}_q \otimes \mathbf{e}_r \otimes \mathbf{e}_s) \\ &= A_{ijkl}B_{pqrs}(\mathbf{e}_i \otimes \mathbf{e}_j \otimes \mathbf{e}_k \otimes \mathbf{e}_l)(\mathbf{e}_p \otimes \mathbf{e}_q) \otimes (\mathbf{e}_r \otimes \mathbf{e}_s) \\ &= A_{ijkl}B_{pqrs}[(\mathbf{e}_i \otimes \mathbf{e}_j \otimes \mathbf{e}_k \otimes \mathbf{e}_l)(\mathbf{e}_p \otimes \mathbf{e}_q)] \otimes (\mathbf{e}_r \otimes \mathbf{e}_s) \\ &= A_{ijkl}B_{pqrs}[\delta_{kp}\delta_{lq}(\mathbf{e}_i \otimes \mathbf{e}_j)] \otimes (\mathbf{e}_r \otimes \mathbf{e}_s) \\ &= A_{ijkl}B_{klrs}(\mathbf{e}_i \otimes \mathbf{e}_j \otimes \mathbf{e}_r \otimes \mathbf{e}_s) \ \Rightarrow\ (\mathbb{AB})_{ijrs} = A_{ijkl}B_{klrs}.\end{aligned} \tag{3.3}$$

Depending upon four indices, a fourth-rank tensor $\mathbb{L}$ cannot be represented by a matrix; however, we will see in Section 3.8 that a matrix representation of a fourth-rank tensor is still possible and that it is currently used in some cases, e.g. in elasticity.

3.3 Conjugation product, transpose, symmetries

For any two tensors $\mathbf{A}, \mathbf{B} \in Lin(\mathcal{V})$, we call the *conjugation product* the tensor $\mathbf{A} \boxtimes \mathbf{B} \in \mathbb{L}\text{in}(\mathcal{V})$ defined by the operation

$$(\mathbf{A} \boxtimes \mathbf{B})\mathbf{L} := \mathbf{A}\mathbf{L}\mathbf{B}^\top \ \forall \mathbf{L} \in Lin(\mathcal{V}).$$

As a consequence, for the dyadic tensors of $\mathcal{B}^2$,

$$(\mathbf{e}_i \otimes \mathbf{e}_j) \boxtimes (\mathbf{e}_k \otimes \mathbf{e}_l) = \mathbf{e}_i \otimes \mathbf{e}_k \otimes \mathbf{e}_j \otimes \mathbf{e}_l \tag{3.4}$$

so that

$$(\mathbf{A} \boxtimes \mathbf{B})_{ijkl} = A_{ik}B_{jl}.$$

Moreover, by the uniqueness of the identity $\mathbb{I}$, $\forall \mathbf{A} \in Lin(\mathcal{V})$,

$$(\mathbf{I} \boxtimes \mathbf{I})\mathbf{A} = \mathbf{I}\mathbf{A}\mathbf{I}^\top = \mathbf{A} \ \Rightarrow \ \mathbb{I} = \mathbf{I} \boxtimes \mathbf{I}.$$

The *transpose* of a fourth-rank tensor $\mathbb{L}$ is the unique tensor $\mathbb{L}^\top$, such that

$$\mathbf{A} \cdot (\mathbb{L}\mathbf{B}) = \mathbf{B} \cdot (\mathbb{L}^\top \mathbf{A}) \ \forall \mathbf{A}, \mathbf{B} \in Lin(\mathcal{V}).$$

By this definition, setting $\mathbf{A} = \mathbf{e}_i \otimes \mathbf{e}_j, \mathbf{B} = \mathbf{e}_k \otimes \mathbf{e}_l$ gives

$$(L^\top)_{ijkl} = L_{klij}.$$

A consequence is that

$$\mathbf{A} \cdot (\mathbb{L}\mathbf{B}) = \mathbf{B} \cdot (\mathbb{L}^\top \mathbf{A}) = \mathbf{A} \cdot (\mathbb{L}^\top)^\top \mathbf{B} \ \Rightarrow \ (\mathbb{L}^\top)^\top = \mathbb{L}.$$

Moreover,

$$\begin{aligned} \mathbf{M} \cdot (\mathbf{A} \otimes \mathbf{B})^\top \mathbf{L} = \mathbf{L} \cdot (\mathbf{A} \otimes \mathbf{B})\mathbf{M} \\ &= \mathbf{L} \cdot \mathbf{A}\mathbf{M} \cdot \mathbf{B} = \mathbf{M} \cdot (\mathbf{B}\mathbf{A} \cdot \mathbf{L}) \\ &= \mathbf{M} \cdot (\mathbf{B} \otimes \mathbf{A})\mathbf{L} \ \Rightarrow \ (\mathbf{A} \otimes \mathbf{B})^\top = \mathbf{B} \otimes \mathbf{A}, \end{aligned}$$

while, cf. Exercise 7, Chapter 2,

$$\begin{aligned} \mathbf{M} \cdot (\mathbf{A} \boxtimes \mathbf{B})^\top \mathbf{L} &= \mathbf{L} \cdot (\mathbf{A} \boxtimes \mathbf{B})\mathbf{M} \\ &= \mathbf{L} \cdot \mathbf{A}\mathbf{M}\mathbf{B}^\top = \mathbf{A}^\top \mathbf{L} \cdot \mathbf{M}\mathbf{B}^\top = \mathbf{M}^\top \mathbf{A}^\top \mathbf{L} \cdot \mathbf{B}^\top \\ &= (\mathbf{M}^\top \mathbf{A}^\top \mathbf{L})^\top \cdot (\mathbf{B}^\top)^\top = \mathbf{L}^\top \mathbf{A}\mathbf{M} \cdot \mathbf{B} = \mathbf{A}\mathbf{M} \cdot \mathbf{L}\mathbf{B} \\ &= \mathbf{M} \cdot \mathbf{A}^\top \mathbf{L}\mathbf{B} = \mathbf{M} \cdot (\mathbf{A}^\top \boxtimes \mathbf{B}^\top)\mathbf{L} \ \Rightarrow \ (\mathbf{A} \boxtimes \mathbf{B})^\top \\ &= \mathbf{A}^\top \boxtimes \mathbf{B}^\top. \end{aligned}$$

The property

$$(\mathbb{A}\mathbb{B})^\top = \mathbb{B}^\top \mathbb{A}^\top$$

can be proved in the same manner as the analogous property of the second-rank tensors.

A tensor $\mathbb{L} \in \mathbb{L}\mathrm{in}(\mathcal{V})$ is *symmetric* $\iff$ $\mathbb{L} = \mathbb{L}^\top$. It is then evident that

$$\mathbb{L} = \mathbb{L}^\top \Rightarrow \ L_{ijkl} = L_{klij},$$

which are relations called *major symmetries.* These symmetries number 36 on the whole so that a symmetric fourth-rank tensor has 45 independent components. Moreover,

$$\begin{aligned} \mathbf{A} \boxtimes \mathbf{B} = (\mathbf{A} \boxtimes \mathbf{B})^\top = \mathbf{A}^\top \boxtimes \mathbf{B}^\top \ &\iff \ \mathbf{A} = \mathbf{A}^\top, \mathbf{B} = \mathbf{B}^\top, \\ \mathbf{A} \otimes \mathbf{B} = (\mathbf{A} \otimes \mathbf{B})^\top = \mathbf{B} \otimes \mathbf{A} \ &\iff \ \mathbf{B} = \lambda \mathbf{A}, \ \lambda \in \mathbb{R}. \end{aligned}$$

Let us now consider the case of a $\mathbb{L} \in \mathbb{L}\mathrm{in}(\mathcal{V})$, such that

$$\mathbb{L}\mathbf{A} = (\mathbb{L}\mathbf{A})^\top \ \ \forall \mathbf{A} \in Lin(\mathcal{V}).$$

Then, by Eq. (3.2),

$$L_{ijkl} = L_{jikl},$$

which are relations called *left minor symmetries*: A tensor $\mathbb{L}$ having the left minor symmetries has values in $Sym(\mathcal{V})$. On the whole, the left minor symmetries number 27. Finally, consider the case of a $\mathbb{L} \in \mathbb{L}\mathrm{in}(\mathcal{V})$, such that

$$\mathbb{L}\mathbf{A} = \mathbb{L}(\mathbf{A}^\top) \quad \forall \mathbf{A} \in Lin(\mathcal{V});$$

then, again by Eq. (3.2), we get

$$L_{ijkl} = L_{ijlk},$$

which are relations called *minor right symmetries*, whose total number is also 27. It is immediate to recognize that if $\mathbb{L}$ has the minor right symmetries, then

$$\mathbb{L}\mathbf{W} = \mathbf{O} \quad \forall \mathbf{W} \in Skw(\mathcal{V}).$$

We say that *a tensor has minor symmetries* if it has both the right and left minor symmetries; the total number of minor symmetries is 45 because, as can be easily checked, some of the left and right minor symmetries are the same, so finally a tensor with the minor symmetries has 36 independent components.

If $\mathbb{L} \in \mathbb{L}\mathrm{in}(\mathcal{V})$ has major and minor symmetries, then the number of independent symmetry relations is actually 60 (some minor and major symmetries coincide), so in such a case, $\mathbb{L}$ depends upon 21 independent components only. This is the case of the elasticity tensor.

Finally, the six *Cauchy–Poisson symmetries*[1] are those of the type

$$L_{ijkl} = L_{ikjl}.$$

A tensor having major, minor, and Cauchy–Poisson symmetries is *completely symmetric*, i.e. swapping any couple of indices gives an identical component. In that case, the number of independent components is only 15.

[1]The Cauchy–Poisson symmetries have played an important role in a celebrated diatribe of the 19th century in elasticity, that between the so-called *rari-* and *multi-constant* theories.

3.4 Trace and scalar product of fourth-rank tensors

We can introduce the scalar product between fourth-rank tensors in the same way we did for second-rank tensors. We first introduce the concept of *trace for fourth-rank tensors* once again using the dyad (here, the tensor dyad):

$$\text{tr}_4\mathbf{A}\otimes\mathbf{B} := \mathbf{A}\cdot\mathbf{B}.$$

The easy proof that $\text{tr}_4 : \mathbb{L}\text{in}(\mathcal{V}) \to \mathbb{R}$ is a linear form is based upon the properties of scalar product of second-rank tensors, and it is left to the reader. An immediate result is that

$$\text{tr}_4\mathbf{A}\otimes\mathbf{B} = A_{ij}B_{ij}.$$

Then, using the canonical decomposition, we have that

$$\begin{aligned}\text{tr}_4\mathbb{L} &= \text{tr}_4(L_{ijkl}(\mathbf{e}_i\otimes\mathbf{e}_j)\otimes(\mathbf{e}_k\otimes\mathbf{e}_l)) = L_{ijkl}(\mathbf{e}_i\otimes\mathbf{e}_j)\cdot(\mathbf{e}_k\otimes\mathbf{e}_l)\\ &= L_{ijkl}\delta_{ik}\delta_{jl} = L_{ijij}\end{aligned}$$

and that

$$\begin{aligned}\text{tr}_4\mathbb{L}^\top &= \text{tr}_4(L_{klij}(\mathbf{e}_i\otimes\mathbf{e}_j)\otimes(\mathbf{e}_k\otimes\mathbf{e}_l)) = L_{klij}(\mathbf{e}_i\otimes\mathbf{e}_j)\cdot(\mathbf{e}_k\otimes\mathbf{e}_l)\\ &= L_{klij}\delta_{ik}\delta_{jl} = L_{ijij} = \text{tr}_4\mathbb{L}.\end{aligned}$$

Then, we define the *scalar product of fourth-rank tensors* as

$$\mathbb{A}\cdot\mathbb{B} := \text{tr}_4(\mathbb{A}^\top\mathbb{B}).$$

By the properties of tr_4, the scalar product is a positive definite symmetric bilinear form:

$$\begin{aligned}\alpha\mathbb{A}\cdot\beta\mathbb{B} &= \text{tr}_4(\alpha\mathbb{A}^\top\beta\mathbb{B}) = \alpha\beta\text{tr}_4(\mathbb{A}^\top\mathbb{B}) = \alpha\beta\mathbb{A}\cdot\mathbb{B},\\ \mathbb{A}\cdot\mathbb{B} &= \text{tr}_4(\mathbb{A}^\top\mathbb{B}) = \text{tr}_4(\mathbb{A}^\top\mathbb{B})^\top = \text{tr}_4(\mathbb{B}^\top\mathbb{A}) = \mathbb{B}\cdot\mathbb{A},\\ \mathbb{A}\cdot\mathbb{A} &= \text{tr}_4(\mathbb{A}^\top\mathbb{A}) = (\mathbb{A}^\top\mathbb{A})_{ijij} = A_{klij}A_{klij} > 0\ \forall\mathbb{A}\in\mathbb{L}in(\mathcal{V}),\\ \mathbb{A}\cdot\mathbb{A} &= 0 \iff \mathbb{A} = \mathbb{O}.\end{aligned}$$

By components,

$$\begin{aligned}\mathbb{A}\cdot\mathbb{B} &= \text{tr}_4((A_{klij}\mathbf{e}_i\otimes\mathbf{e}_j\otimes\mathbf{e}_k\otimes\mathbf{e}_l)(B_{pqrs}\mathbf{e}_p\otimes\mathbf{e}_q\otimes\mathbf{e}_r\otimes\mathbf{e}_s))\\ &= \text{tr}_4(A_{klij}B_{pqrs}\delta_{kp}\delta_{lq}(\mathbf{e}_i\otimes\mathbf{e}_j)\otimes(\mathbf{e}_r\otimes\mathbf{e}_s))\\ &= A_{klij}B_{pqrs}\delta_{kp}\delta_{lq}(\mathbf{e}_i\otimes\mathbf{e}_j)\cdot(\mathbf{e}_r\otimes\mathbf{e}_s)\\ &= A_{klij}B_{pqrs}\delta_{kp}\delta_{lq}\delta_{ir}\delta_{js} = A_{klij}B_{klij}.\end{aligned}$$

The rule for computing the scalar product is hence always the same as was already seen for vectors and second-rank tensors: All the indexes are to be saturated.

In complete analogy with vectors and second-rank tensors, we say that $\mathbb{A}$ is *orthogonal to* $\mathbb{B} \iff$

$$\mathbb{A} \cdot \mathbb{B} = 0,$$

and we define the *norm of* $\mathbb{L}$ as

$$|\mathbb{L}| := \sqrt{\mathbb{L} \cdot \mathbb{L}} = \sqrt{\text{tr}_4 \mathbb{L}^\top \mathbb{L}} = \sqrt{L_{ijkl} L_{ijkl}}.$$

3.5 Projectors and identities

For the spherical part of any $\mathbf{A} \in Sym(\mathcal{V})$, we can write

$$\mathbf{A}^{sph} := \frac{1}{3} \text{tr}\mathbf{A}\ \mathbf{I} = \frac{1}{3} \mathbf{I} \cdot \mathbf{A}\ \mathbf{I} = \frac{1}{3} (\mathbf{I} \otimes \mathbf{I}) \mathbf{A} = \mathbb{S}^{sph} \mathbf{A},$$

where

$$\mathbb{S}^{sph} := \frac{1}{3} \mathbf{I} \otimes \mathbf{I}$$

is the *spherical projector*, i.e. the fourth-rank tensor that extracts from any $\mathbf{A} \in Lin(\mathcal{V})$ its spherical part. Moreover,

$$\mathbf{A}^{dev} := \mathbf{A} - \mathbf{A}^{sph} = \mathbb{I}\mathbf{A} - \mathbb{S}^{sph} \mathbf{A} = \mathbb{D}^{dev} \mathbf{A},$$

where

$$\mathbb{D}^{dev} := \mathbb{I} - \mathbb{S}^{sph}$$

is the *deviatoric projector*, i.e. the fourth-rank tensor that extracts from any $\mathbf{A} \in Lin(\mathcal{V})$ its deviatoric part. It is worth noting that

$$\mathbb{I} = \mathbb{S}^{sph} + \mathbb{D}^{dev}.$$

Moreover, about the components of $\mathbb{S}^{sph}$,

$$\begin{aligned} S^{sph}_{ijkl} &= (\mathbf{e}_i \otimes \mathbf{e}_j) \cdot \frac{1}{3} (\mathbf{I} \otimes \mathbf{I})(\mathbf{e}_k \otimes \mathbf{e}_l) = \frac{1}{3} \mathbf{I} \cdot (\mathbf{e}_i \otimes \mathbf{e}_j)\ \mathbf{I} \cdot (\mathbf{e}_k \otimes \mathbf{e}_l) \\ &= \frac{1}{3} \text{tr}(\mathbf{e}_i \otimes \mathbf{e}_j) \text{tr}(\mathbf{e}_k \otimes \mathbf{e}_l) = \frac{1}{3} \delta_{ij} \delta_{kl} \ \rightarrow \ \mathbb{S}^{sph} \\ &= \frac{1}{3} \delta_{ij} \delta_{kl} (\mathbf{e}_i \otimes \mathbf{e}_j \otimes \mathbf{e}_k \otimes \mathbf{e}_l). \end{aligned}$$

We remark that

$$\mathbb{S}^{sph} = (\mathbb{S}^{sph})^\top.$$

We introduce now the tensor $\mathbb{I}^s$, *restriction of* $\mathbb{I}$ *to* $\mathbf{A} \in Sym(\mathcal{V})$. It can be introduced as follows: $\forall \mathbf{A} \in Sym(\mathcal{V})$,

$$\mathbf{A} = \frac{1}{2}(\mathbf{A} + \mathbf{A}^\top)$$

and

$$\mathbf{A} = \mathbb{I}\mathbf{A} = \frac{1}{2}(\mathbb{I}\mathbf{A} + \mathbb{I}\mathbf{A}^\top) = \frac{1}{2}(I_{ijkl}A_{kl} + I_{ijkl}A_{lk})(\mathbf{e}_i \otimes \mathbf{e}_j \otimes \mathbf{e}_k \otimes \mathbf{e}_l);$$

because $\mathbf{A} = \mathbf{A}^\top$, there is insensitivity to the swap of indexes k and l, so

$$\begin{aligned}\mathbf{A} &= \frac{1}{2}(I_{ijkl}A_{kl} + I_{ijlk}A_{lk})(\mathbf{e}_i \otimes \mathbf{e}_j \otimes \mathbf{e}_k \otimes \mathbf{e}_l)\\ &= \frac{1}{2}(\delta_{ik}\delta_{jl} + \delta_{il}\delta_{jk})A_{kl}(\mathbf{e}_i \otimes \mathbf{e}_j \otimes \mathbf{e}_k \otimes \mathbf{e}_l).\end{aligned}$$

Then, if we admit the interchangeability of indexes k and l, i.e. if we postulate the existence of the minor right symmetries for $\mathbb{I}$, then $\mathbb{I} = \mathbb{I}^s$, with

$$\mathbb{I}^s = \frac{1}{2}(\delta_{ik}\delta_{jl} + \delta_{il}\delta_{jk})(\mathbf{e}_i \otimes \mathbf{e}_j \otimes \mathbf{e}_k \otimes \mathbf{e}_l).$$

It is apparent that

$$I^s_{ijkl} = I^s_{klij},$$

i.e. $\mathbb{I}^s = (\mathbb{I}^s)^\top$ but also that

$$I^s_{ijkl} = \frac{1}{2}(\delta_{il}\delta_{jk} + \delta_{ik}\delta_{jl}) = I^s_{jikl},$$

i.e. $\mathbb{I}^s$ has also the minor left symmetries; in other words, $\mathbb{I}^s$ has the major and minor symmetries, like an elasticity tensor, while this is not the case for $\mathbb{I}$. In fact,

$$I_{ijkl} = I_{jilk} = \delta_{ik}\delta_{jl} \neq \delta_{il}\delta_{jk} = I_{jikl} = I_{ijlk}.$$

Because $\mathbb{S}^{sph}$ and $\mathbb{D}^{dev}$ operate on $Sym(\mathcal{V})$, it is immediate to recognize that it is also

$$\mathbb{D}^{dev} = \mathbb{I}^s - \mathbb{S}^{sph} \Rightarrow \mathbb{I}^s = \mathbb{S}^{sph} + \mathbb{D}^{dev}.$$

It is worth noting that

$$(\mathbb{D}^{dev})^\top = (\mathbb{I}^s - \mathbb{S}^{sph})^\top = (\mathbb{I}^s)^\top - (\mathbb{S}^{sph})^\top = \mathbb{I}^s - \mathbb{S}^{sph} = \mathbb{D}^{dev}.$$

We can now determine the components of $\mathbb{D}^{dev}$:

$$D^{dev}_{ijkl} = I^s_{ijkl} - S^{sph}_{ijkl} = \frac{1}{2}(\delta_{ik}\delta_{jl} + \delta_{il}\delta_{jk}) - \frac{1}{3}\delta_{ij}\delta_{kl} \rightarrow$$

$$\mathbb{D}^{dev} = \left[\frac{1}{2}(\delta_{ik}\delta_{jl} + \delta_{il}\delta_{jk}) - \frac{1}{3}\delta_{ij}\delta_{kl}\right](\mathbf{e}_i \otimes \mathbf{e}_j \otimes \mathbf{e}_k \otimes \mathbf{e}_l).$$

We remark that the result (2.9) implies that $\mathbb{S}^{sph}$ and $\mathbb{D}^{dev}$ are *orthogonal projectors*, i.e. they project the same $\mathbf{A} \in Sym(\mathcal{V})$ into two orthogonal subspaces of $\mathcal{V}$, $Sph(\mathcal{V})$ and $Dev(\mathcal{V})$.

The tensor $\mathbb{T}^{trp} \in \mathbb{L}\text{in}(\mathcal{V})$ defined by the operation

$$\mathbb{T}^{trp}\mathbf{A} = \mathbf{A}^\top \quad \forall \mathbf{A} \in Lin(\mathcal{V})$$

is the *transposition projector* whose components are

$$T^{trp}_{ijkl} = (\mathbf{e}_i \otimes \mathbf{e}_j) \cdot \mathbb{T}^{trp}(\mathbf{e}_k \otimes \mathbf{e}_l) = (\mathbf{e}_i \otimes \mathbf{e}_j) \cdot (\mathbf{e}_l \otimes \mathbf{e}_k) = \delta_{il}\delta_{jk}.$$

The following operation defines the *symmetry projector* $\mathbb{S}^{sym} \in \mathbb{L}\text{in}(\mathcal{V})$:

$$\mathbb{S}^{sym}\mathbf{A} = \frac{1}{2}(\mathbf{A} + \mathbf{A}^\top) \ \forall \mathbf{A} \in Lin(\mathcal{V}),$$

while the *antisymmetry projector* $\mathbb{W}^{skw} \in \mathbb{L}\text{in}(\mathcal{V})$ is defined by

$$\mathbb{W}^{skw}\mathbf{A} = \frac{1}{2}(\mathbf{A} - \mathbf{A}^\top) \ \forall \mathbf{A} \in Lin(\mathcal{V}).$$

Also, $\mathbb{S}^{sym}$ and $\mathbb{W}^{skw}$ are orthogonal projectors because they project the same $\mathbf{A} \in Lin(\mathcal{V})$ into two orthogonal subspaces of $Lin(\mathcal{V})$: $Sym(\mathcal{V})$ and $Skw(\mathcal{V})$, see Exercise 10, Chapter 2.

We prove now two properties of the projectors: $\forall \mathbf{A} \in Lin(\mathcal{V})$,

$$(\mathbb{S}^{sym} + \mathbb{W}^{skw})\mathbf{A} = \frac{1}{2}(\mathbf{A} + \mathbf{A}^\top) + \frac{1}{2}(\mathbf{A} - \mathbf{A}^\top) = \mathbf{A}$$
$$= \mathbb{I}\mathbf{A} \Rightarrow \mathbb{S}^{sym} + \mathbb{W}^{skw} = \mathbb{I}. \tag{3.5}$$

Then,

$$(\mathbb{S}^{sym} - \mathbb{W}^{skw})\mathbf{A} = \frac{1}{2}(\mathbf{A} + \mathbf{A}^\top) - \frac{1}{2}(\mathbf{A} - \mathbf{A}^\top) = \mathbf{A}^\top$$
$$= \mathbb{T}^{trp}\mathbf{A} \Rightarrow \mathbb{S}^{sym} - \mathbb{W}^{skw} = \mathbb{T}^{trp}. \tag{3.6}$$

3.6 Orthogonal conjugator

For any $\mathbf{U} \in Orth(\mathcal{V})$, we define its *orthogonal conjugator* $\mathbb{U} \in \mathbb{L}\mathrm{in}(\mathcal{V})$ as

$$\mathbb{U} := \mathbf{U} \boxtimes \mathbf{U}.$$

Theorem 22 (Orthogonality of $\mathbb{U}$). *The orthogonal conjugator is an orthogonal tensor of $\mathbb{L}in(\mathcal{V})$, i.e. it preserves the scalar product between tensors:*

$$\mathbb{U}\mathbf{A} \cdot \mathbb{U}\mathbf{B} = \mathbf{A} \cdot \mathbf{B} \quad \forall \mathbf{A}, \mathbf{B} \in Lin(\mathcal{V}).$$

Proof. By the assertion in Exercise 12 of Chapter 2 and because $\mathbf{U} \in Orth(\mathcal{V})$, we have

$$\begin{aligned}\mathbb{U}\mathbf{A} \cdot \mathbb{U}\mathbf{B} &= (\mathbf{U} \boxtimes \mathbf{U})\mathbf{A} \cdot (\mathbf{U} \boxtimes \mathbf{U})\mathbf{B} = \mathbf{U}\mathbf{A}\mathbf{U}^\top \cdot \mathbf{U}\mathbf{B}\mathbf{U}^\top \\ &= \mathbf{U}^\top\mathbf{U}\mathbf{A}\mathbf{U}^\top \cdot \mathbf{B}\mathbf{U}^\top = \mathbf{A}\mathbf{U}^\top \cdot \mathbf{B}\mathbf{U}^\top = \mathbf{A}\mathbf{U}^\top\mathbf{U} \cdot \mathbf{B} = \mathbf{A} \cdot \mathbf{B}. \quad \square\end{aligned}$$

Just as for tensors of $Orth(\mathcal{V})$, we also have

$$\mathbb{U}\mathbb{U}^\top = \mathbb{U}^\top\mathbb{U} = \mathbb{I}.$$

In fact, see the assertion in Exercise 4:

$$\mathbb{U}\mathbb{U}^\top = (\mathbf{U} \boxtimes \mathbf{U})(\mathbf{U}^\top \boxtimes \mathbf{U}^\top) = \mathbf{U}\mathbf{U}^\top \boxtimes \mathbf{U}\mathbf{U}^\top = \mathbf{I} \boxtimes \mathbf{I} = \mathbb{I}. \tag{3.7}$$

The orthogonal conjugators also have some properties in relation with projectors.

Theorem 23. *$\mathbb{S}^{sph}$ is unaffected by any orthogonal conjugator, while $\mathbb{D}^{dev}$ commutes with any orthogonal conjugator.*

Proof. For any $\mathbf{L} \in Sym(\mathcal{V})$ and $\mathbf{U} \in Orth(\mathcal{V})$,

$$\begin{aligned}\mathbb{U}\mathbb{S}^{sph}\mathbf{L} &= (\mathbf{U} \boxtimes \mathbf{U})\left(\frac{1}{3}\mathbf{I} \otimes \mathbf{I}\right)\mathbf{L} = \frac{1}{3}(\mathrm{tr}\mathbf{L})(\mathbf{U} \boxtimes \mathbf{U})\mathbf{I} = \frac{1}{3}(\mathrm{tr}\mathbf{L})\mathbf{U}\mathbf{I}\mathbf{U}^\top \\ &= \frac{1}{3}(\mathrm{tr}\mathbf{L})\mathbf{I} = \frac{1}{3}\mathbf{I} \cdot \mathbf{L}\ \mathbf{I} = \frac{1}{3}(\mathbf{I} \otimes \mathbf{I})\mathbf{L} = \mathbb{S}^{sph}\mathbf{L}.\end{aligned}$$

Moreover,

$$\begin{aligned}\mathbb{S}^{sph}\mathbb{U}\mathbf{L} &= \left(\frac{1}{3}\mathbf{I} \otimes \mathbf{I}\right)(\mathbf{U} \boxtimes \mathbf{U})\mathbf{L} = \frac{1}{3}(\mathbf{I} \otimes \mathbf{I})(\mathbf{U}\mathbf{L}\mathbf{U}^\top) = \frac{1}{3}(\mathbf{I} \cdot \mathbf{U}\mathbf{L}\mathbf{U}^\top)\mathbf{I} \\ &= \frac{1}{3}\mathrm{tr}(\mathbf{U}\mathbf{L}\mathbf{U}^\top)\mathbf{I} = \frac{1}{3}\mathrm{tr}(\mathbf{U}^\top\mathbf{U}\mathbf{L})\mathbf{I} = \frac{1}{3}(\mathrm{tr}\mathbf{L})\mathbf{I} = \frac{1}{3}\mathbf{I} \cdot \mathbf{L}\mathbf{I} \\ &= \frac{1}{3}(\mathbf{I} \otimes \mathbf{I})\mathbf{L} = \mathbb{S}^{sph}\mathbf{L}.\end{aligned}$$

Thus, we have proved that

$$\mathbb{S}^{sph}\mathbb{U} = \mathbb{U}\mathbb{S}^{sph} = \mathbb{S}^{sph},$$

i.e. that the spherical projector $\mathbb{S}^{sph}$ is unaffected by any orthogonal conjugator. Furthermore,

$$\mathbb{D}^{dev}\mathbb{U}\mathbf{L} = (\mathbb{I}^s - \mathbb{S}^{sph})\mathbb{U}\mathbf{L} = \mathbb{I}^s\mathbb{U}\mathbf{L} - \mathbb{S}^{sph}\mathbb{U}\mathbf{L} = \mathbb{U}\mathbf{L} - \mathbb{S}^{sph}\mathbf{L} = (\mathbb{U} - \mathbb{S}^{sph})\mathbf{L}$$

and

$$\mathbb{U}\mathbb{D}^{dev}\mathbf{L} = \mathbb{U}(\mathbb{I}^s - \mathbb{S}^{sph})\mathbf{L} = \mathbb{U}\mathbb{I}^s\mathbf{L} - \mathbb{U}\mathbb{S}^{sph}\mathbf{L} = \mathbb{U}\mathbf{L} - \mathbb{S}^{sph}\mathbf{L} = (\mathbb{U} - \mathbb{S}^{sph})\mathbf{L}$$

so that

$$\mathbb{D}^{dev}\mathbb{U} = \mathbb{U}\mathbb{D}^{dev}.$$

□

3.7 Rotations and symmetries

We ponder now how to rotate a fourth-rank tensor, i.e. what are the components of

$$\mathbb{L} = L_{ijkl}\mathbf{e}_i \otimes \mathbf{e}_j \otimes \mathbf{e}_k \otimes \mathbf{e}_l$$

in a basis $\mathcal{B}' = \{\mathbf{e}'_1, \mathbf{e}'_2, \mathbf{e}'_3\}$ obtained by rotating the basis $\mathcal{B} = \{\mathbf{e}_1, \mathbf{e}_2, \mathbf{e}_3\}$ by the rotation $\mathbb{R} = R_{ij}\mathbf{e}_i \otimes \mathbf{e}_j, \mathbb{R} \in Orth(\mathcal{V})^+$. The procedure is exactly the same as already seen for vectors and second-rank tensors:

$$\begin{aligned}\mathbb{L} &= L_{ijkl}\mathbf{e}_i \otimes \mathbf{e}_j \otimes \mathbf{e}_k \otimes \mathbf{e}_l = L_{ijkl}R^\top_{pi}\mathbf{e}'_p \otimes R^\top_{qj}\mathbf{e}'_q \otimes R^\top_{rk}\mathbf{e}'_r \otimes R^\top_{sl}\mathbf{e}'_s \\ &= R^\top_{pi}R^\top_{qj}R^\top_{rk}R^\top_{sl}L_{ijkl}\mathbf{e}'_p \otimes \mathbf{e}'_q \otimes \mathbf{e}'_r \otimes \mathbf{e}'_s,\end{aligned}$$

i.e.

$$L'_{pqrs} = R^\top_{pi}R^\top_{qj}R^\top_{rk}R^\top_{sl}L_{ijkl}.$$

We see clearly that the components of $\mathbb{L}$ in the basis $\mathcal{B}'$ are a linear combination of those in $\mathcal{B}$, the coefficients of the linear combination being fourth powers of the director cosines, R_{ij}s. The introduction of the orthogonal conjugator[2] of the rotation $\mathbf{R}$,

$$\mathbb{R} = \mathbf{R} \boxtimes \mathbf{R},$$

[2]Here, the symbol $\mathbb{R}$ indicates the orthogonal conjugator of $\mathbf{R}$ and not the set of real numbers.

allows us to give a compact expression for the rotation of second- and fourth-rank tensors (for completeness, we also recall that of a vector $\mathbf{w}$):

$$\mathbf{w}' = \mathbf{R}^\top \mathbf{w},$$
$$\mathbf{L}' = \mathbf{R}^\top \mathbf{L} \mathbf{R} = (\mathbf{R}^\top \boxtimes \mathbf{R}^\top)\mathbf{L} = \mathbb{R}^\top \mathbf{L},$$
$$\mathbb{L}' = (\mathbf{R}^\top \boxtimes \mathbf{R}^\top)\mathbb{L}(\mathbf{R} \boxtimes \mathbf{R}) = \mathbb{R}^\top \mathbb{L}\mathbb{R}.$$

Checking the above relations with the orthogonal conjugator $\mathbb{R}$ is left to the reader. It is worth noting that, actually, these transformations are valid not only for $\mathbf{R} \in Orth(\mathcal{V})^+$ but more generally for any $\mathbf{U} \in Orth(\mathcal{V})$, i.e. also for symmetries.

If by $\mathbf{U}$, we denote the tensor of the change of basis under any orthogonal transformation, i.e. if we put $\mathbf{U} = \mathbf{R}^\top$ for the rotations, then the above relations become

$$\mathbf{w}' = \mathbf{U}\mathbf{w},$$
$$\mathbf{L}' = \mathbf{U}\mathbf{L}\mathbf{U}^\top = (\mathbf{U} \boxtimes \mathbf{U})\mathbf{L} = \mathbb{U}\mathbf{L}, \tag{3.8}$$
$$\mathbb{L}' = (\mathbf{U} \boxtimes \mathbf{U})\mathbb{L}(\mathbf{U} \boxtimes \mathbf{U})^\top = \mathbb{U}\mathbb{L}\mathbb{U}^\top.$$

Finally, we say that $\mathbf{L} \in Lin(\mathcal{V})$ or $\mathbb{L} \in \mathbb{L}\mathrm{in}(\mathcal{V})$ is *invariant under an orthogonal transformation* $\mathbf{U}$ if

$$\mathbf{U}\mathbf{L}\mathbf{U}^\top = \mathbf{L}, \quad \mathbb{U}\mathbb{L}\mathbb{U}^\top = \mathbb{L};$$

right multiplying both terms by $\mathbf{U}$ or by $\mathbb{U}$ and through Eq. (3.7), we get that $\mathbf{L}$ or $\mathbb{L}$ are invariant under $\mathbf{U}$ $\iff$

$$\mathbf{U}\mathbf{L} = \mathbf{L}\mathbf{U}, \quad \mathbb{U}\mathbb{L} = \mathbb{L}\mathbb{U},$$

i.e. $\iff$ $\mathbf{L}$ and $\mathbf{U}$ or $\mathbb{L}$ and $\mathbb{U}$ commute. This relation allows, for example, the analysis of material symmetries in anisotropic elasticity.

If a tensor is invariant under *any* orthogonal transformation, i.e. if the previous equations hold true $\forall \mathbf{U} \in Orth(\mathcal{V})$, then the tensor is said to be *isotropic*. A general result[3] is that a fourth-rank tensor $\mathbb{L}$ is isotropic $\iff$ there exist two scalar functions λ, μ such that

$$\mathbb{L}\mathbf{A} = 2\mu\mathbf{A} + \lambda \mathrm{tr}\mathbf{A}\ \mathbf{I} \quad \forall \mathbf{A} \in Sym(\mathcal{V}).$$

The reader is referred to the book by Gurtin (see references) for the proof of this result and for a deeper insight into isotropic functions.

[3] Actually, this is quite a famous result in classical elasticity, the *Lamé's equation*, defining an isotropic elastic material.

3.8 The Kelvin formalism

As already mentioned, though fourth-rank tensors cannot be organized in and represented by a matrix, nevertheless a matrix formalism for these operators exists. Such a formalism is due to Kelvin[4], and it is strictly related to the theory of elasticity, i.e. it concerns Cauchy's stress tensor $\boldsymbol{\sigma}$, the strain tensor $\boldsymbol{\varepsilon}$, and the elasticity tensor $\mathbb{E}$. The relation between $\boldsymbol{\sigma}$ and $\boldsymbol{\varepsilon}$ is given by the celebrated (generalized) *Hooke's law*:

$$\boldsymbol{\sigma} = \mathbb{E}\boldsymbol{\varepsilon}.$$

Both $\boldsymbol{\sigma}, \boldsymbol{\varepsilon} \in Sym(\mathcal{V})$ while $\mathbb{E} = \mathbb{E}^\top$, and it also has the minor symmetries, so $\mathbb{E}$ has only 21 independent components.[5] In the Kelvin formalism, the six independent components of $\boldsymbol{\sigma}$ and $\boldsymbol{\varepsilon}$ are organized into column vectors and renumbered as follows:

$$\{\sigma\} = \begin{Bmatrix} \sigma_1 = \sigma_{11} \\ \sigma_2 = \sigma_{22} \\ \sigma_3 = \sigma_{33} \\ \sigma_4 = \sqrt{2}\sigma_{23} \\ \sigma_5 = \sqrt{2}\sigma_{31} \\ \sigma_6 = \sqrt{2}\sigma_{12} \end{Bmatrix}, \quad \{\varepsilon\} = \begin{Bmatrix} \varepsilon_1 = \varepsilon_{11} \\ \varepsilon_2 = \varepsilon_{22} \\ \varepsilon_3 = \varepsilon_{33} \\ \varepsilon_4 = \sqrt{2}\varepsilon_{23} \\ \varepsilon_5 = \sqrt{2}\varepsilon_{31} \\ \varepsilon_6 = \sqrt{2}\varepsilon_{12} \end{Bmatrix}.$$

The elasticity tensor $\mathbb{E}$ is reduced to a 6×6 matrix $[E]$ as a consequence of the minor symmetries induced by the symmetry of $\boldsymbol{\sigma}$ and $\boldsymbol{\varepsilon}$; this matrix is symmetric because $\mathbb{E} = \mathbb{E}^\top$:

$$[E] = \begin{bmatrix} E_{11} = E_{1111} & E_{12} = E_{1122} & E_{13} = E_{1133} & E_{14} = \sqrt{2}E_{1123} & E_{15} = \sqrt{2}E_{1131} & E_{16} = \sqrt{2}E_{1112} \\ E_{12} = E_{1122} & E_{22} = E_{2222} & E_{23} = E_{2233} & E_{24} = \sqrt{2}E_{2223} & E_{25} = \sqrt{2}E_{2231} & E_{26} = \sqrt{2}E_{2212} \\ E_{13} = E_{1133} & E_{23} = E_{2233} & E_{33} = E_{3333} & E_{34} = \sqrt{2}E_{3323} & E_{35} = \sqrt{2}E_{3331} & E_{36} = \sqrt{2}E_{3312} \\ E_{14} = \sqrt{2}E_{1123} & E_{24} = \sqrt{2}E_{2223} & E_{34} = \sqrt{2}E_{3323} & E_{44} = 2E_{2323} & E_{45} = 2E_{2331} & E_{46} = 2E_{2312} \\ E_{15} = \sqrt{2}E_{1131} & E_{25} = \sqrt{2}E_{2231} & E_{35} = \sqrt{2}E_{3331} & E_{45} = 2E_{2331} & E_{55} = 2E_{3131} & E_{56} = 2E_{3112} \\ E_{16} = \sqrt{2}E_{1112} & E_{26} = \sqrt{2}E_{2212} & E_{36} = \sqrt{2}E_{3312} & E_{46} = 2E_{2312} & E_{56} = 2E_{3112} & E_{66} = 2E_{1212} \end{bmatrix}.$$

In this way, the matrix product

$$\{\sigma\} = [E]\{\varepsilon\} \tag{3.9}$$

[4]W. Thomson (Lord Kelvin): Elements of a Mathematical Theory of Elasticity. *Philos. Trans. R. Soc.*, 146, 481–498, 1856. Later, Voigt (W. Voigt: *Lehrbuch der Kristallphysik*. B. G. Taubner, Leipzig, 1910) gave another, similar matrix formalism for tensors, more widely known than the Kelvin one but less effective.

[5]Actually, the Kelvin formalism can also be extended without major difficulties to tensors that do not possess all the symmetries.

is equivalent to the tensor form of the Hooke's law, and all the operations can be done by the aid of classical matrix algebra,[6] e.g. the computation of the inverse of $\mathbb{E}$, the *compliance tensor.*

An important operation is the expression of tensor $\mathbb{U}$ in Eq. (3.8) in the Kelvin formalism; some tedious but straightforward operations give the result:

$$[U]=\begin{bmatrix} U_{11}^2 & U_{12}^2 & U_{13}^2 & \sqrt{2}U_{12}U_{13} & \sqrt{2}U_{13}U_{11} & \sqrt{2}U_{11}U_{12}\\ U_{21}^2 & U_{22}^2 & U_{23}^2 & \sqrt{2}U_{22}U_{23} & \sqrt{2}U_{23}U_{21} & \sqrt{2}U_{21}U_{22}\\ U_{31}^2 & U_{32}^2 & U_{33}^2 & \sqrt{2}U_{32}U_{33} & \sqrt{2}U_{33}U_{31} & \sqrt{2}U_{31}U_{32}\\ \sqrt{2}U_{21}U_{31} & \sqrt{2}U_{22}U_{32} & \sqrt{2}U_{23}U_{33} & U_{23}U_{32}+U_{22}U_{33} & U_{33}U_{21}+U_{31}U_{23} & U_{31}U_{22}+U_{32}U_{21}\\ \sqrt{2}U_{31}U_{11} & \sqrt{2}U_{32}U_{12} & \sqrt{2}U_{33}U_{13} & U_{32}U_{13}+U_{33}U_{12} & U_{31}U_{13}+U_{33}U_{11} & U_{31}U_{12}+U_{32}U_{11}\\ \sqrt{2}U_{11}U_{21} & \sqrt{2}U_{12}U_{22} & \sqrt{2}U_{13}U_{23} & U_{12}U_{23}+U_{13}U_{22} & U_{11}U_{23}+U_{13}U_{21} & U_{11}U_{22}+U_{12}U_{21} \end{bmatrix}.$$

With some work, it can be checked that

$$[U][U]^\top = [U]^\top[U] = [I],$$

i.e. that $[U]$ is an orthogonal matrix in $\mathbb{R}^6$. Of course,

$$[R] = [U]^\top$$

is the matrix that in the Kelvin formalism represents the tensor $\mathbf{R} = \mathbf{U}^\top$. The change of basis for $\boldsymbol{\sigma}$ and $\boldsymbol{\varepsilon}$ are hence done through the relations

$$\{\sigma'\} = [U]\{\sigma\}, \quad \{\varepsilon'\} = [U]\{\varepsilon\},$$

which when applied to Eq. (3.9) give

$$\{\sigma\} = [E]\{\varepsilon\} \ \rightarrow \ [U]^\top\{\sigma'\} = [E][U]^\top\{\varepsilon'\} \ \rightarrow \ \{\sigma'\} = [U][E][U]^\top\{\varepsilon'\},$$

i.e. in the basis $\mathcal{B}'$,

$$\{\sigma'\} = [E']\{\varepsilon'\},$$

where

$$[E'] = [U][E][U]^\top = [R]^\top[E][R]$$

is the matrix representing $\mathbb{E}$ in $\mathcal{B}'$ in the Kelvin formalism. Though it is possible to give the expression of the components of $[E']$, they are so long that they are omitted here.

[6]Mehrabadi and Cowin have shown that the Kelvin formalism transforms second- and fourth-rank tensors on $\mathbb{R}^3$ into vectors and second-rank tensors on $\mathbb{R}^6$ (M. M. Mehrabadi and S. C. Cowin: Eigentensors of linear anisotropic elastic materials. *Q. J. Mech. Appl. Math.*, 43, 15–41, 1990).

3.9 The polar formalism for plane tensors

The Cartesian representation of tensors makes use of quantities that are basis-dependent, and the change of basis implies algebraic transformations rather complicate. The question of representing tensors using other quantities than Cartesian components is hence of importance. In particular, it should be interesting to represent a tensor making use of only invariants of the tensor itself and of angles, which is the simplest geometrical way to determine a direction.

In the case of plane tensors, this has been done by Verchery[7] who introduced the so-called *polar formalism.* This is basically a mathematical technique to find the invariants of a tensor of any rank. Here, we give just a short insight into the polar formalism of fourth-rank tensors, omitting the proof of the results.[8]

The Cartesian components of a plane fourth-rank tensor $\mathbb{T}$ in a frame rotated through an angle θ can be expressed as

$$\begin{aligned}
T_{1111} &= T_0 + 2T_1 + R_0 \cos 4(\varPhi_0 - \theta) + 4R_1 \cos 2(\varPhi_1 - \theta),\\
T_{1112} &= R_0 \sin 4(\varPhi_0 - \theta) + 2R_1 \sin 2(\varPhi_1 - \theta),\\
T_{1122} &= -T_0 + 2T_1 - R_0 \cos 4(\varPhi_0 - \theta),\\
T_{1212} &= T_0 - R_0 \cos 4(\varPhi_0 - \theta),\\
T_{1222} &= -R_0 \sin 4(\varPhi_0 - \theta) + 2R_1 \sin 2(\varPhi_1 - \theta),\\
T_{2222} &= T_0 + 2T_1 + R_0 \cos 4(\varPhi_0 - \theta) - 4R_1 \cos 2(\varPhi_1 - \theta).
\end{aligned}$$

In the above equations, T_0, T_1, R_0, and R_1 are tensor invariants, with all of them nonnegative, while $\varPhi_0$ and $\varPhi_1$ are angles whose difference, $\varPhi_0 - \varPhi_1$, is also a tensor invariant, so fixing one of the two polar angles corresponds to fixing a frame. In particular, the tensor invariants have a direct physical meaning (e.g. for the elasticity tensor, they are linked to material symmetries and strain energy decomposition). We remark also that the change of frame is extremely simple in the polar formalism: It is sufficient to subtract the angle θ formed by the new frame from the two polar angles.

---

[7]G. Verchery: *Les invariants des tenseurs d'ordre 4 du type de l'élasticité*, Proc. Colloque EUROMECH 115, 1979.

[8]A detailed presentation of the method can be found in the work by Vannucci: *Anisotropic Elasticity*, Springer, 2018.

The Cartesian expression of the polar invariants can be found by inverting the previous expressions:

$$\begin{aligned}
T_0 &= \frac{1}{8}(T_{1111} - 2T_{1122} + 4T_{1212} + T_{2222}),\\
T_1 &= \frac{1}{8}(T_{1111} + 2T_{1122} + T_{2222}),\\
R_0 &= \frac{1}{8}\sqrt{(T_{1111} - 2T_{1122} - 4T_{1212} + T_{2222})^2 + 16(T_{1112} - T_{1222})^2},\\
R_1 &= \frac{1}{8}\sqrt{(T_{1111} - T_{2222})^2 + 4(T_{1112} + T_{1222})^2},\\
\tan 4\Phi_0 &= \frac{4(T_{1112} - T_{1222})}{T_{1111} - 2T_{1122} - 4T_{1212} + T_{2222}},\\
\tan 2\Phi_1 &= \frac{2(T_{1112} + T_{1222})}{T_{1111} - T_{2222}}.
\end{aligned}$$

3.10 Exercises

1. Prove Eq. (3.4).
2. Prove that
$$(\mathbb{A}\mathbb{B})^\top = \mathbb{B}^\top \mathbb{A}^\top.$$
3. Prove that
$$\mathbf{A} \otimes \mathbf{B}\mathbb{L} = \mathbf{A} \otimes \mathbb{L}^\top \mathbf{B}.$$
4. Prove that
$$(\mathbf{A} \boxtimes \mathbf{B})(\mathbf{C} \boxtimes \mathbf{D}) = \mathbf{AC} \boxtimes \mathbf{BD}.$$
5. Prove Eq. (3.3) using the result of the previous exercise.
6. Prove that
$$(\mathbf{A} \otimes \mathbf{B})(\mathbf{C} \boxtimes \mathbf{D}) = \mathbf{A} \otimes ((\mathbf{C}^\top \boxtimes \mathbf{D}^\top)\mathbf{B}).$$
7. Prove that
$$(\mathbf{A} \boxtimes \mathbf{B})(\mathbf{C} \otimes \mathbf{D}) = ((\mathbf{A} \boxtimes \mathbf{B})\mathbf{C}) \otimes \mathbf{D}.$$
8. Let $\mathbf{p} \in \mathcal{S}$ and $\mathbf{P} = \mathbf{p} \otimes \mathbf{p}$, then prove that
$$\mathbf{P} \boxtimes \mathbf{P} = \mathbf{P} \otimes \mathbf{P}.$$
9. Prove that, $\forall \mathbb{A} \in \mathbb{L}\mathrm{in}(\mathcal{V})$,
$$\mathbb{I}\mathbb{A} = \mathbb{A}\mathbb{I} = \mathbb{A}.$$

10. Show that

$$(\mathbf{A} \otimes \mathbf{B}) \cdot (\mathbf{C} \otimes \mathbf{D}) = \mathbf{A} \cdot \mathbf{C}\ \mathbf{B} \cdot \mathbf{D}.$$

11. Show that

$$\mathbb{S}^{sph} = \frac{\mathbf{I}}{|\mathbf{I}|} \otimes \frac{\mathbf{I}}{|\mathbf{I}|}.$$

12. Show that

$$dim(Sph(\mathcal{V})) = 1, \quad dim(Dev(\mathcal{V})) = 5.$$

13. Show the following properties of $\mathbb{S}^{sph}$ and $\mathbb{D}^{dev}$:

$$\mathbb{S}^{sph}\mathbb{S}^{sph} = \mathbb{S}^{sph},$$

$$\mathbb{D}^{dev}\mathbb{D}^{dev} = \mathbb{D}^{dev},$$

$$\mathbb{S}^{sph}\mathbb{D}^{dev} = \mathbb{D}^{dev}\mathbb{S}^{sph} = \mathbb{O}.$$

14. Prove the results in Eqs. (3.5) and (3.6) using the components.
15. Show that

$$\mathbb{S}^{sph} \cdot \mathbb{S}^{sph} = 1,$$

$$\mathbb{D}^{dev} \cdot \mathbb{D}^{dev} = 5,$$

$$\mathbb{S}^{sph} \cdot \mathbb{D}^{dev} = 0.$$

16. Make explicit the orthogonal conjugator $\mathbb{S}_R$ of the tensor $\mathbf{S}_R$ in Eq. (2.49).
17. Using the polar formalism, it can be proved that the material symmetry conditions in plane elasticity are all condensed into the equation

$$R_0 R_1 \sin 4(\Phi_0 - \Phi_1) = 0;$$

determine the different types of possible elastic symmetries.

Chapter 4

Tensor Analysis: Curves

4.1 Curves of points, vectors and tensors

The scalar products in $\mathcal{V}, Lin(\mathcal{V})$, and $\mathbb{L}\text{in}(\mathcal{V})$ allow us to define a *norm*, the *Euclidean norm*, so they automatically endow these spaces with a *metric*, i.e. we are able to measure and calculate a distance between two elements of such a space and in $\mathcal{E}$. This allows us to generalize the concepts of continuity and differentiability already known in $\mathbb{R}$, whose definition intrinsically makes use of a distance between real quantities.

Let $\pi_n = \{p_n \in \mathcal{E}, n \in \mathbb{N}\}$ be a sequence of points in $\mathcal{E}$. We say that π_n *converges to* $p \in \mathcal{E}$ if

$$\lim_{n\to\infty} d(p_n - p) = 0.$$

A similar definition can be given for sequences of vectors or tensors of any rank. Through this definition of convergence, we can now make the concepts of continuity and curve precise.

Let $[a, b]$ be an interval of $\mathbb{R}$; the function

$$p = p(t) : [a, b] \to \mathcal{E}$$

is *continuous* at $t \in [a, b]$ if for each sequence $\{t_n \in [a, b], n \in \mathbb{N}\}$ that converges to t, the sequence π_n defined by $p_n = p(t_n)\ \forall n \in \mathbb{N}$ converges to $p(t) \in \mathcal{E}$. The function $p = p(t)$ is a *curve in* $\mathcal{E} \iff$ it is continuous $\forall t \in [a, b]$. In the same way, we can define a curve of vectors and tensors:

$$\mathbf{v}=\mathbf{v}(t):[a,b]\rightarrow\mathcal{V},$$

$$\mathbf{L}=\mathbf{L}(t):[a,b]\rightarrow Lin(\mathcal{V}),$$

$$\mathbb{L}=\mathbb{L}(t):[a,b]\rightarrow\mathbb{L}\mathrm{in}(\mathcal{V}).$$

Mathematically, a curve is a function that lets correspond to a real value t (the *parameter*) in a given interval, an element of a space: $\mathcal{E},\mathcal{V},Lin(\mathcal{V})$, or $\mathbb{L}(\mathcal{V})$.

4.2 Differentiation of curves

Let $\mathbf{v}=\mathbf{v}(t):[a,b]\rightarrow\mathcal{V}$ be a curve of vectors and $g=g(t):[a,b]\rightarrow\mathbb{R}$ a scalar function. We say that $\mathbf{v}$ is *of the order o with respect to g in* $t_0 \iff$

$$\lim_{t\rightarrow t_0}\frac{|\mathbf{v}(t)|}{|g(t)|}=0,$$

and we write

$$\mathbf{v}(t)=o(g(t)) \text{ for } t\rightarrow t_0.$$

A similar definition can be given for a curve of tensors of any rank. We then say that the curve $\mathbf{v}$ is *differentiable* in $t_0\in]a,b[\iff\exists\mathbf{v}'\in\mathcal{V}$ such that

$$\mathbf{v}(t)-\mathbf{v}(t_0)=(t-t_0)\mathbf{v}'(t_0)+o(t-t_0).$$

We call $\mathbf{v}'(t_0)$ the *derivative*[1] *of* $\mathbf{v}$ at t_0. Applying the definition of derivative to $\mathbf{v}'$, we define the *second derivative* $\mathbf{v}''$ of $\mathbf{v}$ and recursively all the derivatives of higher orders. We say that $\mathbf{v}$ is of *class* C^n if it is continuous with its derivatives up to the order n; if $n\geq 1$, $\mathbf{v}$ is said to be *smooth*. A curve $\mathbf{v}(t)$ of class C^n is said to be *regular* if $\mathbf{v}'\neq\mathbf{o}\ \forall t$. Similar definitions can be given for curves in $\mathcal{E}, Lin(\mathcal{V})$, and $\mathbb{L}\mathrm{in}(\mathcal{V})$, thus defining derivatives of points and tensors. We remark that the derivative of a curve in $\mathcal{E}$, defined as a difference of points, is a curve in $\mathcal{V}$ (we say, in short, that *the derivative of a point is a vector*). For what concerns tensors, the derivative of a tensor of rank r is a tensor of the same rank.

Let $\mathbf{u},\mathbf{v}$ be curves in $\mathcal{V}$, $\mathbf{L},\mathbf{M}$ curves in $Lin(\mathcal{V})$, $\mathbb{L},\mathbb{M}$ curves in $\mathbb{L}\mathrm{in}(\mathcal{V})$, and α a scalar function, with all of them defined and at least of class C^1

[1] The derivative is also written as $\dfrac{d\mathbf{v}}{dt}$, $\mathbf{v}_{,t}$ or also as $\dot{\mathbf{v}}$, with the last symbol being usually reserved, in physics, to the case where t is time. For the sake of brevity, we omit to indicate the derivative of $\mathbf{v}$ at t_0 as $\mathbf{v}'(t_0)$, writing simply $\mathbf{v}'$.

on $[a, b]$. The same definition of the derivative of a curve gives the following results, whose proof is left to the reader:

$$(\mathbf{u}+\mathbf{v})' = \mathbf{u}' + \mathbf{v}',$$
$$(\alpha\mathbf{v})' = \alpha'\mathbf{v} + \alpha\mathbf{v}',$$
$$(\mathbf{u}\cdot\mathbf{v})' = \mathbf{u}'\cdot\mathbf{v} + \mathbf{u}\cdot\mathbf{v}',$$
$$(\mathbf{u}\times\mathbf{v})' = \mathbf{u}'\times\mathbf{v} + \mathbf{u}\times\mathbf{v}',$$
$$(\mathbf{u}\otimes\mathbf{v})' = \mathbf{u}'\otimes\mathbf{v} + \mathbf{u}\otimes\mathbf{v}',$$
$$(\mathbf{L}+\mathbf{M})' = \mathbf{L}' + \mathbf{M}',$$
$$(\alpha\mathbf{L})' = \alpha'\mathbf{L} + \alpha\mathbf{L}',$$
$$(\mathbf{L}\mathbf{v})' = \mathbf{L}'\mathbf{v} + \mathbf{L}\mathbf{v}',$$
$$(\mathbf{L}\mathbf{M})' = \mathbf{L}'\mathbf{M} + \mathbf{L}\mathbf{M}',$$
$$(\mathbf{L}\cdot\mathbf{M})' = \mathbf{L}'\cdot\mathbf{M} + \mathbf{L}\cdot\mathbf{M}',$$
$$(\mathbf{L}\otimes\mathbf{M})' = \mathbf{L}'\otimes\mathbf{M} + \mathbf{L}\otimes\mathbf{M}',$$
$$(\mathbf{L}\boxtimes\mathbf{M})' = \mathbf{L}'\boxtimes\mathbf{M} + \mathbf{L}\boxtimes\mathbf{M}',$$
$$(\mathbb{L}+\mathbb{M})' = \mathbb{L}' + \mathbb{M}',$$
$$(\alpha\mathbb{L})' = \alpha'\mathbb{L} + \alpha\mathbb{L}',$$
$$(\mathbb{L}\mathbf{L})' = \mathbb{L}'\mathbf{L} + \mathbb{L}\mathbf{L}',$$
$$(\mathbb{L}\mathbb{M})' = \mathbb{L}'\mathbb{M} + \mathbb{L}\mathbb{M}',$$
$$(\mathbb{L}\cdot\mathbb{M})' = \mathbb{L}'\cdot\mathbb{M} + \mathbb{L}\cdot\mathbb{M}'.$$

We remark that the derivative of any kind of product is made according to the usual rule of the derivative of a product of functions.

Let $\mathcal{R} = \{o; \mathcal{B}\}$ be a reference frame of the euclidean space $\mathcal{E}$, composed of an *origin* o and a basis $\mathcal{B} = \{\mathbf{e}_1, \mathbf{e}_2, \mathbf{e}_3\}$ of $\mathcal{V}$, $\mathbf{e}_i\cdot\mathbf{e}_j = \delta_{ij}\ \forall i, j = 1, 2, 3$, and let us consider a point $p(t) = (p_1(t), p_2(t), p_3(t))$. If the three *coordinates* $p_i(t)$ are three continuous functions over the interval $[t_1, t_2] \in \mathbb{R}$, then, by the definition given above, the mapping $p(t) : [t_1, t_2] \rightarrow \mathcal{E}$ is a curve in $\mathcal{E}$ and the equation

$$p(t) = (p_1(t), p_2(t), p_3(t)) \ \rightarrow \ \begin{cases} p_1 = p_1(t) \\ p_2 = p_2(t) \\ p_3 = p_3(t) \end{cases}$$

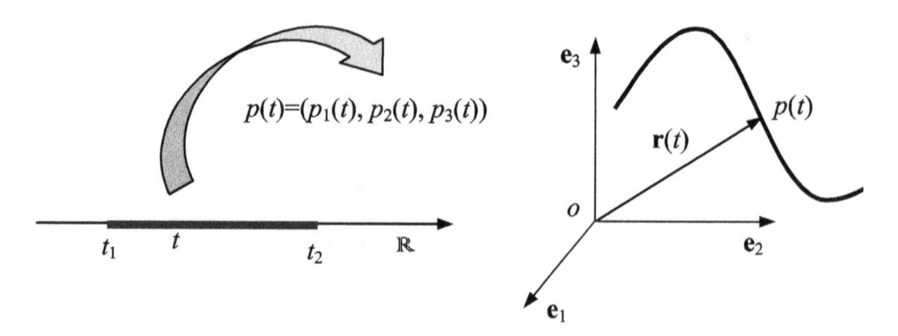

Figure 4.1: Mapping of a curve of points.

is the *parametric point equation of the curve*: To each value of $t \in [t_1, t_2]$, it corresponds to a point of the curve in $\mathcal{E}$, see Fig. 4.1.

The vector function $\mathbf{r}(t) = p(t) - o$ is the *position vector* of point p in $\mathcal{R}$; the equation

$$\mathbf{r}(t) = r_i(t)\mathbf{e}_i = r_1(t)\mathbf{e}_1 + r_2(t)\mathbf{e}_2 + r_3(t)\mathbf{e}_3 \rightarrow \begin{cases} r_1 = r_1(t) \\ r_2 = r_2(t) \\ r_3 = r_3(t) \end{cases}$$

is the *parametric vector equation* of the curve: To each value of $t \in [t_1, t_2]$, there corresponds a vector of $\mathcal{V}$ that determines a point of the curve in $\mathcal{E}$ through the operation $p(t) = o + \mathbf{r}(t)$.

Similarly, if the components $L_{ij}(t)$ are continuous functions of a parameter t, the mapping $\mathbf{L}(t) : [t_1, t_2] \rightarrow Lin(\mathcal{V})$ defined by

$$\mathbf{L}(t) = L_{ij}(t)\mathbf{e}_i \otimes \mathbf{e}_j, \quad i, j = 1, 2, 3,$$

is a curve of tensors. In a similar way, we can give a curve of fourth-rank tensors $\mathbb{L}(t) : [t_1, t_2] \rightarrow \mathbb{L}\text{in}(\mathcal{V})$ by

$$\mathbb{L}(t) = L_{ijkl}(t)\mathbf{e}_i \otimes \mathbf{e}_j \otimes \mathbf{e}_k \otimes \mathbf{e}_l, \quad i, j, k, l = 1, 2, 3.$$

It is noted that the choice of the parameter is *not unique*: The equation $p = p[\tau(t)]$ still represents the same curve $p = p(t)$ through the *change of parameter* $\tau = \tau(t)$.

The definition given above for the derivative of a curve of points $p = p(t)$ in $t = t_0$ is equivalent to the following one[2] (probably more familiar to the

[2]This is also true for the derivatives of vector or tensor curves.

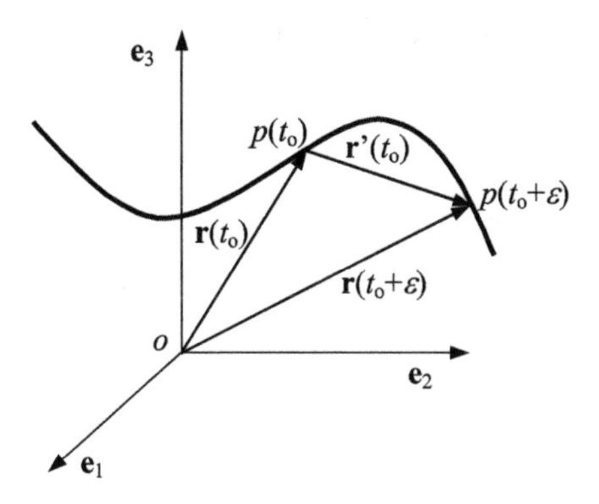

Figure 4.2: Derivative of a curve.

reader):

$$\frac{dp(t)}{dt} = \lim_{\varepsilon \to 0} \frac{p(t_0 + \varepsilon) - p(t_0)}{\varepsilon},$$

represented in Fig. 4.2, where it is apparent that $\mathbf{r}'(t) = \dfrac{dp(t)}{dt}$ is a vector.

An important case is that of a vector $\mathbf{v}(t)$ whose norm $v(t)$ is constant $\forall t$:

$$(v^2)' = (\mathbf{v} \cdot \mathbf{v})' = \mathbf{v}' \cdot \mathbf{v} + \mathbf{v} \cdot \mathbf{v}' = 2\mathbf{v}' \cdot \mathbf{v} = 0. \tag{4.1}$$

The derivative of such a vector is orthogonal to it $\forall t$. The contrary is also true, as is immediately apparent.

Finally, using the above rules and assuming that the reference frame $\mathcal{R}$ is independent of t, we get easily that

$$\begin{aligned}
p'(t) &= p_i'(t)\ \mathbf{e}_i,\\
\mathbf{v}'(t) &= v_i'(t)\ \mathbf{e}_i,\\
\mathbf{L}'(t) &= L_{ij}'(t)\ \mathbf{e}_i \otimes \mathbf{e}_j,\\
\mathbb{L}'(t) &= L_{ijkl}'(t)\ \mathbf{e}_i \otimes \mathbf{e}_j \otimes \mathbf{e}_k \otimes \mathbf{e}_l,
\end{aligned} \tag{4.2}$$

i.e. the derivative of a curve of points, vectors, or tensors is simply calculated by differentiating the coordinates of the components. Using this result, it is immediate to prove that

$$\begin{aligned}
(\mathbf{L}^\top)' &= \mathbf{L}'^\top,\\
(\mathbb{L}^\top)' &= \mathbb{L}'^\top,
\end{aligned}$$

while for any invertible tensor $\mathbf{L}$, we have (we state the following results without proof[3])

$$(\mathbf{L}^{-1})' = -\mathbf{L}^{-1}\mathbf{L}'\mathbf{L}^{-1},$$

$$(\det \mathbf{L})' = \det \mathbf{L}\ \mathrm{tr}(\mathbf{L}'\mathbf{L}^{-1}) = \det \mathbf{L}\ \mathbf{L}^{\top\prime} \cdot \mathbf{L}^{-1} = \det \mathbf{L}\ \mathbf{L}' \cdot \mathbf{L}^{-\top}.$$

Let $\mathbf{Q}(t) : \mathbb{R} \to Orth(\mathcal{V})$ be a differentiable function. We call *spin tensor* the tensor $\mathbf{S}(t)$ defined as

$$\mathbf{S}(t) := \mathbf{Q}'(t)\mathbf{Q}^\top(t).$$

Then, we have the following.[4]

Theorem 24 (Characterization of the spin tensor). $\mathbf{S}(t) \in$ *Skw*$(\mathcal{V})$ $\forall t \in \mathbb{R}$.

Proof. As $\mathbf{Q}(t) \in Orth\mathcal{V}\ \ \forall t$, then

$$\mathbf{Q}\mathbf{Q}^\top = \mathbf{I} \Rightarrow (\mathbf{Q}\mathbf{Q}^\top)' = \mathbf{Q}'\mathbf{Q}^\top + \mathbf{Q}\mathbf{Q}^{\top\prime} = \mathbf{I}' = \mathbf{O} \Rightarrow \mathbf{Q}\mathbf{Q}^{\top\prime} = -\mathbf{Q}'\mathbf{Q}^\top,$$

so

$$\mathbf{S}^\top = (\mathbf{Q}'\mathbf{Q}^\top)^\top = \mathbf{Q}\mathbf{Q}^{\top\prime} = -\mathbf{Q}'\mathbf{Q}^\top = -\mathbf{S}.$$ □

4.3 Integral of a curve of vectors and length of a curve

We define the *integral of a curve of vectors* $\mathbf{r}(t)$ between a and $b \in [t_1, t_2]$ the curve that is obtained by integrating each component of the curve:

$$\int_a^b \mathbf{r}(t)\ dt = \int_a^b r_i(t)\ dt\ \mathbf{e}_i.$$

If the curve is regular, we can generalize the second fundamental theorem of the integral calculus:

$$\mathbf{r}(t) = \mathbf{r}(a) + \int_a^t \mathbf{r}'(t^*)\ dt^*.$$

Because

$$\mathbf{r}(t) = p(t) - o, \quad \mathbf{r}'(t) = (p(t) - o)' = p'(t),$$

[3]The interested reader can find these proofs in the text by Gurtin, see the suggested texts.

[4]The spin tensor and the following result are of importance in kinematics: If t is time and $\mathbf{Q}(t) \in Orth\mathcal{V}^+$, then the axial vector of $\mathbf{S}(t)$ is $\boldsymbol{\omega}(t)$, the *angular velocity*.

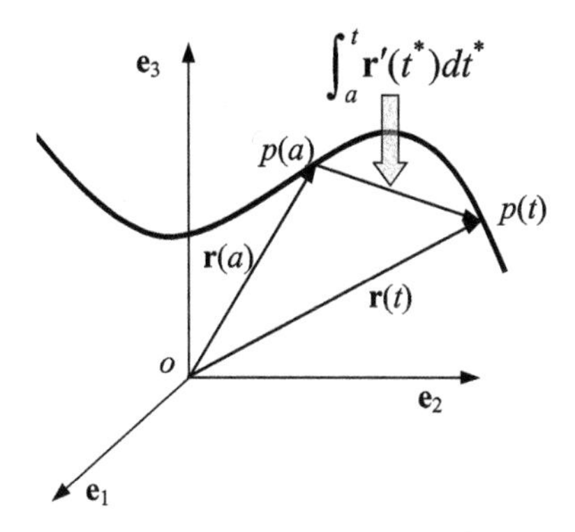

Figure 4.3: Integral of a vector curve.

we also get

$$p(t) = p(a) + \int_a^t p'(t^*)\ dt^*.$$

The integral of a vector function is the generalization of the vector sum, see Fig. 4.3.

Let $\mathbf{r}(t) : [a, b] \to \mathcal{E}$ be a regular curve, σ a partition of $[a, b]$ of the type $a = t_0 < t_1 < \cdots < t_n = b$, and

$$\sigma_{max} = \max_{i=1,\ldots,n} |t_i - t_{i-1}|.$$

The length ℓ_σ of the polygonal line whose vertices are the points $\mathbf{r}(t_i)$ is hence

$$\ell_\sigma = \sum_{i=1}^{n} |\mathbf{r}(t_i) - \mathbf{r}(t_{i-1})|.$$

We define the *length of the curve* $\mathbf{r}(t)$ the (positive) number

$$\ell := \sup_\sigma \ell_\sigma.$$

Theorem 25. *Let* $\mathbf{r}(t) : [a, b] \Rightarrow \mathcal{E}$ *be a regular curve, then*

$$\ell = \int_a^b |\mathbf{r}'(t)| dt.$$

Proof. By the fundamental theorem of calculus,

$$\mathbf{r}(t_i) - \mathbf{r}(t_{i-1}) = \int_{t_{i-1}}^{t_i} \mathbf{r}'(t)dt$$

so that, using Minkowski's inequality,

$$|\mathbf{r}(t_i) - \mathbf{r}(t_{i-1})| = \left| \int_{t_{i-1}}^{t_i} \mathbf{r}'(t)dt \right| \leq \int_{t_{i-1}}^{t_i} |\mathbf{r}'(t)|dt,$$

whence

$$\ell \leq \int_a^b |\mathbf{r}'(t)|dt. \tag{4.3}$$

Because $\mathbf{r}'(t)$ is continuous on $[a, b]$, $\forall \varepsilon > 0\ \exists \delta > 0$ such that $|t - \bar{t}| < \delta \Rightarrow |\mathbf{r}'(t) - \mathbf{r}'(\bar{t})| < \varepsilon$. Let $t \in [t_{i-1}, t_i]$ and $\sigma_{max} < \delta$, which is always possible by the choice of the partition σ; again by Minkowski's inequality,

$$|\mathbf{r}'(t)| \leq |\mathbf{r}'(t) - \mathbf{r}'(t_i)| + |\mathbf{r}'(t_i)| < \varepsilon + |\mathbf{r}'(t_i)|,$$

whence

$$\begin{aligned}
\int_{t_{i-1}}^{t_i} |\mathbf{r}'(t)|dt &< \int_{t_{i-1}}^{t_i} |\mathbf{r}'(t_i)|dt + \varepsilon(t_i - t_{i-1}) = \left| \int_{t_{i-1}}^{t_i} \mathbf{r}'(t_i)dt \right| + \varepsilon(t_i - t_{i-1}) \\
&\leq \left| \int_{t_{i-1}}^{t_i} \mathbf{r}'(t)dt \right| + \left| \int_{t_{i-1}}^{t_i} (\mathbf{r}'(t_i) - \mathbf{r}'(t))dt \right| + \varepsilon(t_i - t_{i-1}) \\
&\leq |\mathbf{r}(t_i) - \mathbf{r}(t_{i-1})| + 2\varepsilon(t_i - t_{i-1}).
\end{aligned}$$

Summing up over all the intervals $[t_{i-1}, t_i]$, we get

$$\int_a^b |\mathbf{r}'(t)|dt \leq \ell_\sigma + 2\varepsilon(b - a) \leq \ell + 2\varepsilon(b - a),$$

and because ε is arbitrary,

$$\int_a^b |\mathbf{r}'(t)|dt \leq \ell,$$

which by Eq. (4.3) implies the thesis. □

Let $t = f(\tau) : [c, d] \to [a, b]$ be a bijective function that operates the change of parameter from t to τ. If $\mathbf{r}_t(t) : [a, b] \to \mathcal{V}$ is a parametric equation of a curve and $\mathbf{r}_\tau : [c, d] \to \mathcal{V}$ is a *re-parameterization* of the same curve, we then have the following.

Theorem 26. *The length of a curve does not depend upon its parameterization.*

Proof. Let $\mathbf{r}_t(t) : [a, b] \to \mathcal{E}$ be a regular curve and $t = f(\tau) : [c, d] \to [a, b]$ be a change of parameter, then $dt = f'(\tau)d\tau$ and

$$\ell = \int_a^b |\mathbf{r}_t'(t)|dt = \int_c^d |\mathbf{r}_t'(f(\tau))f'(\tau)|d\tau = \int_c^d |\mathbf{r}_\tau'(\tau)|d\tau.$$

□

A simple way to determine a point $p(t)$ on a curve is to fix a point p_0 on the curve and to measure the length $s(t)$ of the arc of curve between $p_0 = p(t = 0)$ and $p(t)$. This length $s(t)$ is called *curvilinear abscissa*[5]:

$$s(t) = \int_0^t |\mathbf{r}'(t^*)|dt^* = \int_0^t |(p(t^*) - o)'|dt^*. \tag{4.4}$$

From Eq. (4.4), we get

$$\frac{ds}{dt} = |\mathbf{r}'(t)| > 0$$

so that $s(t)$ is an increasing function of t, and the length of an infinitesimal arc is

$$ds = \sqrt{dr_1^2 + dr_2^2 + dr_3^2}.$$

For a plane curve $y = f(x)$, we can always put $t = x$, which gives the parametric equation

$$p(t) = (t, f(t)),$$

or in vector form,

$$\mathbf{r}(t) = t\ \mathbf{e}_1 + f(t)\ \mathbf{e}_2,$$

from which we obtain

$$\frac{ds}{dt} = |\mathbf{r}'(t)| = |p'(t)| = \sqrt{1 + f'^2(t)}, \tag{4.5}$$

which gives the length of a plane curve between $t = x_0$ and $t = x$ as a function of the abscissa x:

$$s(x) = \int_{x_0}^x \sqrt{1 + f'^2(t)}dt.$$

---

[5]The curvilinear abscissa is also called *arc length* or *natural parameter*.

4.4 The Frenet–Serret basis

We define the *tangent vector* $\boldsymbol{\tau}(t)$ to a regular curve $p = p(t)$ as the vector

$$\boldsymbol{\tau}(t) := \frac{p'(t)}{|p'(t)|}.$$

By the definition of the derivative, this unit vector is always oriented as the increasing values of t; hence, the straight line tangent to the curve in $p_0 = p(t_0)$ has the equation

$$q(\bar{t}) = p(t_0) + \bar{t}\ \boldsymbol{\tau}(t_0).$$

If the curvilinear abscissa s is chosen as a parameter for the curve through the change of parameter $s = s(t)$, we get

$$\boldsymbol{\tau}(t) = \frac{p'(t)}{|p'(t)|} = \frac{p'[s(t)]}{|p'[s(t)]|} = \frac{1}{s'(t)}\frac{dp(s)}{ds}\frac{ds(t)}{dt} = \frac{dp(s)}{ds} \rightarrow \boldsymbol{\tau}(s) = p'(s). \tag{4.6}$$

So, if the parameter of the curve is s, the derivative of the curve is $\boldsymbol{\tau}$, i.e. it is automatically a unit vector. The above equation, in addition, shows that the change of parameter does not change the direction of the tangent because it is only a scalar, the derivative of the parameter's change, that multiplies the vector. Nevertheless, in general, a change of parameter can change the orientation of the curve.

Because the norm of $\boldsymbol{\tau}$ is constant, its derivative is a vector orthogonal to $\boldsymbol{\tau}$, see Eq. (4.1). That is why we call *principal normal vector* to a curve the unit vector

$$\boldsymbol{\nu}(t) := \frac{\boldsymbol{\tau}'(t)}{|\boldsymbol{\tau}'(t)|}. \tag{4.7}$$

$\boldsymbol{\nu}$ is defined only on the points of the curve where $\boldsymbol{\tau}' \neq \mathbf{o}$, which implies that $\boldsymbol{\nu}$ is not defined on the points of a straight line. This simply means that there is not, among the infinite unit normal vectors to a straight line, a normal with special properties, a *principal* one, uniquely linked to $\boldsymbol{\tau}$.

Unlike $\boldsymbol{\tau}$, whose orientation changes with the choice of the parameter, $\boldsymbol{\nu}$ is an *intrinsic* local characteristic of the curve: *It is not affected by the choice of the parameter.* In fact, by the same definition, $\boldsymbol{\nu}$ does not depend upon the reference frame; then, because the direction of $\boldsymbol{\tau}$ is also independent of the parameter's choice, the only factor that could affect $\boldsymbol{\nu}$ is the orientation of the curve, which depends upon the parameter. But a change in the orientation affects, in (4.7), both $\boldsymbol{\tau}$ and the sign of the increment dt so that $\boldsymbol{\tau}'(t) = d\boldsymbol{\tau}/dt$ does not change, nor does $\boldsymbol{\nu}$, which is hence an intrinsic property of the curve.

The vector

$$\boldsymbol{\beta}(t) := \boldsymbol{\tau}(t) \times \boldsymbol{\nu}(t)$$

is called the *binormal vector*; by construction, it is orthogonal to $\boldsymbol{\tau}$ and $\boldsymbol{\nu}$, and it is a unit vector. In addition, it is evident that

$$\boldsymbol{\tau} \times \boldsymbol{\nu} \cdot \boldsymbol{\beta} = 1,$$

so the set $\{\boldsymbol{\tau}, \boldsymbol{\nu}, \boldsymbol{\beta}\}$ forms a positively oriented orthonormal basis that can be defined at any regular point of a curve with $\boldsymbol{\tau}' \neq \mathbf{o}$. Such a basis is called the *Frenet–Serret local basis*, local in the sense that it changes with the position along the curve. The plane $(\boldsymbol{\tau}, \boldsymbol{\nu})$ is the *osculating plane*, the plane $(\boldsymbol{\nu}, \boldsymbol{\beta})$ the *normal plane*, and the plane $(\boldsymbol{\beta}, \boldsymbol{\tau})$ the *rectifying plane*, see Fig. 4.4. The osculating plane is particularly important: If we consider a plane passing through three nonaligned points of the curve, when these points become closer and closer, still remaining on the curve, the plane tends to the osculating plane. The osculating plane at a point of a curve is the plane that better approaches the curve near the point. A plane curve is entirely contained in the osculating plane, which is fixed.

The principal normal $\boldsymbol{\nu}$ is always oriented toward the part of the space with respect to the rectifying plane where the curve is; in particular, for a plane curve, $\boldsymbol{\nu}$ is always directed toward the concavity of the curve. To show that, it is sufficient to prove that the vector $p(t+\varepsilon) - p(t)$ forms with

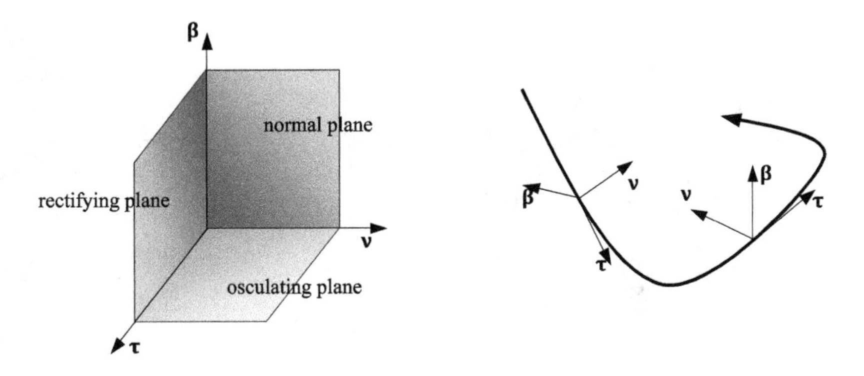

Figure 4.4: The Frenet–Serret basis.

$\boldsymbol{\nu}$ an angle $\psi \leq \pi/2$, i.e. that $(p(t+\varepsilon)-p(t)) \cdot \boldsymbol{\nu} \geq 0$. In fact,

$$p(t+\varepsilon)-p(t)=\varepsilon\, p'(t)+\frac{1}{2}\varepsilon^2 p''(t)+o(\varepsilon^2)$$
$$\Rightarrow (p(t+\varepsilon)-p(t)) \cdot \boldsymbol{\nu}=\frac{1}{2}\varepsilon^2 p''(t)\cdot \boldsymbol{\nu}+o(\varepsilon^2),$$

but

$$p''(t)\cdot\boldsymbol{\nu}=(\boldsymbol{\tau}'|p'|+\boldsymbol{\tau}|p'|')\cdot\boldsymbol{\nu}=(|\boldsymbol{\tau}'||p'|\boldsymbol{\nu}+\boldsymbol{\tau}|p'|')\cdot\boldsymbol{\nu}=|\boldsymbol{\tau}'||p'|$$

so that, to within infinitesimal quantities of order $o(\varepsilon^2)$, we obtain

$$(p(t+\varepsilon)-p(t))\cdot\boldsymbol{\nu}=\frac{1}{2}\varepsilon^2|\boldsymbol{\tau}'||p'|\geq 0.$$

4.5 Curvature of a curve

It is important, in several situations, to evaluate how much a curve moves away from a straight line in the neighborhood of a point. To do that, we calculate the angle formed by the tangents at two close points, determined by the curvilinear abscissa s and $s+\varepsilon$, and we measure the angle $\chi(s,\varepsilon)$ that they form, see Fig. 4.5.

We then define the *curvature* of the curve in $p=p(s)$ as the limit

$$c(s)=\lim_{\varepsilon\rightarrow 0}\left|\frac{\chi(s,\varepsilon)}{\varepsilon}\right|.$$

The curvature is hence a nonnegative scalar that measures the rapidity of variation in the direction of the curve per unit length of the curve (that is why $c(s)$ is defined as a function of the curvilinear abscissa); by the same definition, the curvature is an *intrinsic property* of the curve, i.e.

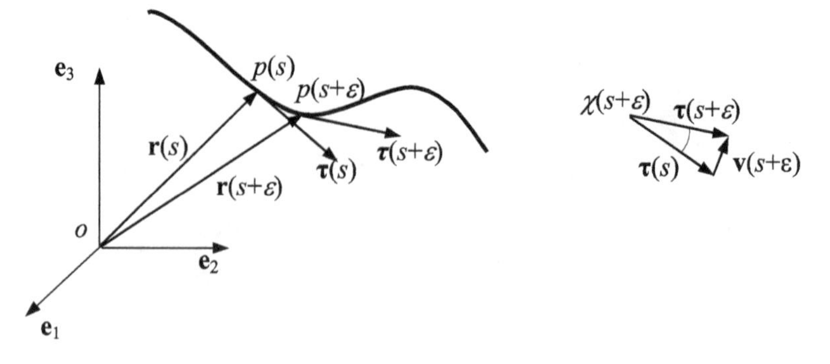

Figure 4.5: Curvature of a curve.

independent of the parameter's choice. For a straight line, the curvature is identically null everywhere.

The curvature is linked to the second derivative of the curve; referring to Fig. 4.5, it is

$$\begin{aligned} c(s) &= \lim_{\varepsilon\to 0}\left|\frac{\chi(s,\varepsilon)}{\varepsilon}\right| = \lim_{\varepsilon\to 0}\left|\frac{\sin\chi(s,\varepsilon)}{\varepsilon}\right| = \lim_{\varepsilon\to 0}\left|\frac{2}{\varepsilon}\sin\frac{\chi(s,\varepsilon)}{2}\right| \\ &= \lim_{\varepsilon\to 0}\left|\frac{\mathbf{v}(s,\varepsilon)}{\varepsilon}\right| = \lim_{\varepsilon\to 0}\left|\frac{\boldsymbol{\tau}(s+\varepsilon)-\boldsymbol{\tau}(s)}{\varepsilon}\right| = |\boldsymbol{\tau}'(s)| = |p''(s)|. \end{aligned}$$

Another formula for the calculation of $c(s)$ can be obtained if we consider that

$$\frac{d\boldsymbol{\tau}[s(t)]}{dt} = \frac{d\boldsymbol{\tau}}{ds}\frac{ds}{dt} = \frac{d\boldsymbol{\tau}}{ds}|p'(t)| \ \rightarrow\ \frac{d\boldsymbol{\tau}}{ds} = \frac{1}{|p'(t)|}\frac{d\boldsymbol{\tau}}{dt}$$

so that

$$c(s) = |\boldsymbol{\tau}'(s)| = \frac{1}{|p'(t)|}\left|\frac{d\boldsymbol{\tau}}{dt}\right| = \frac{|\boldsymbol{\tau}'(t)|}{|p'(t)|}. \tag{4.8}$$

A better formula can be obtained using the *complementary projector* onto $\boldsymbol{\tau}$, i.e. the tensor $\mathbf{I}-\boldsymbol{\tau}\otimes\boldsymbol{\tau}$, introduced in Exercise 2, Chapter 2:

$$\begin{aligned} \frac{d\boldsymbol{\tau}}{ds} &= \frac{1}{|p'(t)|}\frac{d\boldsymbol{\tau}}{dt} = \frac{1}{|p'(t)|}\frac{d}{dt}\frac{p'(t)}{|p'(t)|} = \frac{1}{|p'|}\frac{p''|p'| - p'\dfrac{p''\cdot p'}{|p'|}}{|p'|^2} \\ &= \frac{p'' - \boldsymbol{\tau}\, p''\cdot\boldsymbol{\tau}}{|p'|^2} = (\mathbf{I}-\boldsymbol{\tau}\otimes\boldsymbol{\tau})\frac{p''}{|p'|^2}. \end{aligned}$$

Consequently,

$$c(s) = \left|\frac{d\boldsymbol{\tau}(s)}{ds}\right| = \frac{1}{|p'|^2}|(\mathbf{I}-\boldsymbol{\tau}\otimes\boldsymbol{\tau})p''|.$$

Now, we use Eq. (2.29) with $\mathbf{w}=\boldsymbol{\tau}$; denoting by $\mathbf{W}_\tau$ the axial tensor of $\boldsymbol{\tau}$, then

$$\mathbf{W}_\tau\mathbf{W}_\tau = -\frac{1}{2}|\mathbf{W}_\tau|^2(\mathbf{I}-\boldsymbol{\tau}\otimes\boldsymbol{\tau}),$$

whence

$$\mathbf{I}-\boldsymbol{\tau}\otimes\boldsymbol{\tau} = -2\frac{\mathbf{W}_\tau\mathbf{W}_\tau}{|\mathbf{W}_\tau|^2} = -\mathbf{W}_\tau\mathbf{W}_\tau,$$

because if $\boldsymbol{\tau} = (\tau_1, \tau_2, \tau_3)$, then

$$|\mathbf{W}_\tau|^2 = \mathbf{W}_\tau \cdot \mathbf{W}_\tau = \begin{bmatrix} 0 & -\tau_3 & \tau_2 \\ \tau_3 & 0 & -\tau_1 \\ -\tau_2 & \tau_1 & 0 \end{bmatrix} \cdot \begin{bmatrix} 0 & -\tau_3 & \tau_2 \\ \tau_3 & 0 & -\tau_1 \\ -\tau_2 & \tau_1 & 0 \end{bmatrix}$$
$$= 2(\tau_1^2 + \tau_2^2 + \tau_3^2) = 2.$$

So, because $\mathbf{W}_\tau \in Skw(\mathcal{V})$,

$$\mathbf{W}_\tau\ \mathbf{u} = \boldsymbol{\tau} \times \mathbf{u} \quad \forall \mathbf{u} \in \mathcal{V}.$$

Finally, using Eq. (2.33), the orthogonality property of cross product, Eq. (2.31), and Eq. (2.35), we get

$$|(\mathbf{I} - \boldsymbol{\tau} \otimes \boldsymbol{\tau})p''| = |-\mathbf{W}_\tau \mathbf{W}_\tau p''| = |-\mathbf{W}_\tau(\boldsymbol{\tau} \times p'')| = |-\boldsymbol{\tau} \times (\boldsymbol{\tau} \times p'')|$$
$$= |\boldsymbol{\tau} \times (\boldsymbol{\tau} \times p'')| = |\boldsymbol{\tau} \times p''| = \frac{|p' \times p''|}{|p'|}$$

so that, finally,

$$c = \frac{|p' \times p''|}{|p'|^3}. \tag{4.9}$$

Applying this last formula to a plane curve $p(t) = (x(t), y(t))$, we get

$$c = \frac{|x'y'' - x''y'|}{(x'^2 + y'^2)^{\frac{3}{2}}},$$

and if the curve is given in the form $y = y(x)$ so that the parameter $t = x$, then we obtain

$$c = \frac{|y''|}{(1 + y'^2)^{\frac{3}{2}}}.$$

This last formula shows that if $|y'| \ll 1$, then

$$c \simeq |y''|.$$

This result is fundamental to the linearized (infinitesimal) theory of beams, plates, and shells.

4.6 The Frenet–Serret formula

From Eq. (4.7) for $t = s$ and Eq. (4.8), we get

$$\frac{d\boldsymbol{\tau}}{ds} = c\boldsymbol{\nu}, \tag{4.10}$$

which is the *first Frenet–Serret formula*, giving the variation in $\boldsymbol{\tau}$ per unit length of the curve. Such a variation is a vector whose norm is the curvature and that has as direction that of $\boldsymbol{\nu}$. We remark that, because $t = s$, by Eq. (4.6), it is also that

$$p''(s) = c(s)\boldsymbol{\nu}(s). \tag{4.11}$$

Let us now consider the variation in $\boldsymbol{\beta}$ per unit length of the curve; because $\boldsymbol{\beta}$ is a unit vector, we have

$$\frac{d\boldsymbol{\beta}}{ds} \cdot \boldsymbol{\beta} = 0$$

and

$$\boldsymbol{\beta} \cdot \boldsymbol{\tau} = 0 \ \Rightarrow \ \frac{d(\boldsymbol{\beta} \cdot \boldsymbol{\tau})}{ds} = \frac{d\boldsymbol{\beta}}{ds} \cdot \boldsymbol{\tau} + \boldsymbol{\beta} \cdot \frac{d\boldsymbol{\tau}}{ds} = 0.$$

Through Eq. (4.10) and because $\boldsymbol{\beta} \cdot \boldsymbol{\nu} = 0$, we get

$$\frac{d\boldsymbol{\beta}}{ds} \cdot \boldsymbol{\tau} = -c\boldsymbol{\beta} \cdot \boldsymbol{\nu} = 0$$

so that $\dfrac{d\boldsymbol{\beta}}{ds}$ is necessarily parallel to $\boldsymbol{\nu}$. We then set

$$\frac{d\boldsymbol{\beta}}{ds} = \vartheta \boldsymbol{\nu},$$

which is the *second Frenet–Serret formula*. The scalar $\vartheta(s)$ is called the *torsion of the curve* in $p = p(s)$. So, we see that the variation in $\boldsymbol{\beta}$ per unit length is a vector parallel to $\boldsymbol{\nu}$ and proportional to the torsion of the curve.

We can now find the variation in $\boldsymbol{\nu}$ per unit length of the curve:

$$\frac{d\boldsymbol{\nu}}{ds} = \frac{d(\boldsymbol{\beta} \times \boldsymbol{\tau})}{ds} = \frac{d\boldsymbol{\beta}}{ds} \times \boldsymbol{\tau} + \boldsymbol{\beta} \times \frac{d\boldsymbol{\tau}}{ds} = \vartheta \ \boldsymbol{\nu} \times \boldsymbol{\tau} + c \ \boldsymbol{\beta} \times \boldsymbol{\nu},$$

so finally

$$\frac{d\boldsymbol{\nu}}{ds} = -c \ \boldsymbol{\tau} - \vartheta \ \boldsymbol{\beta},$$

which is the *third Frenet–Serret formula*: the variation in $\boldsymbol{\nu}$ per unit length of the curve is a vector of the rectifying plane.

The three formulae of Frenet–Serret (discovered independently by J. F. Frenet in 1847 and by J. A. Serret in 1851) can be condensed into the symbolic matrix product

$$\begin{Bmatrix} \boldsymbol{\tau}' \\ \boldsymbol{\nu}' \\ \boldsymbol{\beta}' \end{Bmatrix} = \begin{bmatrix} 0 & c & 0 \\ -c & 0 & -\vartheta \\ 0 & \vartheta & 0 \end{bmatrix} \begin{Bmatrix} \boldsymbol{\tau} \\ \boldsymbol{\nu} \\ \boldsymbol{\beta} \end{Bmatrix}.$$

The matrix in the equation above is called the *matrix of Cartan*, and it is skew.

4.7 The torsion of a curve

We have introduced the torsion of a curve in the previous section, with the second formula of Frenet–Serret. The torsion measures the deviation of a curve from flatness: If a curve is planar, it belongs to the osculating plane, and $\boldsymbol{\beta}$, which is perpendicular to the osculating pane, is hence a constant vector. So, its derivative is null, and by the second Frenet–Serret formula, $\vartheta = 0$.

Conversely, if $\vartheta = 0$ everywhere, $\boldsymbol{\beta}$ is a constant vector, and hence, the osculating plane does not change and the curve is planar. So, we have that *a curve is planar if and only if the torsion is null* $\forall p(s)$.

Using the Frenet–Serret formulae in the expression of $p'''(s)$, we get a formula for the torsion:

$$\begin{aligned}
p'(t) &= |p'|\boldsymbol{\tau} = \frac{dp}{ds}\frac{ds}{dt} = s'\boldsymbol{\tau} \;\Rightarrow\; |p'| = s' \;\rightarrow \\
p''(t) &= s''\boldsymbol{\tau} + s'\boldsymbol{\tau}' = s''\boldsymbol{\tau} + s'^2\frac{d\boldsymbol{\tau}}{ds} = s''\boldsymbol{\tau} + c\, s'^2\boldsymbol{\nu} \;\rightarrow \\
p'''(t) &= s'''\boldsymbol{\tau} + s''\boldsymbol{\tau}' + (c\, s'^2)'\boldsymbol{\nu} + c\, s'^2\boldsymbol{\nu}' \\
&= s'''\boldsymbol{\tau} + s''s'\frac{d\boldsymbol{\tau}}{ds} + (c\, s'^2)'\boldsymbol{\nu} + c\, s'^3\frac{d\boldsymbol{\nu}}{ds} \\
&= s'''\boldsymbol{\tau} + s''s'c\boldsymbol{\nu} + (c\, s'^2)'\boldsymbol{\nu} - c\, s'^3(c\boldsymbol{\tau} + \vartheta\boldsymbol{\beta}) \\
&= (s''' - c^2 s'^3)\boldsymbol{\tau} + (s''s'c + c's'^2 + 2c\, s's'')\boldsymbol{\nu} - c\, s'^3\vartheta\boldsymbol{\beta},
\end{aligned}$$

whence, through Eq. (4.9),

$$\begin{aligned}
p' \times p'' \cdot p''' &= s'\boldsymbol{\tau} \times (s''\boldsymbol{\tau} + c\, s'^2\boldsymbol{\nu}) \cdot [(s''' - c^2 s'^3)\boldsymbol{\tau} \\
&\quad + (s''s'c + c's'^2 + 2c\, s's'')\boldsymbol{\nu} - c\, s'^3\vartheta\boldsymbol{\beta}] \\
&= -c^2 s'^6\vartheta = -c^2|p'|^6\vartheta = -\frac{|p' \times p''|^2}{|p'|^6}|p'|^6\vartheta
\end{aligned}$$

so that, finally,

$$\vartheta = -\frac{p' \times p'' \cdot p'''}{|p' \times p''|^2}.$$

We remark that while the curvature is linked to the second derivative of the curve, the torsion is also a function of the third derivative.

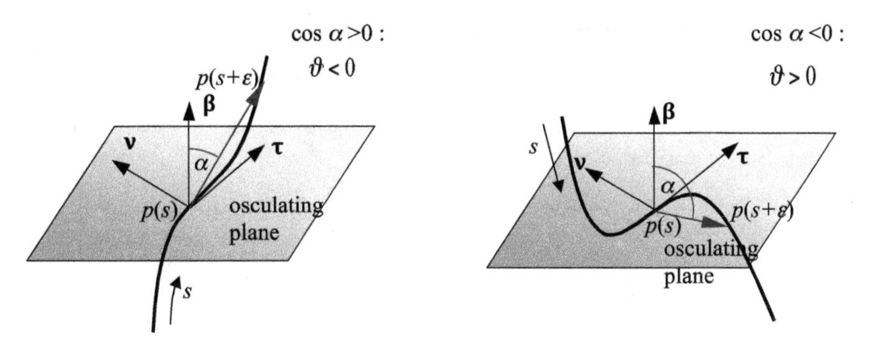

Figure 4.6: Torsion of a curve.

Unlike curvature, which is intrinsically positive, the torsion can be negative. In fact, again using the Frenet–Serret formulae,

$$
\begin{aligned}
p(s+\varepsilon)-p(s) &= \varepsilon\, p' + \frac{1}{2}\varepsilon^2 p'' + \frac{1}{6}\varepsilon^3 p''' + o(\varepsilon^3)\\
&= \varepsilon\boldsymbol{\tau} + \frac{1}{2}\varepsilon^2 c\boldsymbol{\nu} + \frac{1}{6}\varepsilon^3 (c\boldsymbol{\nu})' + o(\varepsilon^3)\\
&= \varepsilon\boldsymbol{\tau} + \frac{1}{2}\varepsilon^2 c\boldsymbol{\nu} + \frac{1}{6}\varepsilon^3 (c'\boldsymbol{\nu} - c^2\boldsymbol{\tau} - c\,\vartheta\boldsymbol{\beta}) + o(\varepsilon^3)\\
&\Rightarrow (p(s+\varepsilon)-p(s))\cdot\boldsymbol{\beta} = -\frac{1}{6}\varepsilon^3 c\,\vartheta + o(\varepsilon^3).
\end{aligned}
$$

The above dot product determines if the point $p(s+\varepsilon)$ is located, with respect to the osculating plane, on the side of $\boldsymbol{\beta}$ or on the opposite one, see Fig. 4.6: If following the curve for increasing values of s, $\varepsilon > 0$, the point passes into the semi-space of $\boldsymbol{\beta}$ from the opposite one, because $1/6\, c\,\varepsilon^3 > 0$, it will be $\vartheta < 0$, while in the opposite case, it will be $\vartheta > 0$.

This result is intrinsic, i.e. it does not depend upon the choice of the parameter, hence of the positive orientation of the curve; in fact, $\boldsymbol{\nu}$ is intrinsic, but changing the orientation of the curve, $\boldsymbol{\tau}$, and hence $\boldsymbol{\beta}$, change in orientation.

4.8 Osculating sphere and circle

The *osculating sphere*[6] to a curve at a point p is a sphere to which the curve tends to adhere to the neighborhood of p. Mathematically, if q_s is the

[6]The word osculating comes from the latin word *osculo*, which means to kiss; an osculating sphere or circle or plane is a geometric object that is very close to the curve, as close as two lovers are in a kiss.

center of the sphere relative to the point $p(s)$, then

$$|p(s+\varepsilon)-q_s|^2=|p(s)-q_s|^2+o(\varepsilon^3).$$

Using this definition, discarding the terms of order $o(\varepsilon^3)$ and using the Frenet–Serret formulae, we get

$$\begin{aligned}|p(s+\varepsilon)-q_s|^2&=|p(s)-q_s+\varepsilon p'+\frac{1}{2}\varepsilon^2p''+\frac{1}{6}\varepsilon^3p'''+o(\varepsilon^3)|^2\\&=|p(s)-q_s+\varepsilon\boldsymbol{\tau}+\frac{1}{2}\varepsilon^2c\,\boldsymbol{\nu}+\frac{1}{6}\varepsilon^3(c\boldsymbol{\nu})'+o(\varepsilon^3)|^2\\&=|p(s)-q_s|^2+2\varepsilon(p(s)-q_s)\cdot\boldsymbol{\tau}+\varepsilon^2+\varepsilon^2c(p(s)-q_s)\cdot\boldsymbol{\nu}\\&\quad+\frac{1}{3}\varepsilon^3(p(s)-q_s)\cdot(c'\boldsymbol{\nu}-c^2\boldsymbol{\tau}-c\,\vartheta\boldsymbol{\beta})+o(\varepsilon^3),\end{aligned}$$

which gives

$$\begin{aligned}(p(s)-q_s)\cdot\boldsymbol{\tau}&=0,\\(p(s)-q_s)\cdot\boldsymbol{\nu}&=-\frac{1}{c}=-\rho,\\(p(s)-q_s)\cdot\boldsymbol{\beta}&=-\frac{c'}{c^2\vartheta}=\frac{\rho'}{\vartheta},\end{aligned}$$

and finally,

$$q_s=p+\rho\,\boldsymbol{\nu}-\frac{\rho'}{\vartheta}\boldsymbol{\beta},\tag{4.12}$$

so the center of the sphere belongs to the normal plane; the sphere is not defined for a plane curve. The quantity ρ is the *radius of curvature* of the curve, which is defined as

$$\rho=\frac{1}{c}.$$

The radius of the osculating sphere is

$$\rho_s=|p-q_s|=\sqrt{\rho^2+\left(\frac{\rho'}{\vartheta}\right)^2}.$$

The intersection between the osculating sphere and the osculating plane at the same point p is the *osculating circle*. This circle has the property of sharing the same tangent in p with the curve, and its radius is the radius of curvature, ρ. From Eq. (4.12), we get the position of the osculating circle center q:

$$q=p+\rho\,\boldsymbol{\nu}.\tag{4.13}$$

An example can be seen in Fig. 4.7, where the osculating plane, circle, and sphere are shown for a point p of a conical helix.

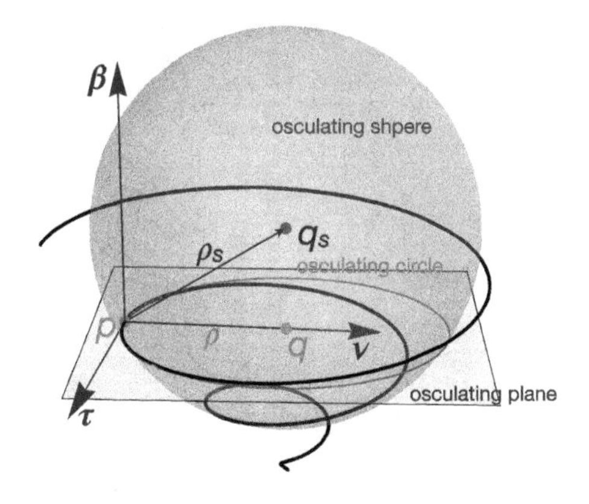

Figure 4.7: Osculating plane, circle, and sphere for a point p of a conical helix.

The osculating circle is a diametral circle of the osculating sphere only when $q = q_s$, so if and only if

$$\frac{\rho'}{\vartheta} = -\frac{c'}{c^2\vartheta} = 0,$$

i.e. when the curvature is constant.

4.9 Evolute, involute and envelopes of plane curves

For any plane curve $\gamma(s)$, the center of the osculating circle q describes a curve $\boldsymbol{\delta}(\sigma)$ that is called the *evolute* of $\gamma(s)$ (s and σ are curvilinear abscissa). A point q of the evolute is then given by Eq. (4.13). We call *involute* of a curve $\gamma(s)$ a curve $\boldsymbol{\mu}(\sigma)$ whose evolute is $\gamma(s)$. We call the *envelope* of a family of plane curves $\varphi(s,\kappa), \kappa \in \mathbb{R}$ being a parameter, a curve that is tangent in each of its points to the curve of $\varphi(s,\kappa)$ passing through that point.

Let us consider the evolute $\boldsymbol{\delta}(\sigma)$ of a curve $\gamma(s)$; the tangent to $\boldsymbol{\delta}(\sigma)$ is the vector, cf. Eq. (4.13),

$$\boldsymbol{\tau}_\delta = \frac{dq}{d\sigma} = \frac{dq}{ds}\frac{ds}{d\sigma}.$$

But, cf. again Eq. (4.13) and the Frenet–Serret formulae,

$$\frac{dq}{ds} = \frac{dp}{ds} + \frac{d\rho}{ds}\boldsymbol{\nu} + \rho\frac{d\boldsymbol{\nu}}{ds} = \boldsymbol{\tau} + \frac{d\rho}{ds}\boldsymbol{\nu} - \rho\, c\, \boldsymbol{\tau} = \frac{d\rho}{ds}\boldsymbol{\nu},$$

so

$$\boldsymbol{\tau}_\delta = \frac{dq}{d\sigma} = \frac{d\rho}{ds}\frac{ds}{d\sigma}\boldsymbol{\nu}.$$

Because

$$\left|\frac{dq}{d\sigma}\right| = |\boldsymbol{\nu}| = 1,$$

then

$$\frac{d\rho}{ds}\frac{ds}{d\sigma} = 1 \Rightarrow \frac{d\rho}{ds} = \frac{d\sigma}{ds}$$

and

$$\boldsymbol{\tau}_\delta = \boldsymbol{\nu}.$$

The evolute, $\boldsymbol{\delta}(\sigma)$, of $\boldsymbol{\gamma}(s)$ is hence the envelope of its principal normals $\boldsymbol{\nu}(s)$.

This result helps us in finding the equation of the involute $\boldsymbol{\mu}(\sigma)$ of a curve $\boldsymbol{\gamma}(s)$; let $p = p(s)$ be a point of $\boldsymbol{\gamma}(s)$; then, if $b \in \boldsymbol{\mu}(\sigma)$, it must be that

$$(b - p) \cdot \boldsymbol{\nu} = 0,$$

where $\boldsymbol{\nu}$ is the principal normal to $\boldsymbol{\gamma}(s)$ in p because $\boldsymbol{\gamma}(s)$ is the evolute of $\boldsymbol{\mu}(\sigma)$, which implies, for the last result, that $\boldsymbol{\tau} = \boldsymbol{\nu}_\mu$, with $\boldsymbol{\tau}$ the tangent to $\boldsymbol{\gamma}(s)$ in p and $\boldsymbol{\nu}_\mu$ the principal normal to $\boldsymbol{\mu}(\sigma)$ in b, see Fig. 4.8.

Therefore,

$$b(s) - p(s) = f(s)\boldsymbol{\tau}(s) \rightarrow b(s) = p(s) + f(s)\boldsymbol{\tau}(s),$$

with $f = f(s)$ a scalar function of s; we remark that $b = b(s)$, i.e. the arc length s of $\boldsymbol{\gamma}(s)$ is the parameter also for $\boldsymbol{\mu}(s)$, but in general, $\sigma \neq s$. Upon differentiation, we get

$$b'(s) = (1 + f'(s))\boldsymbol{\tau}(s) + f(s)c(s)\boldsymbol{\nu}(s).$$

Then, because $b'(s) = |b'(s)|\boldsymbol{\tau}_\mu$ is orthogonal to $\boldsymbol{\nu}_\mu = \boldsymbol{\tau}$, it is parallel to $\boldsymbol{\nu}$, so it must be that

$$1 + f'(s) = 0 \Rightarrow f(s) = a - s, \quad a \in \mathbb{R}.$$

Finally, the equation of the involute $\boldsymbol{\mu}(s)$ to $\boldsymbol{\gamma}(s)$ is

$$b(s) = p(s) + (a - s)\boldsymbol{\tau}(s),$$

and we remark that the involute is not unique.

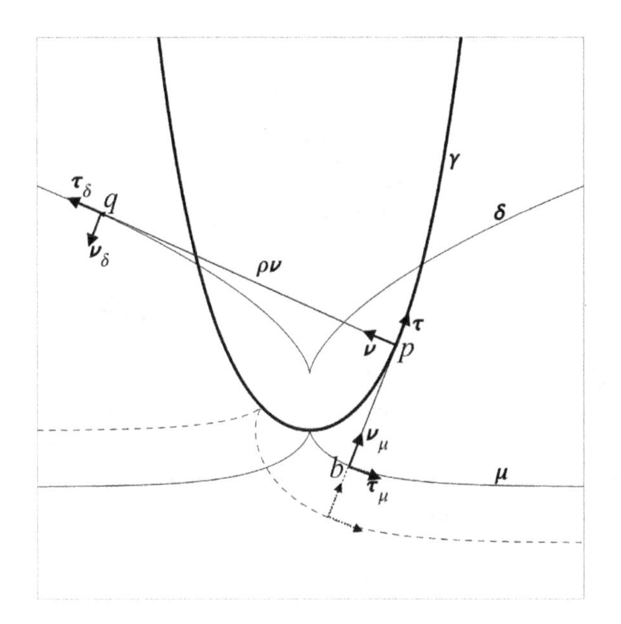

Figure 4.8: Evolute, $\boldsymbol{\delta}$, and involutes for $a = 0$, denoted by $\boldsymbol{\mu}$, and $a = 1$, dashed, of a catenary γ.

4.10 The theorem of Bonnet

The curvature, $c(s)$, and the torsion, $\vartheta(s)$, are the only differential parameters that completely describe a curve. In other words, given two functions $c(s)$ and $\vartheta(s)$, then a curve exists with such a curvature and torsion (we remark that there are no conditions bounding these parameters). This is proved in the following.

Theorem 27 (Bonnet's theorem). *Given two scalar functions $c(s) \in C^1$ and $\vartheta(s) \in C^0$, there always exists a unique curve $\gamma \in C^3$ whose curvilinear abscissa is s, curvature $c(s)$, and torsion $\vartheta(s)$.*

Proof. Let

$$\mathbf{e} = \begin{pmatrix} \boldsymbol{\tau} \\ \boldsymbol{\nu} \\ \boldsymbol{\beta} \end{pmatrix}$$

be the column vector whose elements are the vectors of the Frenet–Serret basis. Then,

$$\frac{d\mathbf{e}(s)}{ds} = \mathbf{C}(s)\mathbf{e}(s), \tag{4.14}$$

with

$$\mathbf{C}(s)=\begin{bmatrix}0 & c(s) & 0\\ -c(s) & 0 & -\vartheta(s)\\ 0 & \vartheta(s) & 0\end{bmatrix}$$

the matrix of Cartan. Adding the initial condition

$$\mathbf{e}(0)=\begin{pmatrix}\mathbf{e}_1\\ \mathbf{e}_2\\ \mathbf{e}_3\end{pmatrix},$$

we have a Cauchy problem for the basis $\mathbf{e}(0)$. As known, such a problem admits a unique solution, i.e. we can associate to $c(s)$ and $\vartheta(s)$ a family of bases $\mathbf{e}(s)$ (that are orthonormal because if one of them were not so, the Cartan's matrix should not be skew). Call $\boldsymbol{\tau}(s)$ the first vector of the basis $\mathbf{e}(s)$ and define the function

$$p(s):=p_0+\int_0^s \boldsymbol{\tau}(s^*)ds^*;$$

$p(s)$ is the curve we are looking for (it depends upon an arbitrary point p_0, i.e. upon an inessential rigid displacement). In fact, because $|\boldsymbol{\tau}|=1$, s is the curvilinear abscissa of the curve. Then, it is sufficient to write the Frenet–Serret equations identifying them with system (4.14). □

4.11 Canonic equations of a curve

We call the *canonic equations* of a curve at a point p_0 the equations of the curve referred to the Frenet–Serret basis in p_0. For this purpose, we expand the curve in a Taylor series of initial point p_0:

$$p(s)=p_0+s\ p'(0)+\frac{1}{2}s^2p''(0)+\frac{1}{6}s^3p'''(0)+o(s^3).$$

In the Frenet–Serret basis,

$$p'(0)=\boldsymbol{\tau}(0),\quad p''(0)=c(0)\boldsymbol{\nu}(0),\quad p'''(0)=\left.\frac{dc\boldsymbol{\nu}}{ds}\right|_{s=0}$$
$$=c'(0)\boldsymbol{\nu}(0)-c^2(0)\boldsymbol{\tau}(0)-c(0)\vartheta(0)\boldsymbol{\beta}(0),$$

so

$$p(s)=p_0+s\ \boldsymbol{\tau}(0)+\frac{1}{2}s^2c(0)\boldsymbol{\nu}(0)+\frac{1}{6}s^3(-c^2(0)\boldsymbol{\tau}(0)+c'(0)\boldsymbol{\nu}(0)$$
$$-c(0)\vartheta(0)\boldsymbol{\beta}(0))+o(s^3).$$

The coordinates of a point $p(s)$ close to p_0 in the basis $(\boldsymbol{\tau}(0), \boldsymbol{\nu}(0), \boldsymbol{\beta}(0))$ are hence

$$p_1(s) = s - \frac{1}{6}c^2(0)s^3 + o(s^3),$$
$$p_2(s) = \frac{1}{2}c(0)s^2 + \frac{1}{6}c'(0)s^3 + o(s^3),$$
$$p_3(s) = -\frac{1}{6}c(0)\vartheta(0)s^3 + o(s^3).$$

The projections of the curve onto the planes of the Frenet–Serret basis hence have, close to p_0 (i.e. retaining the first non-null term in the expressions above), the following equations:

- On the osculating plane,

$$\begin{cases} p_1(s) = s, \\ p_2(s) = \frac{1}{2}c(0)s^2, \end{cases}$$

  or, eliminating s,

$$p_2 = \frac{1}{2}c(0)p_1^2,$$

  which is the equation of a parabola.
- On the rectifying plane,

$$\begin{cases} p_1(s) = s, \\ p_3(s) = -\frac{1}{6}c(0)\vartheta(0)s^3, \end{cases}$$

  or, eliminating s,

$$p_3 = -\frac{1}{6}c(0)\vartheta(0)p_1^3,$$

  which is the equation of a cubic parabola.
- On the normal plane,

$$\begin{cases} p_2(s) = \frac{1}{2}c(0)s^2, \\ p_3(s) = -\frac{1}{6}c(0)\vartheta(0)s^3, \end{cases}$$

  or, eliminating s,

$$p_3^2 = \frac{2}{9}\frac{\vartheta^2(0)}{c(0)}p_2^3,$$

  which is the equation of a semicubic parabola, with a cusp at the origin, hence a singular point, though the curve $p(s)$ is regular.

4.12 Exercises

1. Using the same definition of the derivative of a curve, prove the relations in Section (4.2).
2. Prove the relations in Eq. (4.2).
3. The curve whose polar equation is

$$r = a\,\theta, \quad a \in \mathbb{R},$$

is an *Archimedes' spiral*, Fig. 4.9(a). Find its curvature $c(\theta)$ and its length $\ell(\theta)$, and prove that any straight line passing through the origin is divided by the spiral in segments of constant length $2\pi\, a$ (that is why the Archimedes' spiral is used to record disks).
4. The curve whose polar equation is

$$r = ae^{b\theta}, \quad a, b \in \mathbb{R},$$

is the *logarithmic spiral.* Prove that the origin is an asymptotic point of the curve, find its curvature $c(\theta)$ and its length $\ell(\theta)$, and show that the length of the segments in which a straight line by the origin is divided by two consecutive intersections with the spiral varies as a geometrical progression. Then, prove its *equiangular property*: The angle α between $p(\theta) - o$ and $\boldsymbol{\tau}(\theta)$ is constant. Finally, show that the evolute of the logarithmic spiral is a logarithmic spiral itself (and hence that its involute is still a logarithmic spiral, that's why Jc. Bernoulli coined for this curve the Latin sentence *eadem mutata resurgo.*)

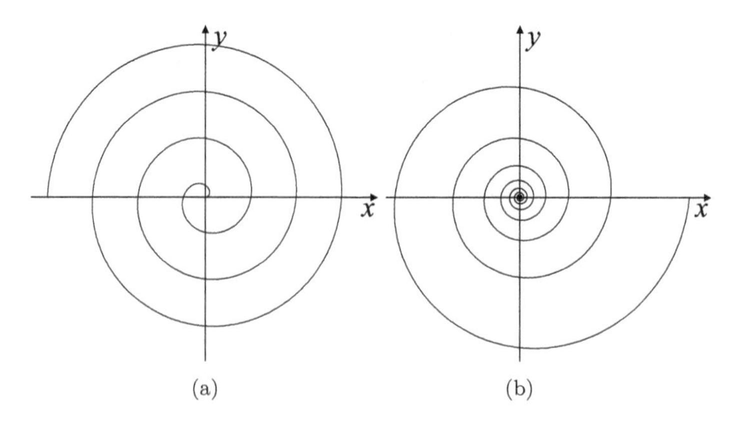

Figure 4.9: The (a) Archimedes' and (b) logarithmic spirals.

5. The curve whose parametric equation is

$$p(\theta) = a(\cos\theta + \theta\sin\theta)\mathbf{e}_1 + a(\sin\theta - \theta\cos\theta)\mathbf{e}_2,$$

with θ the angle formed by $p(\theta) - o$ and with the axis x_1, is the *involute of the circle*, see Fig. 4.10. Find its curvature $c(\theta)$ and its length $\ell(\theta)$, and prove that its evolute is exactly the circle of center o and radius a (that is why the involute of the circle is used to profile gears).

6. The curve whose parametric equation is

$$p(\theta) = a\cos\omega\theta\mathbf{e}_1 + a\sin\omega\theta\mathbf{e}_2 + b\omega\theta\mathbf{e}_3$$

is a *circular helix*, i.e. a *helix* that winds on a circular cylinder of radius a, see Fig. 4.11. Show that the angle φ formed by the helix and any generatrix of the cylinder is constant (a property that defines a helix in the general case). Then, find its length $\ell(\theta)$, its curvature $c(\theta)$, torsion $\vartheta(\theta)$, and the pitch d, i.e. the distance, on a same generatrix, between two successive intersections with the helix. Prove then the *Bertrand's theorem*: A curve is a cylindrical helix if and only if the ratio $c/\vartheta = const$. Finally, prove that for the above circular helix, there are two constants A and B such that

$$p' \times p'' = A\mathbf{u}(\theta) + B\mathbf{e}_3,$$

with

$$\mathbf{u} = \sin\omega\theta\mathbf{e}_1 - \cos\omega\theta\mathbf{e}_2;$$

then, find A and B.

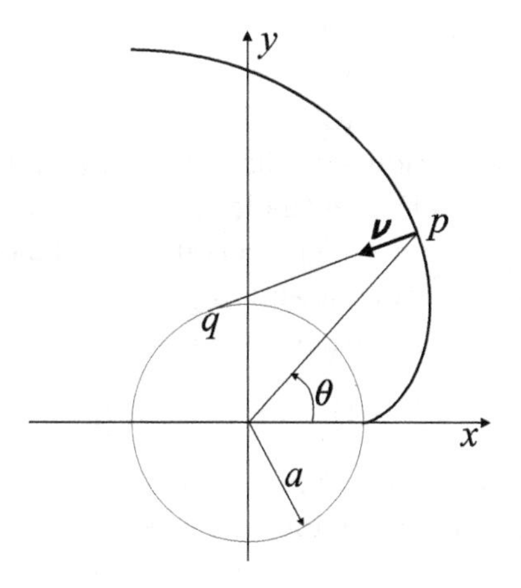

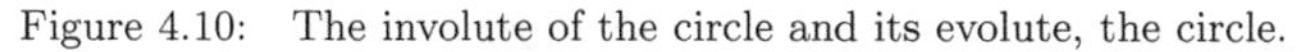
Figure 4.10: The involute of the circle and its evolute, the circle.

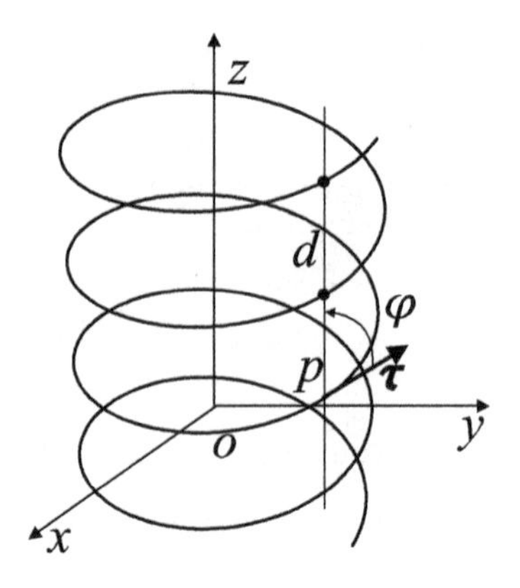

Figure 4.11: The circular helix.

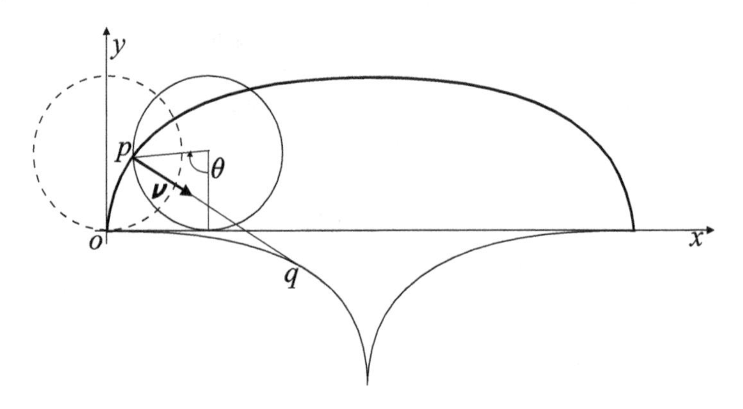

Figure 4.12: The cycloid and its evolute.

7. Find the equation of the *cycloid*, i.e. of the curve that is the trace of a point of a circle of radius r rolling without slipping on a horizontal axis, see Fig. 4.12. Calculate the length of the cycloid for a complete round of the circle, determine its curvature, and show that the evolute of the cycloid is the cycloid itself (Huygens, 1659).
8. The planar curve whose parametric equation is

$$p(t) = t\mathbf{e}_1 + \cosh t\mathbf{e}_2$$

is the *catenary* (Jc. Bernoulli, 1690; Jn. Bernoulli, Leibniz, and Huygens, 1691). It is the equilibrium curve of a heavy, perfectly flexible, and inextensible cable. Calculate the curvature of the catenary and the equation of its evolute and of its involutes (see Fig. 4.8).

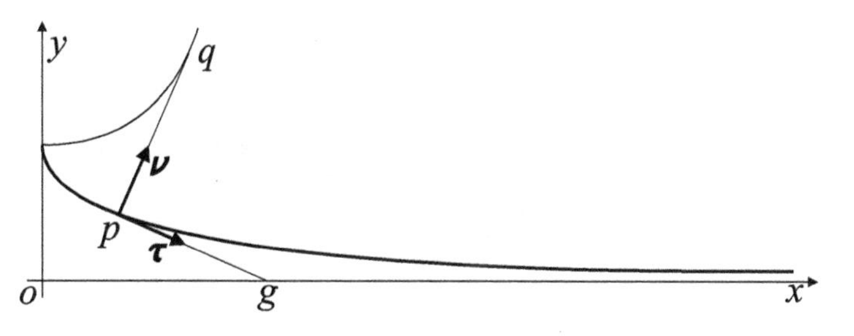

Figure 4.13: The tractrix and its evolute.

9. The planar curve whose parametric equation is

$$p(t) = \left(\cos t + \ln \tan \frac{t}{2}\right) \mathbf{e}_1 + \sin t \mathbf{e}_2$$

is the *tractrix* (Perrault, 1670; Newton, 1676; Huygens, 1693). This is the curve along which an object moves, under the influence of friction, when pulled on a horizontal plane by a line segment attached to a tractor that moves at a right angle to the initial line between the object and the puller at an infinitesimal speed, see Fig. 4.13. Show that the length of the tangent to the tractrix between the points on the tractrix itself and the axis x is constant $\forall t$, calculate the length of the curve between t_1 and t_2, calculate the curvature of the tractrix, and finally show that its evolute is the catenary.

10. For the curve whose cylindrical equation is

$$\begin{cases} r = 1, \\ z = \sin \theta, \end{cases}$$

find the highest curvature and determine whether or not it is planar.

11. Let $p = p(t)$ be the path of a moving particle of mass m, with t being the time. Define the velocity and the acceleration of p as, respectively, the first and second derivative of p with respect to t. Decompose these two vectors in the Frenet–Serret basis and interpret physically the result. Recalling the second Newton's principle of mechanics, what about the forces on p?

Chapter 5

Tensor Analysis: Fields

5.1 Scalar, vector and tensor fields

Let $\Omega \subset \mathcal{E}$ and $\mathbf{f} : \Omega \to \mathcal{V}$. We say that $\mathbf{f}$ is *continuous at* $p \in \Omega \iff \forall$ sequence $\pi_n = \{p_n \in \Omega, n \in \mathbb{N}\}$ that converges to $p \in \mathcal{E}$, the sequence $\{\mathbf{v}_n = \mathbf{f}(p_n), n \in \mathbb{N}\}$ converges to $\mathbf{f}(p)$ in $\mathcal{V}$. The function $\mathbf{f}(p) : \Omega \to \mathcal{V}$ is a *vector field on* Ω if it is continuous at each $p \in \Omega$. In the same way, we can define a *scalar field* $\varphi(p) : \Omega \to \mathbb{R}$ and a *tensor field* $\mathbf{L}(p) : \Omega \to Lin(\mathcal{V})$.

A *deformation* is any continuous and bijective function $f(p) : \Omega \to \mathcal{E}$, i.e. any transformation of a region $\Omega \subset \mathcal{E}$ into another region of $\mathcal{E}$; bijectivity imposes that to any point $p \in \Omega$ corresponds one and only point in the transformed region, and vice versa, which is the mathematical condition expressing the physical constraint of mass conservation.

Finally, the basic difference between fields/deformations and curves is that a field or a deformation is defined over a subset of $\mathcal{E}$, not of $\mathbb{R}$. In practice, this implies that the components of the field/deformation are functions of three variables, the coordinates x_i of a point $p \in \Omega$.

5.2 Differentiation of fields, differential operators

Let $\psi(p)$ be a scalar, vector, or tensor field or also a deformation; we define the *directional derivative* of $\psi(p)$ in the direction of $\mathbf{e} \in \mathcal{S}$ the limit

$$\frac{d\psi(p)}{d\mathbf{e}} := \lim_{\alpha \to 0} \frac{\psi(p + \alpha\mathbf{e}) - \psi(p)}{\alpha}, \quad \alpha \in \mathbb{R}.$$

The directional derivative measures the rate of variation of $\psi(p)$ in the direction of $\mathbf{e}$. In the particular case of $\mathbf{e} = \mathbf{e}_i,\ i = 1, 2, 3$, i.e. of the directions of the basis $\{\mathbf{e}_1, \mathbf{e}_2, \mathbf{e}_3\}$ of $\mathcal{V}$, then

$$\frac{d\psi(p)}{d\mathbf{e}_i} = \lim_{\alpha\to 0} \frac{\psi(p+\alpha\mathbf{e}_i) - \psi(p)}{\alpha}$$

is the *partial derivative of* ψ *with respect to* x_i, for example, if $i = 1$, then

$$\frac{d\psi(p)}{d\mathbf{e}_1} = \lim_{\alpha\to 0} \frac{\psi(x_1+\alpha, x_2, x_3) - \psi(x_1, x_2, x_3)}{\alpha}.$$

The partial derivative with respect to x_i is usually indicated as $\dfrac{\partial\psi}{\partial x_i}$ or also as $\psi_{,i}$.

Let $\mathbf{v}(p) : \Omega \to \mathcal{V}$; we say that $\mathbf{v}$ is *differentiable in* $p_0 \in \Omega \iff \exists \ \text{grad}\mathbf{v} \in Lin(\mathcal{V})$ such that

$$\mathbf{v}(p_0 + \mathbf{u}) = \mathbf{v}(p) + \text{grad}\mathbf{v}(p)\ \mathbf{u} + o(u)$$

when $\mathbf{u} \to \mathbf{o}$. If $\mathbf{v}$ is differentiable $\forall p \in \Omega$, grad$\mathbf{v}$ defines a tensor field on Ω called the *gradient of* $\mathbf{v}$. It is also possible to define higher order differential operators using higher order tensors, but this will not be done here. If $\mathbf{v}$ is continuous with grad$\mathbf{v}$ $\forall p \in \Omega$, then $\mathbf{v}$ is of class C^1 (*smooth*).

Let $\mathbf{v}$ be a vector field of class C^1 on Ω. Then, the *divergence* of $\mathbf{v}$ is the scalar field defined by

$$\text{div}\mathbf{v} := \text{tr}(\text{grad}\mathbf{v}),$$

while curl$\mathbf{v}$ is the unique vector field that satisfies the relation

$$(\text{grad}\mathbf{v} - \text{grad}\mathbf{v}^\top)\mathbf{u} = (\text{curl}\mathbf{v}) \times \mathbf{u} \quad \forall \mathbf{u} \in \mathcal{V}.$$

The *divergence of a tensor field* $\mathbf{L}$ is the unique vector field div$\mathbf{L}$ that satisfies

$$(\text{div}\mathbf{L}) \cdot \mathbf{u} = \text{div}(\mathbf{L}^\top \mathbf{u}) \quad \forall \mathbf{u} = const. \in \mathcal{V}.$$

Let $\varphi(p) : \Omega \to \mathbb{R}$ be a scalar field over Ω. Similar to the case of vector fields, we say that φ is *differentiable at* $p_0 \in \Omega \iff \exists \ \text{grad}\varphi \in \mathcal{V}$ such that

$$\varphi(p + \mathbf{u}) = \varphi(p) + \text{grad}\varphi(p) \cdot \mathbf{u} + o(u)$$

when $\mathbf{u} \to \mathbf{o}$. If φ is differentiable $\forall p \in \Omega$, gradφ defines a vector field on Ω called the *gradient of* φ. If gradφ is differentiable, its gradient is the tensor $\text{grad}^{II}\varphi$ called the *second gradient* or *Hessian*. It is immediate to show that under the continuity assumption,

$$\text{grad}^{II}\varphi = (\text{grad}^{II}\varphi)^\top.$$

A *level set* of a scalar field $\varphi(p)$ is the set $\mathcal{S}_L$ such that

$$\varphi(p) = const. \quad \forall p \in \mathcal{S}_L.$$

Considering hence two points p and $p + \mathbf{u}$ of the same $\mathcal{S}_L$, then by the definition of differentiability of $\varphi(p)$ itself, we see that gradφ is a vector that is orthogonal to $\mathcal{S}_L$ at p. The curves of $\mathcal{E}$ that are tangent to gradφ $\forall p \in \Omega$ are the *streamlines of* φ; they have the property to be orthogonal to any $\mathcal{S}_L$ of φ $\forall p \in \Omega$.

gradφ allows us to calculate the directional derivative of φ along any direction $\mathbf{n} \in \mathcal{S}$ as

$$\frac{d\varphi}{d\mathbf{n}} = \text{grad}\varphi \cdot \mathbf{n}.$$

The highest variation of φ is hence in the direction of gradφ, and $|\text{grad}\varphi|$ is the value of this variation; we also remark that gradφ is a vector directed along the increasing values of φ.

Similarly, for a vector field $\mathbf{v}$, the directional derivative along any direction $\mathbf{n} \in \mathcal{S}$ can be computed as

$$\frac{d\mathbf{v}}{d\mathbf{n}} = \text{grad}\mathbf{v}\ \mathbf{n}.$$

Let ψ be a scalar of vector field of class C^2 at least. Then, the *laplacian* $\Delta\psi$ *of* ψ is defined by

$$\Delta\psi := \text{div}(\text{grad}\psi).$$

By the linearity of the trace, and hence of the divergence, we see easily that the laplacian of a vector field is the vector field whose components are the laplacian of each corresponding component of the field. A field is said to be *harmonic* on Ω if its laplacian is null $\forall p \in \Omega$.

The definitions given above for the differentiable field, gradient, and class C^1 can be repeated *verbatim* for a deformation $f(p) : \Omega \to \mathcal{E}$.

5.3 Properties of the differential operators

The differential operators, gradient, divergence, curl, and laplacian, have some interesting properties that are useful for calculations; they are introduced in this section.

Theorem 28 (Gradient of products). *Let* φ, ψ *be scalar and* $\mathbf{u}, \mathbf{v}, \mathbf{w}$ *be vector fields, with all of them differentiable. Then:*

$$\begin{aligned} &\text{(i) } \text{grad}(\varphi\psi) = \varphi\ \text{grad}\psi + \psi\ \text{grad}\varphi,\\ &\text{(ii) } \text{grad}(\varphi\mathbf{v}) = \varphi\ \text{grad}\mathbf{v} + \mathbf{v} \otimes \text{grad}\varphi, \qquad (5.1)\\ &\text{(iii) } \text{grad}(\mathbf{v} \cdot \mathbf{w}) = (\text{grad}\mathbf{w})^\top \mathbf{v} + (\text{grad}\mathbf{v})^\top \mathbf{w}. \end{aligned}$$

Proof. The proof is based upon the definition of gradient itself[1]:

(i)
$$(\varphi\psi)(p+\mathbf{u}) = \varphi\psi + \mathrm{grad}(\varphi\psi)\cdot\mathbf{u} + o(u),$$

but also,

$$\begin{aligned}(\varphi\psi)(p+\mathbf{u}) &= \varphi(p+\mathbf{u})\psi(p+\mathbf{u}) = (\varphi + \mathrm{grad}\varphi\cdot\mathbf{u} + o(u))\\ &\quad\times(\psi + \mathrm{grad}\psi\cdot\mathbf{u} + o(u))\\ &= \varphi\psi + \varphi\ \mathrm{grad}\psi\cdot\mathbf{u} + \psi\ \mathrm{grad}\varphi\cdot\mathbf{u} + o(u)\\ &= \varphi\psi + (\varphi\ \mathrm{grad}\psi + \psi\ \mathrm{grad}\varphi)\cdot\mathbf{u} + o(u),\end{aligned}$$

so by comparison

$$\mathrm{grad}(\varphi\psi) = \varphi\mathrm{grad}\psi + \psi\mathrm{grad}\varphi.$$

(ii) In the same way,

$$(\varphi\mathbf{v})(p+\mathbf{u}) = \varphi\mathbf{v} + \mathrm{grad}(\varphi\mathbf{v})\mathbf{u} + o(u) = \varphi\mathbf{v} + \mathrm{grad}(\varphi\mathbf{v})\mathbf{u} + o(u),$$

but also,

$$\begin{aligned}(\varphi\mathbf{v})(p+\mathbf{u}) &= \varphi(p+\mathbf{u})\mathbf{v}(p+\mathbf{u}) = (\varphi + \mathrm{grad}\varphi\cdot\mathbf{u} + o(u))\\ &\quad\times(\mathbf{v} + \mathrm{grad}\mathbf{v}\ \mathbf{u} + o(u))\\ &= \varphi\mathbf{v} + \varphi\mathrm{grad}\mathbf{v}\ \mathbf{u} + \mathrm{grad}\varphi\cdot\mathbf{u}\ \mathbf{v} + o(u)\\ &= \varphi\mathbf{v} + (\varphi\ \mathrm{grad}\mathbf{v} + \mathbf{v}\otimes\mathrm{grad}\mathbf{v})\mathbf{u} + o(u),\end{aligned}$$

so comparing the two results, we get

$$\mathrm{grad}(\varphi\mathbf{v}) = \varphi\ \mathrm{grad}\mathbf{v} + \mathbf{v}\otimes\mathrm{grad}\mathbf{v}.$$

(iii) In the same way,

$$(\mathbf{v}\cdot\mathbf{w})(p+\mathbf{u}) = \mathbf{v}\cdot\mathbf{w} + \mathrm{grad}(\mathbf{v}\cdot\mathbf{w})\mathbf{u} + o(u),$$

but also,

$$\begin{aligned}(\mathbf{v}\cdot\mathbf{w})(p+\mathbf{u}) &= \mathbf{v}(p+\mathbf{u})\cdot\mathbf{w}(p+\mathbf{u}) = (\mathbf{v} + \mathrm{grad}\mathbf{v}\ \mathbf{u} + o(u))\\ &\quad\cdot(\mathbf{w} + \mathrm{grad}\mathbf{w}\ \mathbf{u} + o(u))\\ &= \mathbf{v}\cdot\mathbf{w} + \mathbf{v}\cdot(\mathrm{grad}\mathbf{w}\ \mathbf{u}) + (\mathrm{grad}\mathbf{v}\ \mathbf{u})\cdot\mathbf{w} + o(u)\\ &= \mathbf{v}\cdot\mathbf{w} + ((\mathrm{grad}\mathbf{w})^\top\mathbf{v} + (\mathrm{grad}\mathbf{v})^\top\mathbf{w})\cdot\mathbf{u} + o(u),\end{aligned}$$

whence, by comparison of the two results,

$$\mathrm{grad}(\mathbf{v}\cdot\mathbf{w}) = (\mathrm{grad}\mathbf{w})^\top\mathbf{v} + (\mathrm{grad}\mathbf{v})^\top\mathbf{w}.$$ □

[1]For the sake of brevity, we omit to indicate the point p, e.g. we simply write φ for $\varphi(p)$ and gradφ for grad$\varphi(p)$.

Another important result[2] relating the gradient and the curl of a vector field is as follows.

Theorem 29. *If* $\mathbf{v}$ *is a differentiable vector field, then*

$$(\text{grad}\mathbf{v})\mathbf{v} = (\text{curl}\mathbf{v}) \times \mathbf{v} + \frac{1}{2}\text{grad}\mathbf{v}^2.$$

Proof.

$$\begin{aligned}(\text{curl}\mathbf{v}) \times \mathbf{v} &= (\text{grad}\mathbf{v} - (\text{grad}\mathbf{v}^\top))\mathbf{v} = (\text{grad}\mathbf{v})\mathbf{v} - (\text{grad}\mathbf{v})^\top\mathbf{v} \\ &= (\text{grad}\mathbf{v})\mathbf{v} - \frac{1}{2}((\text{grad}\mathbf{v})^\top\mathbf{v} + (\text{grad}\mathbf{v})^\top\mathbf{v}),\end{aligned}$$

and by property (iii) of the previous theorem,

$$(\text{grad}\mathbf{v})^\top\mathbf{v} + (\text{grad}\mathbf{v})^\top\mathbf{v} = \text{grad}(\mathbf{v} \cdot \mathbf{v}) = \text{grad}\mathbf{v}^2$$

so that

$$(\text{curl}\mathbf{v}) \times \mathbf{v} = (\text{grad}\mathbf{v})\mathbf{v} - \frac{1}{2}\text{grad}\mathbf{v}^2,$$

whence we obtain the thesis. □

The proof of the following properties of the gradient are left to the reader as an exercise:

$$\begin{aligned}\text{grad}(\mathbf{v} \cdot \mathbf{w}) &= (\text{grad}\mathbf{w})\mathbf{v} + (\text{grad}\mathbf{v})\mathbf{w} + \mathbf{v} \times \text{curl}\mathbf{w} + \mathbf{w} \times \text{curl}\mathbf{v}, \\ \text{grad}(\mathbf{u} \cdot \mathbf{v}\ \mathbf{w}) &= (\mathbf{u} \cdot \mathbf{v})\text{grad}\mathbf{w} + (\mathbf{w} \otimes \mathbf{u})\text{grad}\mathbf{v} + (\mathbf{w} \otimes \mathbf{v})\text{grad}\mathbf{u}, \\ \text{grad}\mathbf{v} \cdot \text{grad}\mathbf{v}^\top &= \text{div}((\text{grad}\mathbf{v})\mathbf{v} - (\text{div}\mathbf{v})\mathbf{v}) + (\text{div}\mathbf{v})^2.\end{aligned} \tag{5.2}$$

Theorem 30 (Divergence of products). *Let* $\varphi, \mathbf{u}, \mathbf{v}, \mathbf{w}, \mathbf{L}$ *be differentiable scalar, vector, or tensor fields. Then:*

$$\begin{aligned}&\text{(i) } \text{div}(\varphi\mathbf{v}) = \varphi\text{div}\mathbf{v} + \mathbf{v} \cdot \text{grad}\varphi, \\ &\text{(ii) } \text{div}(\mathbf{v} \otimes \mathbf{w}) = \mathbf{v}\text{div}\mathbf{w} + (\text{grad}\mathbf{v})\mathbf{w}, \\ &\text{(iii) } \text{div}(\varphi\mathbf{L}) = \varphi\text{div}\mathbf{L} + \mathbf{L}\text{grad}\varphi, \\ &\text{(iv) } \text{div}(\mathbf{L}^\top\mathbf{v}) = \mathbf{L} \cdot \text{grad}\mathbf{v} + \mathbf{v} \cdot \text{div}\mathbf{L}, \\ &\text{(v) } \text{div}(\mathbf{v} \times \mathbf{w}) = \mathbf{w} \cdot \text{curl}\mathbf{v} - \mathbf{v} \cdot \text{curl}\mathbf{w}.\end{aligned}$$

---

[2]This result is fundamental to fluid mechanics, as it allows us to get an interesting form of the Navier–Stokes equations.

Proof. (i) Using the definition of divergence and property (ii) of Theorem 28, we get

$$\text{div}(\varphi\mathbf{v}) = \text{tr}(\text{grad}(\varphi\mathbf{v})) = \text{tr}(\varphi\ \text{grad}\mathbf{v} + \mathbf{v} \otimes \text{grad}\varphi)$$

$$= \varphi\ \text{tr}(\text{grad}\mathbf{v}) + \text{tr}(\mathbf{v} \otimes \text{grad}\varphi) = \varphi\ \text{div}\mathbf{v} + \mathbf{v} \cdot \text{grad}\varphi.$$

(ii) By the definition of divergence of a tensor, $\forall \mathbf{a} = const. \in \mathcal{V}$, and using the previous property along with property (iii) of Theorem 28:

$$(\text{div}(\mathbf{v} \otimes \mathbf{w}) \cdot \mathbf{a} = \text{div}((\mathbf{v} \otimes \mathbf{w})^\top \mathbf{a}) = \text{div}(\mathbf{w} \otimes \mathbf{v}\ \mathbf{a}) = \text{div}(\mathbf{v} \cdot \mathbf{a}\ \mathbf{w})$$

$$= \mathbf{v} \cdot \mathbf{a}\text{div}\mathbf{w} + \mathbf{w} \cdot \text{grad}(\mathbf{a} \cdot \mathbf{v})$$

$$= \text{div}\mathbf{w}\ \mathbf{v} \cdot \mathbf{a} + \mathbf{w} \cdot (\text{grad}\mathbf{v})^\top \mathbf{a} + \mathbf{w} \cdot (\text{grad}\mathbf{a})^\top \mathbf{v}$$

$$= (\mathbf{v}\text{div}\mathbf{w} + \text{grad}\mathbf{v}\ \mathbf{w}) \cdot \mathbf{a}.$$

(iii) By the definition of divergence of a tensor, $\forall \mathbf{a} = const. \in \mathcal{V}$, and using property (i), along with property (iii) of Theorem 28:

$$\text{div}(\varphi\mathbf{L}) \cdot \mathbf{a} = \text{div}((\varphi\mathbf{L})^\top \mathbf{a}) = \text{div}(\varphi\mathbf{L}^\top \mathbf{a}) = \varphi\text{div}(\mathbf{L}^\top \mathbf{a}) + \mathbf{L}^\top \mathbf{a} \cdot \text{grad}\varphi$$

$$= \varphi\text{div}\mathbf{L} \cdot \mathbf{a} + \mathbf{a} \cdot \mathbf{L}\text{grad}\varphi = (\varphi\text{div}\mathbf{L} + \mathbf{L}\text{grad}\varphi) \cdot \mathbf{a}.$$

(iv) By the definition of divergence of a tensor and using the previous property:

$$\text{div}(\mathbf{L}^\top \mathbf{v}) = \text{div}(\mathbf{L}^\top v_j \mathbf{e}_j) = \text{div}((v_j \mathbf{L}^\top)\mathbf{e}_j) = \text{div}(v_j \mathbf{L}^\top)^\top \cdot \mathbf{e}_j$$

$$= \text{div}(v_j \mathbf{L}) \cdot \mathbf{e}_j = v_j \text{div}\mathbf{L} \cdot \mathbf{e}_j + \mathbf{L}\ \text{grad}v_j \cdot \mathbf{e}_j$$

$$= \mathbf{v} \cdot \text{div}\mathbf{L} + (L_{pq}\mathbf{e}_p \otimes \mathbf{e}_q(\text{grad}v_j)_m \mathbf{e}_m) \cdot \mathbf{e}_j$$

$$= \mathbf{v} \cdot \text{div}\mathbf{L} + L_{pq} v_{j,m} \delta_{qm} \delta_{jp} = \mathbf{v} \cdot \text{div}\mathbf{L} + \mathbf{L} \cdot \text{grad}\mathbf{v}.$$

(v) This property can be proved making use of the expression of the cross product with the Ricci's alternator, given in Eq. (2.30):

$$\text{curl}\mathbf{v} = \epsilon_{ijk} v_{k,j} \mathbf{e}_i$$

and

$$\mathbf{v} \times \mathbf{w} = \epsilon_{ijk} v_j w_k \mathbf{e}_i,$$

whence

$$\text{div}(\mathbf{v} \times \mathbf{w}) = \text{div}(\epsilon_{ijk} v_j w_k \mathbf{e}_i) = \epsilon_{ijk}(v_j w_k)_{,i} = \epsilon_{ijk} v_{j,i} w_k + \epsilon_{ijk} w_{k,i} v_j.$$

Moreover,

$$\mathbf{w} \cdot \text{curl}\mathbf{v} = w_m \mathbf{e}_m \cdot \epsilon_{pqr} v_{r,q} \mathbf{e}_p = \epsilon_{pqr} v_{r,q} w_m \delta_{pm} = \epsilon_{pqr} v_{r,q} w_p = \epsilon_{qrp} v_{r,q} w_p$$

and

$$\mathbf{v} \cdot \text{curl}\mathbf{w} = v_m \mathbf{e}_m \cdot \epsilon_{pqr} w_{r,q} \mathbf{e}_p = \epsilon_{pqr} w_{r,q} v_m \delta_{pm} = \epsilon_{pqr} w_{r,q} v_p = -\epsilon_{qpr} w_{r,q} v_p,$$

so finally, comparing the last three results (all the subscripts are dummy indexes, so their denomination is inessential),

$$\mathrm{div}(\mathbf{v} \times \mathbf{w}) = \mathbf{w} \cdot \mathrm{curl}\mathbf{v} - \mathbf{v} \cdot \mathrm{curl}\mathbf{w}. \qquad \square$$

The divergence has also the following properties:

$$\begin{aligned} \mathrm{div}(\mathrm{grad}\mathbf{v}^\top) &= \mathrm{grad}(\mathrm{div}\mathbf{v}), \\ \mathrm{div}((\mathrm{grad}\mathbf{v})\mathbf{v}) &= \mathrm{grad}\mathbf{v} \cdot \mathrm{grad}\mathbf{v}^\top + \mathbf{v} \cdot \mathrm{grad}(\mathrm{div}\mathbf{v}), \\ \mathrm{div}(\varphi\mathbf{L}\mathbf{v}) &= \varphi\mathbf{L}^\top \cdot \mathrm{grad}\mathbf{v} + \varphi\mathbf{v} \cdot \mathrm{div}\mathbf{L}^\top + \mathbf{L}\mathbf{v} \cdot \mathrm{grad}\varphi, \end{aligned} \tag{5.3}$$

whose proofs can be a good exercise for the reader.

The relations of the gradient and divergence with the curl are given by the following.

Theorem 31. *Let* φ *and* $\mathbf{v}$ *be the scalar and vector fields of class* C^2*. Then:*

$$\text{(i) } \mathrm{div}(\mathrm{curl}\mathbf{v}) = 0,$$

$$\text{(ii) } \mathrm{curl}(\mathrm{grad}\varphi) = \mathbf{o}.$$

Proof. (i) Using again the Ricci's alternator to represent the cross product,

$$\mathrm{div}(\mathrm{curl}\mathbf{v}) = \mathrm{div}(\epsilon_{ijk}v_{k,j}\mathbf{e}_i) = \epsilon_{ijk}v_{k,j}\mathrm{div}\mathbf{e}_i + \epsilon_{ijk}v_{k,ji} = \epsilon_{ijk}v_{k,ji}$$

$$= v_{3,21} + v_{1,32} + v_{2,13} - v_{2,31} - v_{3,12} - v_{1,23} = 0.$$

(ii) In a similar manner,

$$\mathrm{curl}(\mathrm{grad}\varphi) = \epsilon_{ijk}\varphi_{,kj}\mathbf{e}_i = \varphi_{32} + \varphi_{,13} + \varphi_{,21} - \varphi_{,23} - \varphi_{,31} - \varphi_{,12} = 0. \qquad \square$$

The following theorem gives an interesting relation between the curl of a vector and the divergence of its axial tensor.

Theorem 32 (Curl of an axial vector). *Let* $\mathbf{w}$ *be a differentiable vector field and* $\mathbf{W}$ *its axial tensor field. Then,*

$$\mathrm{curl}\mathbf{w} = -\mathrm{div}\mathbf{W}.$$

Proof. Using properties (iv) and (v) of Theorem 30 and because $\mathbf{W} = -\mathbf{W}^\top, \forall\mathbf{a} = const. \in \mathcal{V}$, we get

$$\begin{aligned} \mathrm{div}(\mathbf{w} \times \mathbf{a}) &= \mathbf{a} \cdot \mathrm{curl}\mathbf{w} - \mathbf{w} \cdot \mathrm{curl}\mathbf{a} = \mathbf{a} \cdot \mathrm{curl}\mathbf{w}, \\ \mathrm{div}(\mathbf{W}\mathbf{a}) &= \mathrm{div}(-\mathbf{W}^\top\mathbf{a}) = -\mathbf{W} \cdot \mathrm{grad}\mathbf{a} - \mathbf{a} \cdot \mathrm{div}\mathbf{W} = -\mathbf{a} \cdot \mathrm{div}\mathbf{W}. \end{aligned}$$

Now, because $\forall\mathbf{a}, \mathbf{w} \times \mathbf{a} = \mathbf{W}\mathbf{a} \Rightarrow \mathrm{div}(\mathbf{w} \times \mathbf{a}) = \mathrm{div}(\mathbf{W}\mathbf{a})$, we get the thesis. $\square$

The way the curl of a curl[3] is computed is given by the following theorem.

Theorem 33 (Curl of a curl). *Let* $\mathbf{v}$ *be a vector field of class* $\geq C^2$. *Then,*

$$\text{curl}(\text{curl}\mathbf{v}) = \text{grad}(\text{div}\mathbf{v}) - \Delta\mathbf{v}.$$

Proof. Using properties (iv) and (v) of Theorem 30, along with the first of Eq. (5.3), $\forall \mathbf{a} = const. \in \mathcal{V}$, we get

$$\text{div}((\text{curl}\mathbf{v}) \times \mathbf{a}) = \mathbf{a} \cdot \text{curl}(\text{curl}\mathbf{v}) - \text{curl}\mathbf{v} \cdot \text{curl}\mathbf{a} = \mathbf{a} \cdot \text{curl}(\text{curl}\mathbf{v}),$$

and by the definition of curl and laplacian,

$$\begin{aligned}
\text{div}((\text{curl}\mathbf{v}) \times \mathbf{a}) &= \text{div}((\text{grad}\mathbf{v} - (\text{grad}\mathbf{v})^\top)\mathbf{a}) = \text{div}(\text{grad}\mathbf{v}\; a) \\
&\quad - \text{div}((\text{grad}\mathbf{v})^\top \mathbf{a}) \\
&= (\text{grad}\mathbf{v})^\top \cdot \text{grad}\mathbf{a} + \mathbf{a} \cdot \text{div}(\text{grad}\mathbf{v})^\top - \text{grad}\mathbf{v} \cdot \text{grad}\mathbf{a} \\
&\quad - \mathbf{a} \cdot \text{div}(\text{grad}\mathbf{v}) \\
&= \mathbf{a} \cdot (\text{div}(\text{grad}\mathbf{v})^\top - \text{div}(\text{grad}\mathbf{v})) = \mathbf{a} \cdot (\text{grad}(\text{div}\mathbf{v}) - \Delta\mathbf{v}),
\end{aligned}$$

whence, by comparison,

$$\text{curl}(\text{curl}\mathbf{v}) = \text{grad}(\text{div}\mathbf{v}) - \Delta\mathbf{v}.$$

□

The proof of the following properties of the curl can be obtained using the above results, and it is a good exercise:

$$\begin{aligned}
\text{curl}(\varphi\mathbf{v}) &= \varphi\text{curl}\mathbf{v} + \text{grad}\varphi \times \mathbf{v}, \\
\text{curl}(\mathbf{v} \times \mathbf{w}) &= (\text{grad}\mathbf{v})\mathbf{w} - (\text{grad}\mathbf{w})\mathbf{v} + \mathbf{v}\text{div}\mathbf{w} - \mathbf{w}\text{div}\mathbf{v}.
\end{aligned} \tag{5.4}$$

Finally, we have a theorem also for the laplacian of a product.

Theorem 34 (Laplacian of products). *Let* $\varphi, \psi, \mathbf{u}, \mathbf{v}$ *be the scalar and vector fields of class* $\geq C^2$. *Then:*

$$\text{(i) } \Delta(\varphi\psi) = 2\text{grad}\varphi \cdot \text{grad}\psi + \varphi\Delta\psi + \psi\Delta\varphi,$$

$$\text{(ii) } \Delta(\mathbf{v} \cdot \mathbf{w}) = 2\text{grad}\mathbf{v} \cdot \text{grad}\mathbf{w} + \mathbf{v} \cdot \Delta\mathbf{w} + \mathbf{w} \cdot \Delta\mathbf{v}.$$

[3]This relation is useful in fluid mechanics for writing the vorticity equation.

Proof. (i) Using property (i) of Theorems 28 and 30, we get

$$\begin{aligned}\Delta(\varphi\psi) &= \mathrm{div}(\mathrm{grad}(\varphi\psi)) = \mathrm{div}(\varphi\ \mathrm{grad}\psi + \psi\ \mathrm{grad}\varphi) = \mathrm{div}(\varphi\ \mathrm{grad}\psi) \\ &\quad + \mathrm{div}(\psi\ \mathrm{grad}\varphi) \\ &= \varphi\ \mathrm{div}(\mathrm{grad}\psi) + \mathrm{grad}\psi \cdot \mathrm{grad}\varphi + \psi\ \mathrm{div}(\mathrm{grad}\varphi) + \mathrm{grad}\varphi \cdot \mathrm{grad}\psi \\ &= 2\mathrm{grad}\varphi \cdot \mathrm{grad}\psi + \varphi\ \Delta\psi + \psi\ \Delta\varphi.\end{aligned}$$

(ii) Using properties (iii) of Theorem 28 and (iv) of Theorem 30, we obtain

$$\begin{aligned}\Delta(\mathbf{v}\cdot\mathbf{w}) &= \mathrm{div}(\mathrm{grad}(\mathbf{v}\cdot\mathbf{w})) = \mathrm{div}((\mathrm{grad}\mathbf{w})^\top\mathbf{v} + (\mathrm{grad}\mathbf{v})^\top\mathbf{w}) \\ &= \mathrm{div}((\mathrm{grad}\mathbf{w})^\top\mathbf{v}) + \mathrm{div}((\mathrm{grad}\mathbf{v})^\top\mathbf{w}) \\ &= \mathrm{grad}\mathbf{w}\cdot\mathrm{grad}\mathbf{v} + \mathbf{v}\cdot\mathrm{div}(\mathrm{grad}\mathbf{w}) + \mathrm{grad}\mathbf{v}\cdot\mathrm{grad}\mathbf{w} \\ &\quad + \mathbf{w}\cdot\mathrm{div}(\mathrm{grad}\mathbf{v}) \\ &= 2\mathrm{grad}\mathbf{v}\cdot\mathrm{grad}\mathbf{w} + \mathbf{v}\cdot\Delta\mathbf{w} + \mathbf{w}\cdot\Delta\mathbf{v}.\end{aligned}$$

□

5.4 Theorems on fields

We recall here some classical theorems on fields and operators.

Theorem 35 (Harmonic fields). *If* $\mathbf{v}(p)$ *is a vector field of class* $\geq \mathrm{C}^2$ *such that*

$$\mathrm{div}\mathbf{v} = 0, \qquad \mathrm{curl}\mathbf{v} = \mathbf{o},$$

then $\mathbf{v}$ *is harmonic:* $\Delta\mathbf{v} = \mathbf{o}$.

Proof. By the definition of curl,

$$\mathrm{curl}\mathbf{v} = \mathbf{o} \Rightarrow\ \mathrm{grad}\mathbf{v} - (\mathrm{grad}\mathbf{v})^\top = \mathbf{o} \Rightarrow \mathrm{div}(\mathrm{grad}\mathbf{v} - (\mathrm{grad}\mathbf{v})^\top) = \mathbf{o},$$

and through Eq. $(5.3)_1$, the definition of laplacian, and because by hypothesis $\mathrm{div}\mathbf{v} = 0$, we have

$$\mathrm{div}(\mathrm{grad}\mathbf{v} - (\mathrm{grad}\mathbf{v})^\top) = \Delta\mathbf{v} - \mathrm{grad}(\mathrm{div}\mathbf{v}) = \Delta\mathbf{v}.$$

□

We state now without proof a lemma[4] that, basically, allows us to transform a volume integral on a domain Ω to a surface integral on the boundary surface $\partial\Omega$.

Theorem 36 (Divergence lemma). *Let* $\mathbf{v}(p)$ *be a vector field of class* $\geq \mathrm{C}^1$ *on a regular region* $\Omega \subset \mathcal{E}$. *Then,*

$$\int_{\partial\Omega} \mathbf{v}\otimes\mathbf{n}\ dA = \int_\Omega \mathrm{grad}\mathbf{v}\ dV.$$

[4]In the following, $\partial\Omega$ indicates the boundary of Ω.

This lemma is fundamental for proving the three forms of the Gauss theorem, which is of paramount importance in many fields of mathematical physics.

Theorem 37 (Divergence or Gauss theorem). *Let $\varphi, \mathbf{v}, \mathbf{L}$ be, respectively, a scalar, vector, and tensor field of class $\geq \mathrm{C}^1$ on a regular region $\Omega \subset \mathcal{E}$. Then:*

$$\text{(i)} \int_{\partial\Omega} \varphi\mathbf{n}\, dA = \int_{\Omega} \text{grad}\varphi\, dV,$$

$$\text{(ii)} \int_{\partial\Omega} \mathbf{v}\cdot\mathbf{n}\, dA = \int_{\Omega} \text{div}\mathbf{v}\, dV,$$

$$\text{(iii)} \int_{\partial\Omega} \mathbf{L}\mathbf{n}\, dA = \int_{\Omega} \text{div}\mathbf{L}\, dV.$$

Proof. (i) $\forall \mathbf{a} = const. \in \mathcal{V}$, by the lemma of divergence,

$$\int_{\Omega} \text{grad}(\varphi\mathbf{a}) dV = \int_{\partial\Omega} \varphi\mathbf{a}\otimes\mathbf{n}\, dA = \mathbf{a}\otimes\int_{\partial\Omega} \varphi\mathbf{n}\, dA,$$

but also, by (ii) of Theorem 28,

$$\int_{\Omega} \text{grad}(\varphi\mathbf{a}) dV = \int_{\Omega} (\varphi\, \text{grad}\mathbf{a} + \mathbf{a}\otimes\text{grad}\varphi) dV = \mathbf{a}\otimes\int_{\Omega} \text{grad}\varphi\, dV,$$

whence, by comparison,

$$\int_{\partial\Omega} \varphi\mathbf{n}\, dA = \int_{\Omega} \text{grad}\varphi\, dV.$$

(ii) Still, by the divergence lemma,

$$\text{tr}\int_{\Omega} \text{grad}\mathbf{v}\, dV = \text{tr}\int_{\partial\Omega} \mathbf{v}\otimes\mathbf{n}\, dA = \int_{\partial\Omega} \text{tr}(\mathbf{v}\otimes\mathbf{n}) dA = \int_{\partial\Omega} \mathbf{v}\cdot\mathbf{n}\, dA,$$

but also,

$$\text{tr}\int_{\Omega} \text{grad}\mathbf{v}\, dV = \int_{\Omega} \text{tr}(\text{grad}\mathbf{v}) dV = \int_{\Omega} \text{div}\mathbf{v}\, dV,$$

so by comparison,

$$\int_{\partial\Omega} \mathbf{v}\cdot\mathbf{n}\, dA = \int_{\Omega} \text{div}\mathbf{v}\, dV.$$

(iii) $\forall \mathbf{a} = const. \in \mathcal{V}$, by the lemma of divergence, (iv) of Theorem 30 and (ii) just proved,

$$\int_{\Omega} \text{div}(\mathbf{L}^{\top}\mathbf{a})dV = \int_{\partial\Omega} (\mathbf{L}^{\top}\mathbf{a}) \cdot \mathbf{n}\, dA = \int_{\partial\Omega} \mathbf{a} \cdot \mathbf{L}\mathbf{n}\, dA = \mathbf{a} \cdot \int_{\partial\Omega} \mathbf{L}\mathbf{n}\, dA,$$

but also,

$$\int_{\Omega} \text{div}(\mathbf{L}^{\top}\mathbf{a})dV = \int_{\Omega} (\text{div}\mathbf{L}) \cdot \mathbf{a} + \mathbf{L} \cdot \text{grad}\mathbf{a}\, dV = \mathbf{a} \cdot \int_{\Omega} \text{div}\mathbf{L}\, dV,$$

so once again by comparison,

$$\int_{\partial\Omega} \mathbf{L}\mathbf{n}\, dA = \int_{\Omega} \text{div}\mathbf{L}\, dV.$$

□

The following identities follow directly from the Gauss theorem:

$$\begin{aligned} &\int_{\partial\Omega} \mathbf{v} \cdot \mathbf{L}\mathbf{n}\, dA = \int_{\Omega} (\mathbf{v} \cdot \text{div}\mathbf{L} + \mathbf{L} \cdot \text{grad}\mathbf{v})dV, \\ &\int_{\partial\Omega} (\mathbf{L}\mathbf{n}) \otimes \mathbf{v}\, dA = \int_{\Omega} ((\text{div}\mathbf{L}) \otimes \mathbf{v} + \mathbf{L}(\text{grad}\mathbf{v}^{\top}))dV, \\ &\int_{\partial\Omega} (\mathbf{w} \cdot \mathbf{n})\mathbf{v}\, dA = \int_{\Omega} (\mathbf{v}\text{div}\mathbf{w} + (\text{grad}\mathbf{v})\mathbf{w})dV. \end{aligned} \tag{5.5}$$

A direct consequence of the Gauss theorem is the following theorem.

Theorem 38 (Flux theorem). *Let $\mathbf{v}(p)$ be a vector field of class $\geq \mathrm{C}^1$ on an open subset* R *of $\mathcal{E}$. Then,*

$$\text{div}\mathbf{v} = 0 \iff \int_{\partial\Omega} \mathbf{v} \cdot \mathbf{n}\, dA = 0 \quad \forall \Omega \subset \mathrm{R}.$$

Proof. It immediately follows from (ii) of the Gauss theorem. □

Another consequence of the Gauss theorem is the next theorem.

Theorem 39 (Curl theorem). *Let $\mathbf{v}(p)$ be a vector field of class $\geq \mathrm{C}^1$ on a regular region $\Omega \subset \mathcal{E}$. Then,*

$$\int_{\partial\Omega} \mathbf{n} \times \mathbf{v}\, dA = \int_{\Omega} \text{curl}\mathbf{v}\, dV.$$

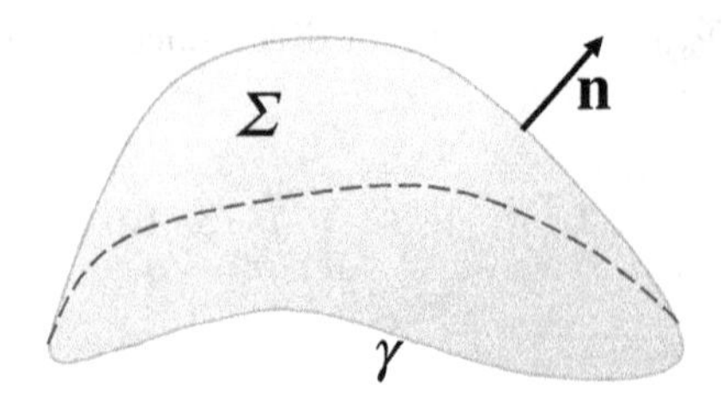

Figure 5.1: Scheme for the Stokes theorem.

Proof. If $\mathbf{V}$ is the axial tensor of $\mathbf{v}$, by Theorem 32 and (iii) of the Gauss theorem,

$$\begin{aligned}\int_{\partial\Omega} \mathbf{n}\times\mathbf{v}\, dA &= -\int_{\partial\Omega} \mathbf{v}\times\mathbf{n}\, dA = -\int_{\partial\Omega} \mathbf{V}\mathbf{n}\, dA\\ &= -\int_{\Omega} \text{div}\mathbf{V}\, dV = \int_{\Omega} \text{curl}\mathbf{v}\, dV.\end{aligned}$$

□

The following classical theorems on fields are recalled here without proof.

Theorem 40 (Potential theorem). *Let* $\mathbf{v}(p)$ *be a vector field of class* $\geq \mathrm{C}^1$ *on a simply connected region* $\Omega \subset \mathcal{E}$*. Then,*

$$\text{curl}\mathbf{v} = \mathbf{o} \iff \mathbf{v} = \text{grad}\varphi,$$

with $\varphi(p)$ *a scalar field of class* $\geq \mathrm{C}^2$*, the potential.*

Theorem 41 (Stokes theorem). *Let* $\mathbf{v}(p)$ *be a vector field of class* $\geq \mathrm{C}^1$ *on a regular region* $\Omega \subset \mathcal{E}$*,* Σ *an open surface whose support is the closed line* γ *and* $\mathbf{n} \in \mathcal{S}$ *the normal to* Σ*, see Fig. 5.1. Then,*

$$\oint_{\gamma} \mathbf{v}\cdot d\ell = \int_{\Sigma} \text{curl}\mathbf{v}\cdot\mathbf{n}\, dA.$$

The parametric equation of γ *must be chosen in such a way that*

$$p'(t_1)\times p'(t_2)\cdot\mathbf{n} > 0 \quad \forall t_2 > t_1.$$

Theorem 42 (Green's formula). *Let* $\varphi(p), \psi(p)$ *be two scalar fields of class* $\geq \mathrm{C}^2$ *on a regular region* $\Omega \subset \mathcal{E}$*, with* $\mathbf{n} \in \mathcal{S}$ *the normal to* $\partial\Omega$*. Then,*

$$\int_{\partial\Omega}\left(\psi\frac{d\varphi}{d\mathbf{n}} - \varphi\frac{d\psi}{d\mathbf{n}}\right) dA = \int_{\Omega}(\psi\,\Delta\varphi - \varphi\,\Delta\psi)dV.$$

5.5 Differential operators in Cartesian coordinates

The Cartesian expression of the differential operators can be found without difficulty by applying the properties of such operators shown previously and considering that the vectors on the Cartesian basis are fixed. Then,[5]

$$
\begin{aligned}
&\mathrm{grad} f = f_{,i}\, \mathbf{e}_i,\\
&\mathrm{grad}\mathbf{v} = v_{i,j}\mathbf{e}_i \otimes \mathbf{e}_j,\\
&\mathrm{div}\mathbf{v} = v_{i,i},\\
&\mathrm{div}\mathbf{L} = L_{ij,j}\mathbf{e}_i,\\
&\Delta f = f_{,ii},\\
&\Delta \mathbf{v} = \Delta v_i \mathbf{e}_i = v_{i,jj}\mathbf{e}_i,\\
&\mathrm{curl}\mathbf{v} = (v_{3,2} - v_{2,3})\mathbf{e}_1 + (v_{1,3} - v_{3,1})\mathbf{e}_2 + (v_{2,1} - v_{1,2})\mathbf{e}_3.
\end{aligned}
\tag{5.6}
$$

The so-called *operator nabla* ∇:,

$$
\nabla := \frac{\partial \cdot}{\partial x_i}\mathbf{e}_i = \frac{\partial \cdot}{\partial x_1}\mathbf{e}_1 + \frac{\partial \cdot}{\partial x_2}\mathbf{e}_2 + \frac{\partial \cdot}{\partial x_3}\mathbf{e}_3, \tag{5.7}
$$

is often used to indicate the differential operators:

$$
\begin{aligned}
\mathrm{grad} f &= \nabla f,\\
\mathrm{div}\mathbf{v} &= \nabla \cdot \mathbf{v},\\
\mathrm{curl}\mathbf{v} &= \nabla \times \mathbf{v},\\
\Delta f &= \nabla^2 f.
\end{aligned}
$$

5.6 Differential operators in cylindrical coordinates

The cylindrical coordinates ρ, θ, z of a point p whose Cartesian coordinates in the (fixed) frame $\{o; \mathbf{e}_1, \mathbf{e}_2, \mathbf{e}_3\}$ are $p = (x_1, x_2, x_3)$ are shown in Fig. 5.2.

[5] In what follows, and also in the following sections, $f, \mathbf{v}, \mathbf{L}$ are, respectively, scalar, vector, and tensor fields.

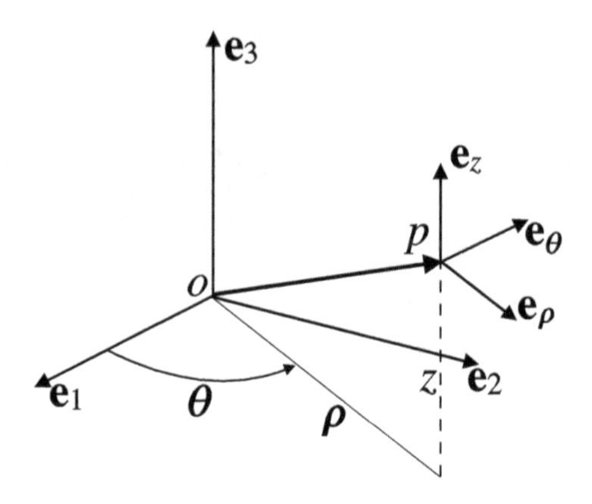

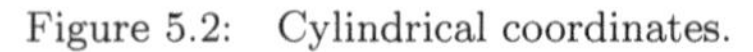
Figure 5.2: Cylindrical coordinates.

They are related together by

$$\begin{aligned}\rho &= \sqrt{x_1^2 + x_2^2},\\ \theta &= \arctan \frac{x_2}{x_1},\\ z &= x_3,\end{aligned} \tag{5.8}$$

or conversely,

$$\begin{aligned}x_1 &= \rho \cos \theta,\\ x_2 &= \rho \sin \theta,\\ x_3 &= z.\end{aligned} \tag{5.9}$$

We note that $\rho \geq 0$ and that the *anomaly* θ is bounded by $0 \leq \theta < 2\pi$. A vector $p - o = x_i \mathbf{e}_i$ in the cylindrical basis is expressed as

$$p - o = \rho \mathbf{e}_\rho + z \mathbf{e}_z,$$

and the rotation tensor transforming the Cartesian basis, $\{\mathbf{e}_1, \mathbf{e}_2, \mathbf{e}_3\}$, into the cylindrical one, $\{\mathbf{e}_\rho, \mathbf{e}_\theta, \mathbf{e}_z\}$, is

$$\mathbf{Q} = \begin{bmatrix} \cos\theta & -\sin\theta & 0 \\ \sin\theta & \cos\theta & 0 \\ 0 & 0 & 1 \end{bmatrix},$$

so the relations between the vectors of the Cartesian and the cylindrical basis are

$$\mathbf{e}_\rho = \cos\theta \mathbf{e}_1 + \sin\theta \mathbf{e}_2,$$
$$\mathbf{e}_\theta = -\sin\theta \mathbf{e}_1 + \cos\theta \mathbf{e}_2,$$
$$\mathbf{e}_z = \mathbf{e}_3,$$

and vice versa:

$$\begin{aligned} \mathbf{e}_1 &= \cos\theta \mathbf{e}_\rho - \sin\theta \mathbf{e}_\theta, \\ \mathbf{e}_2 &= \sin\theta \mathbf{e}_\rho + \cos\theta \mathbf{e}_\theta, \\ \mathbf{e}_3 &= \mathbf{e}_z. \end{aligned} \tag{5.10}$$

The question is: How can we express the differential operators in the (moving) frame $\{p; \mathbf{e}_\rho, \mathbf{e}_\theta, \mathbf{e}_z\}$? To this end, we can proceed as follows: From Eq. (5.8),

$$f_{,i} = f_{,\rho}\frac{\partial \rho}{\partial x_i} + f_{,\theta}\frac{\partial \theta}{\partial x_i} + f_{,z}\frac{\partial z}{\partial x_i} \rightarrow \begin{cases} f_{,1} = f_{,\rho}\dfrac{x_1}{\rho} - f_{,\theta}\dfrac{x_2}{\rho^2}, \\ f_{,2} = f_{,\rho}\dfrac{x_2}{\rho} + f_{,\theta}\dfrac{x_1}{\rho^2}, \\ f_{,3} = f_{,z}. \end{cases} \tag{5.11}$$

So, by Eqs. $(5.6)_1$ and (5.10),

$$\begin{aligned} \operatorname{grad} f = f_{,i}\mathbf{e}_i = {} & \left(f_{,\rho}\frac{x_1}{\rho} - f_{,\theta}\frac{x_2}{\rho^2}\right)(\cos\theta \mathbf{e}_\rho - \sin\theta \mathbf{e}_\theta) \\ & + \left(f_{,\rho}\frac{x_2}{\rho} + f_{,\theta}\frac{x_1}{\rho^2}\right)(\sin\theta \mathbf{e}_\rho + \cos\theta \mathbf{e}_\theta) + f_{,z}\mathbf{e}_z. \end{aligned}$$

Finally, by Eq. (5.9) and through some standard operations, we obtain

$$\operatorname{grad} f = f_{,\rho}\mathbf{e}_\rho + \frac{1}{\rho} f_{,\theta}\, \mathbf{e}_\theta + f_{,z}\mathbf{e}_z.$$

The gradient of a vector field $\mathbf{v}$ can be obtained in a similar way: If we denote by $\mathbf{v}_{\text{Cart}}$ the vector $\mathbf{v}$ expressed by its Cartesian components (v_1, v_2, v_3) and by $\mathbf{v}_{\text{cyl}}$ the same vector expressed through the cylindrical ones, $\{v_\rho, v_\theta, v_z\}$, then, cf. Section 2.11,

$$\mathbf{v}_{\text{Cart}} = \mathbf{Q}\mathbf{v}_{\text{cyl}} \rightarrow \begin{cases} v_1 = v_\rho \cos\theta - v_\theta \sin\theta, \\ v_2 = v_\rho \sin\theta + v_\theta \cos\theta, \\ v_3 = v_z. \end{cases} \tag{5.12}$$

Applying Eq. (5.11) to these components, we get

$$\begin{aligned} v_{i,1} &= v_{i,\rho}\frac{x_1}{\rho} - v_{i,\theta}\frac{x_2}{\rho^2}, \\ v_{i,2} &= v_{i,\rho}\frac{x_2}{\rho} + v_{i,\theta}\frac{x_1}{\rho^2}, \\ v_{i,3} &= v_{i,z}. \end{aligned} \tag{5.13}$$

Injecting these expressions into Eq. $(5.6)_2$ for $v_{i,j}$s and (5.10) for $\mathbf{e}_i$s gives finally[6]

$$\begin{aligned} \text{grad}\mathbf{v} = {} & v_{\rho,\rho}(\mathbf{e}_\rho \otimes \mathbf{e}_\rho) + \frac{1}{\rho}(v_{\rho,\theta} - v_\theta)(\mathbf{e}_\rho \otimes \mathbf{e}_\theta) + v_{\rho,z}(\mathbf{e}_\rho \otimes \mathbf{e}_z) \\ & + v_{\theta,\rho}(\mathbf{e}_\theta \otimes \mathbf{e}_\rho) + \frac{1}{\rho}(v_{\theta,\theta} + v_\rho)(\mathbf{e}_\theta \otimes \mathbf{e}_\theta) + v_{\theta,z}(\mathbf{e}_\theta \otimes \mathbf{e}_z) \\ & + v_{z,\rho}(\mathbf{e}_z \otimes \mathbf{e}_\rho) + \frac{1}{\rho}v_{z,\theta}(\mathbf{e}_z \otimes \mathbf{e}_\theta) + v_{z,z}(\mathbf{e}_z \otimes \mathbf{e}_z), \end{aligned}$$

or, in matrix form,

$$\text{grad}\mathbf{v} = \begin{bmatrix} v_{\rho,\rho} & \frac{1}{\rho}(v_{\rho,\theta} - v_\theta) & v_{\rho,z} \\ v_{\theta,\rho} & \frac{1}{\rho}(v_{\theta,\theta} + v_\rho) & v_{\theta,z} \\ v_{z,\rho} & \frac{1}{\rho}v_{z,\theta} & v_{z,z} \end{bmatrix}.$$

By the definition of divergence, we get immediately

$$\text{div}\mathbf{v} = v_{\rho,\rho} + \frac{1}{\rho}(v_{\theta,\theta} + v_\rho) + v_{z,z}. \tag{5.14}$$

Now, from Eq. (5.6), we see that div$\mathbf{L}$ is the vector whose components are the divergence of the rows of the matrix representing $\mathbf{L}$. So, we need first

[6]Though straightforward, the details of the calculations for this formula, as for the following ones, are particularly long and tedious, which is why they are omitted here; however, they are a good exercise for the reader.

to calculate the Cartesian components of $\mathbf{L}$ as functions of the cylindrical ones, cf. Section 2.11:

$$\mathbf{L}_{Cart} = \mathbf{Q}\mathbf{L}_{cyl}\mathbf{Q}^\top \rightarrow \begin{cases} L_{11} = -\sin\theta(L_{\rho\theta}\cos\theta - L_{\theta\theta}\sin\theta) + \cos\theta(L_{\rho\rho}\cos\theta - L_{\theta\rho}\sin\theta), \\ L_{12} = \cos\theta(L_{\rho\theta}\cos\theta - L_{\theta\theta}\sin\theta) + \sin\theta(L_{\rho\rho}\cos\theta - L_{\theta\rho}\sin\theta), \\ L_{13} = L_{\rho z}\cos\theta - L_{\theta z}\sin\theta, \\ L_{21} = -\sin\theta(L_{\theta\theta}\cos\theta + L_{\rho\theta}\sin\theta) + \cos\theta(L_{\theta\rho}\cos\theta + L_{\rho\rho}\sin\theta), \\ L_{22} = \cos\theta(L_{\theta\theta}\cos\theta + L_{\rho\theta}\sin\theta) + \sin\theta(L_{\theta\rho}\cos\theta + L_{\rho\rho}\sin\theta), \\ L_{23} = L_{\theta z}\cos\theta + L_{\rho z}\sin\theta, \\ L_{31} = L_{z\rho}\cos\theta - L_{z\theta}\sin\theta, \\ L_{32} = L_{z\theta}\cos\theta + L_{z\rho}\sin\theta, \\ L_{33} = L_{zz}, \end{cases}$$

then, applying Eqs. (5.10) and (5.14) in Eq. $(5.6)_3$ for the vectors $\mathbf{v}_i = (L_{i1}, L_{i2}, L_{i3})$, $i = 1, 2, 3$, we get, through long but standard operations and after putting $\theta = 0$ to obtain the components of div$\mathbf{L}$ in the basis $\{\mathbf{e}_\rho, \mathbf{e}_\theta, \mathbf{e}_z\}$,

$$\begin{aligned} \text{div}\mathbf{L} = &\left(\frac{1}{\rho}((\rho L_{\rho\rho})_{,\rho} + L_{\rho\theta,\theta} - L_{\theta\theta}) + L_{\rho z,z}\right)\mathbf{e}_\rho \\ &+ \left(L_{\theta\rho,\rho} + \frac{1}{\rho}(L_{\theta\theta,\theta} + L_{\rho\theta} + L_{\theta\rho}) + L_{\theta z,z}\right)\mathbf{e}_\theta \\ &+ \left(\frac{1}{\rho}((\rho L_{z\rho})_{,\rho} + L_{z\theta,\theta}) + L_{zz,z}\right)\mathbf{e}_z. \end{aligned}$$

To obtain $\Delta f = f_{,ii}$, we need to apply twice Eq. (5.11), which gives

$$\begin{aligned} f_{,11} &= \left(f_{,\rho}\frac{x_1}{\rho} - f_{,\theta}\frac{x_2}{\rho^2}\right)_{,1} = f_{,\rho 1}\frac{x_1}{\rho} + f_{,\rho}\frac{\rho - x_1\rho_{,1}}{\rho^2} - f_{,\theta 1}\frac{x_2}{\rho^2} + f_{,\theta}\frac{2x_2\rho\rho_{,1}}{\rho^4} \\ &= \left(f_{,\rho\rho}\frac{x_1}{\rho} - f_{,\rho\theta}\frac{x_2}{\rho^2}\right)\frac{x_1}{\rho} + f_{,\rho}\frac{\rho^2 - x_1^2}{\rho^3} - \left(f_{,\rho\theta}\frac{x_1}{\rho} - f_{,\theta\theta}\frac{x_2}{\rho^2}\right)\frac{x_2}{\rho^2} + f_{,\theta}\frac{2x_1x_2}{\rho^4} \\ &= f_{,\rho\rho}\cos^2\theta - 2f_{,\rho\theta}\frac{\sin\theta\cos\theta}{\rho} + f_{,\rho}\frac{\sin^2\theta}{\rho} + f_{,\theta\theta}\frac{\sin^2\theta}{\rho^2} + 2f_{,\theta}\frac{\sin\theta\cos\theta}{\rho^2}, \end{aligned}$$

$$
\begin{aligned}
f_{,22} &= \left(f_{,\rho}\frac{x_2}{\rho} + f_{,\theta}\frac{x_1}{\rho^2}\right)_{,2} = f_{,\rho 2}\frac{x_2}{\rho} + f_{,\rho}\frac{\rho - x_2\rho_{,2}}{\rho^2} + f_{,\theta 2}\frac{x_1}{\rho^2} - f_{,\theta}\frac{2x_1\rho\rho_{,2}}{\rho^4} \\
&= \left(f_{,\rho\rho}\frac{x_2}{\rho} + f_{,\rho\theta}\frac{x_1}{\rho^2}\right)\frac{x_2}{\rho} + f_{,\rho}\frac{\rho^2 - x_2^2}{\rho^3} + \left(f_{,\rho\theta}\frac{x_2}{\rho} + f_{,\theta\theta}\frac{x_1}{\rho^2}\right)\frac{x_1}{\rho^2} - f_{,\theta}\frac{2x_1x_2}{\rho^4} \\
&= f_{,\rho\rho}\sin^2\theta + 2f_{,\rho\theta}\frac{\sin\theta\cos\theta}{\rho} + f_{,\rho}\frac{\cos^2\theta}{\rho} + f_{,\theta\theta}\frac{\cos^2\theta}{\rho^2} - 2f_{,\theta}\frac{\sin\theta\cos\theta}{\rho^2},
\end{aligned}
$$

$$
f_{,33} = f_{,zz}.
$$

So, adding these terms together, we finally have

$$
\Delta f = \frac{1}{\rho}(\rho f_{,\rho})_{,\rho} + \frac{1}{\rho^2}f_{,\theta\theta} + f_{,zz}.
$$

The laplacian $\Delta\mathbf{v}$ of a vector field $\mathbf{v}$, Eq. $(5.6)_6$, can be obtained by following the same steps for each one of the components in Eq. (5.12), which gives

$$
\begin{aligned}
\Delta\mathbf{v} = &\left(\frac{1}{\rho}(\rho v_{\rho,\rho})_{,\rho} + \frac{1}{\rho^2}v_{\rho,\theta\theta} + v_{\rho,zz} - \frac{1}{\rho^2}(v_\rho + 2v_{\theta,\theta})\right)\mathbf{e}_\rho \\
&+ \left(\frac{1}{\rho}(\rho v_{\theta,\rho})_{,\rho} + \frac{1}{\rho^2}v_{\theta,\theta\theta} + v_{\theta,zz} - \frac{1}{\rho^2}(v_\theta - 2v_{\rho,\theta})\right)\mathbf{e}_\theta \\
&+ \left(\frac{1}{\rho}(\rho v_{z,\rho})_{,\rho} + \frac{1}{\rho^2}v_{z,\theta\theta} + v_{z,zz}\right)\mathbf{e}_z.
\end{aligned}
$$

Finally, injecting Eqs. (5.10), (5.12), and (5.13) into Eq. $(5.6)_7$ gives

$$
\operatorname{curl}\mathbf{v} = \left(\frac{1}{\rho}v_{z,\theta} - v_{\theta,z}\right)\mathbf{e}_\rho + (v_{\rho,z} - v_{z,\rho})\mathbf{e}_\theta + \left(\frac{1}{\rho}((\rho v_\theta)_{,\rho} - v_{\rho,\theta})\right)\mathbf{e}_z. \tag{5.15}
$$

5.7 Differential operators in spherical coordinates

The spherical coordinates r, φ, θ of a point p, whose Cartesian coordinates in the (fixed) frame $\{o; \mathbf{e}_1, \mathbf{e}_2, \mathbf{e}_3\}$ are $p = (x_1, x_2, x_3)$, are shown in Fig. 5.3.

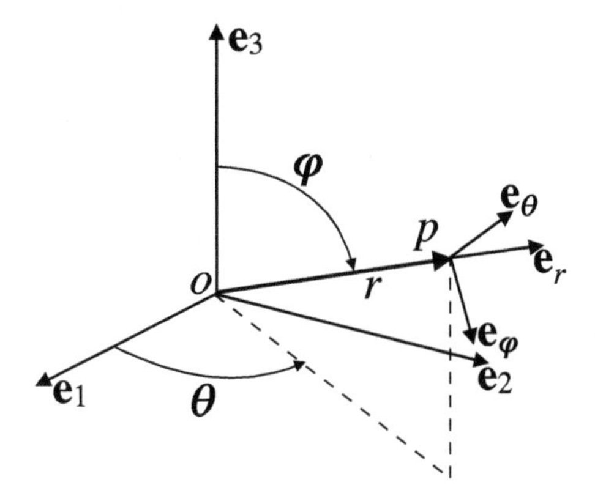

Figure 5.3: Spherical coordinates.

They are related together by

$$r = \sqrt{x_1^2 + x_2^2 + x_3^2},$$

$$\varphi = \arctan \frac{\sqrt{x_1^2 + x_2^2}}{x_3},$$

$$\theta = \arctan \frac{x_2}{x_1},$$

or conversely,

$$x_1 = r \cos\theta \sin\varphi,$$

$$x_2 = r \sin\theta \sin\varphi,$$

$$x_3 = r \cos\varphi.$$

We note that $r \geq 0$ and that the *anomaly* θ is bounded by $0 \leq \theta < 2\pi$ while the *colatitude* φ by $0 \leq \varphi \leq \pi$.

The procedure to determine the expression of the differential operators in spherical coordinates, i.e. in the (moving) frame $\{p; \mathbf{e}_r, \mathbf{e}_\varphi, \mathbf{e}_\theta\}$, is identical to that used for the cylindrical coordinates, but the analytical developments are even more complicated and long, so they are omitted here and only the final formulae are given as follows:

$$\text{grad} f = f_{,r}\mathbf{e}_r + \frac{1}{r} f_{,\varphi}\mathbf{e}_\varphi + \frac{1}{r\sin\varphi} f_{,\theta}\mathbf{e}_\theta,$$

$$
\begin{aligned}
\mathrm{grad}\mathbf{v} = {} & v_{r,r}\mathbf{e}_r \otimes \mathbf{e}_r + \frac{1}{r}(v_{r,\varphi} - v_\varphi)\mathbf{e}_r \otimes \mathbf{e}_\varphi + \frac{1}{r}\left(\frac{1}{\sin\varphi}v_{r,\theta} - v_\theta\right)\mathbf{e}_r \otimes \mathbf{e}_\theta \\
& + v_{\varphi,r}\mathbf{e}_\varphi \otimes \mathbf{e}_r + \frac{1}{r}(v_{\varphi,\varphi} + v_r)\mathbf{e}_\varphi \otimes \mathbf{e}_\varphi \\
& + \frac{1}{r}\left(\frac{1}{\sin\varphi}v_{\varphi,\theta} - v_\theta \cot\varphi\right)\mathbf{e}_\varphi \otimes \mathbf{e}_\theta \\
& + v_{\theta,r}\mathbf{e}_\theta \otimes \mathbf{e}_r + \frac{1}{r}v_{\theta,\varphi}\mathbf{e}_\theta \otimes \mathbf{e}_\varphi \\
& + \frac{1}{r}\left(\frac{1}{\sin\varphi}v_{\theta,\theta} + v_r + v_\varphi \cot\varphi\right)\mathbf{e}_\theta \otimes \mathbf{e}_\theta,
\end{aligned}
$$

or in matrix form,

$$
\mathrm{grad}\mathbf{v} = \begin{bmatrix} v_{r,r} & \frac{1}{r}(v_{r,\varphi} - v_\varphi) & \frac{1}{r}\left(\frac{1}{\sin\varphi}v_{r,\theta} - v_\theta\right) \\ v_{\varphi,r} & \frac{1}{r}(v_{\varphi,\varphi} + v_r) & \frac{1}{r}\left(\frac{1}{\sin\varphi}v_{\varphi,\theta} - v_\theta\cot\varphi\right) \\ v_{\theta,r} & \frac{1}{r}v_{\theta,\varphi} & \frac{1}{r}\left(\frac{1}{\sin\varphi}v_{\theta,\theta} + v_r + v_\varphi\cot\varphi\right) \end{bmatrix},
$$

$$
\mathrm{div}\mathbf{v} = \frac{1}{r^2}(r^2 v_r)_{,r} + \frac{1}{r\sin\varphi}((v_\varphi \sin\varphi)_{,\varphi} + v_{\theta,\theta}),
$$

$$
\begin{aligned}
\mathrm{div}\mathbf{L} = {} & \left(\frac{1}{r^2}(r^2 L_{rr})_{,r} + \frac{1}{r}L_{r\varphi,\varphi} + \frac{1}{r\sin\varphi}L_{r\theta,\theta} - \frac{L_{\varphi\varphi} + L_{\theta\theta}}{r} + \frac{\cot\varphi}{r}L_{r\varphi}\right)\mathbf{e}_r \\
& + \left(\frac{1}{r^2}(r^2 L_{\varphi r})_{,r} + \frac{1}{r}L_{\varphi\varphi,\varphi} + \frac{1}{r\sin\varphi}L_{\varphi\theta,\theta} + \frac{1}{r}L_{r\varphi}\right. \\
& \left. + \frac{\cot\varphi}{r}(L_{\varphi\varphi} - L_{\theta\theta})\right)\mathbf{e}_\varphi \\
& + \left(\frac{1}{r^2}(r^2 L_{\theta r})_{,r} + \frac{1}{r}L_{\theta\varphi,\varphi} + \frac{1}{r\sin\varphi}L_{\theta\theta,\theta}\right. \\
& \left. + \frac{1}{r}L_{r\theta} + \frac{\cot\varphi}{r}(L_{\varphi\theta} + L_{\theta\varphi})\right)\mathbf{e}_\theta,
\end{aligned}
$$

$$
\Delta f = \frac{1}{r^2}(r^2 f_{,r})_{,r} + \frac{1}{r^2\sin\varphi}\left(\frac{f_{,\theta\theta}}{\sin\varphi} + (f_{,\varphi}\sin\varphi)_{,\varphi}\right),
$$

$$
\begin{aligned}
\Delta \mathbf{v} = & \left(v_{r,rr} + \frac{2v_{r,r}}{r} + \frac{v_{r,\varphi\varphi} - 2v_{\varphi,\varphi}}{r^2} + \frac{v_{r,\varphi} - 2v_\varphi}{r^2 \tan\varphi} \right. \\
& \left. + \frac{1}{r^2 \sin\varphi} \left(\frac{v_{r,\theta\theta}}{\sin\varphi} - 2v_{\theta,\theta} \right) - \frac{2v_r}{r^2} \right) \mathbf{e}_r \\
& + \left(v_{\varphi,rr} + \frac{2v_{\varphi,r}}{r} + \frac{v_{\varphi,\varphi\varphi} + 2v_{r,\varphi}}{r^2} + \frac{v_{\varphi,\varphi} - v_\varphi \cot\varphi}{r^2 \tan\varphi} \right. \\
& \left. + \frac{1}{r^2 \sin^2\varphi} (v_{\varphi,\theta\theta} - 2v_{\theta,\theta} \cos\varphi) - \frac{v_\varphi}{r^2} \right) \mathbf{e}_\varphi \\
& + \left(v_{\theta,rr} + \frac{2v_{\theta,r}}{r} + \frac{v_{\theta,\varphi\varphi}}{r^2} + \left(v_{\theta,\varphi} + \frac{2v_{\varphi,\theta}}{\sin\varphi} \right) \frac{1}{r^2 \tan\varphi} \right. \\
& \left. + \frac{1}{r^2 \sin\varphi} \left(\frac{v_{\theta,\theta\theta}}{\sin\varphi} + 2v_{r,\theta} \right) - \frac{v_\theta}{r^2 \sin^2\varphi} \right) \mathbf{e}_\theta, \\
\text{curl}\mathbf{v} = & \left(\frac{1}{r \sin\varphi} ((v_\theta \sin\varphi)_{,\varphi} - v_{\varphi,\theta}) \right) \mathbf{e}_r \\
& + \left(\frac{1}{r \sin\varphi} v_{r,\theta} - \frac{1}{r} (r v_\theta)_{,r} \right) \mathbf{e}_\varphi \\
& + \left(\frac{1}{r} ((r v_\varphi)_{,r} - v_{r,\varphi}) \right) \mathbf{e}_\theta.
\end{aligned}
$$

5.8 Exercises

1. Prove the relations of Eq. (5.2).
2. Prove the properties of the divergence in Eq. (5.3).
3. Prove the properties of the curl in Eq. (5.4).
4. Prove the identities in Eq. (5.5).
5. Prove that
$$\frac{d\varphi}{d\mathbf{n}} = \text{grad}\varphi \cdot \mathbf{n}, \quad \frac{d\mathbf{v}}{d\mathbf{n}} = \text{grad}\mathbf{v}\ \mathbf{n} \quad \forall \mathbf{n} \in \mathcal{S}.$$
6. Prove the results of Eq. (5.6).
7. Consider a rigid body $\mathtt{B}$ and a point $p_0 \in \mathtt{B}$. From the kinematics of rigid bodies, we know that the velocity of another point $p \in \mathtt{B}$ is given by
$$\mathbf{v}(p) = \mathbf{v}(p_0) + \boldsymbol{\omega} \times (p - p_0),$$
with $\boldsymbol{\omega}$ the angular velocity. Prove that
$$\boldsymbol{\omega} = \frac{1}{2}\text{curl}\mathbf{v}, \quad \text{div}\mathbf{v} = 0.$$

8. In the infinitesimal theory of strain, a deformation is *isochoric* when $\operatorname{div}\mathbf{u} = 0$, $\mathbf{u}$ being the corresponding displacement vector. Determine which, among the following ones, are locally or globally isochoric deformations:

   i. $\mathbf{u} = \alpha(x_1, x_2, x_3), \quad \alpha \in \mathbb{R}, \ |\alpha| \ll 1;$
   ii. $\mathbf{u} = \beta(x_2 + x_3, x_1 + x_3, x_1 + x_2), \quad \beta \in \mathbb{R}, \ |\beta| \ll 1;$
   iii. $\mathbf{u} = \gamma(x_1x_2, x_2x_3, x_3x_1), \quad \gamma \in \mathbb{R}, \ |\gamma| \ll 1;$
   iv. $\mathbf{u} = \delta(\sin x_1, -\cos x_2, \sin x_3), \quad \delta \in \mathbb{R}, \ |\delta| \ll 1.$

9. In fluid mechanics, the condition $\operatorname{div}\mathbf{v} = 0$, with $\mathbf{v}$ being the velocity field, characterizes *incompressible flows*. Verify that the following velocity fields, given in cylindrical coordinates, correspond to incompressible flows ($\alpha \in \mathbb{R}$):

   i. source or sink: $\mathbf{v} = \dfrac{\alpha}{\rho}\mathbf{e}_\rho;$
   ii. vortex: $\mathbf{v} = \dfrac{\alpha}{\rho}\mathbf{e}_\theta;$
   iii. doublet: $\mathbf{v} = \dfrac{\alpha}{\rho^2}(\cos\theta\mathbf{e}_\rho + \sin\theta\mathbf{e}_\theta).$

10. A flow with $\operatorname{curl}\mathbf{v} = \mathbf{o}$ is said to be *irrotational*; check that the flows in the previous exercise are irrotational.

Chapter 6

Curvilinear Coordinates

6.1 Introduction

All the developments in the previous chapters are intended for the case where algebraic and differential operators are expressed in a Cartesian frame, i.e. with *rectangular coordinates.* The points of $\mathcal{E}$ are thus referred to a system of coordinates taken along straight lines that are mutually orthogonal and with the same unit along each one of the directions of the frame. Though this is a very important and common case, it is not the only possibility, and in many cases, non-rectangular coordinate frames are used or arise in the mathematical developments (a typical example is that of the geometry of surfaces, see Chapter 7). A non-rectangular coordinate frame is a frame where coordinates can be taken along non-orthogonal directions or along some lines that intersect at right angles but that are not straight lines, or even when both of these cases occur. This situation is often denoted in the literature as that of *curvilinear coordinates*; the transformations to be done to algebraic and differential operators in the case of curvilinear coordinates is the topic of this chapter.

6.2 Curvilinear coordinates, metric tensor

Let us consider an arbitrary origin o of $\mathcal{E}$ and an orthonormal basis $e = \{\mathbf{e}_1, \mathbf{e}_2, \mathbf{e}_3\}$ of $\mathcal{V}$; we indicate the coordinates of a point $p \in \mathcal{E}$ with respect to the frame $\mathcal{R} = \{o; \mathbf{e}_1, \mathbf{e}_2, \mathbf{e}_3\}$ by $x_k : p = (x_1, x_2, x_3)$. Then, we also consider another set of coordinate lines for $\mathcal{E}$, where the position of a point $p \in \mathcal{E}$ with respect to the same arbitrary origin o of $\mathcal{E}$ is now determined by

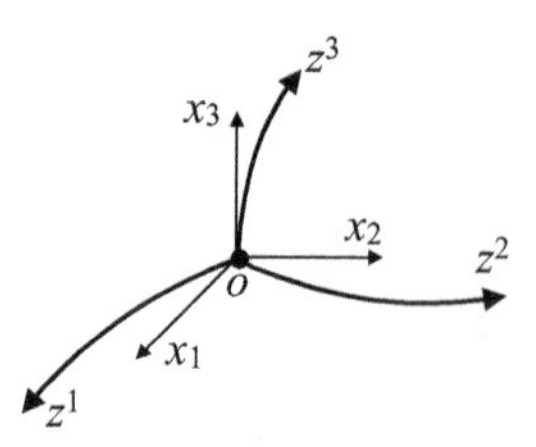

Figure 6.1: Cartesian and curvilinear coordinates.

a set of three numbers $z^j : p = \{z^1, z^2, z^3\}$. Nothing is *a priori* required of coordinates z^j, namely they do not need to be a set of Cartesian coordinates, i.e. referring to an orthonormal basis of $\mathcal{V}$. In principle, the coordinates z^j can be taken along non-straight lines that do not need to be mutually orthogonal at o and also with different units along each line. That is why we call z^js *curvilinear coordinates*, see Fig. 6.1. Any point $p \in \mathcal{E}$ can be identified by either set of coordinates; mathematically, this means that there must be an isomorphism between x_ks and z^js, i.e. invertible relations of the kind

$$z^j = z^j(x_1, x_2, x_3) = z^j(x_k), \quad x_k = x_k(z^1, z^2, z^3) = x_k(z^j) \;\; \forall j, k = 1, 2, 3, \tag{6.1}$$

exist between the two sets of coordinates. The distance between two points $p, q \in \mathcal{E}$ is[1]

$$s = \sqrt{(p-q)\cdot(p-q)} = \sqrt{(x_k^p - x_k^q)(x_k^p - x_k^q)},$$

but this is no longer true for curvilinear coordinates:

$$s \neq \sqrt{(z^{jp} - z^{jq})(z^{jp} - z^{jq})}.$$

However, if $p \to q$, we can define

$$dx_k = x_k^p - x_k^q, \quad dz^j = z^{jp} - z^{jq},$$

so using Eq. $(6.1)_2$,

$$dx_k = \frac{\partial x_k}{\partial z^j} dz^j. \tag{6.2}$$

[1]The distance between two points p and q is still defined as the Euclidean norm of $(p-q)$, i.e. it is independent of the set of coordinates.

The (infinitesimal) distance between p and q will then be

$$ds = \sqrt{dx_k dx_k} = \sqrt{\frac{\partial x_k}{\partial z^j}\frac{\partial x_k}{\partial z^l}dz^j dz^l} = \sqrt{g_{jl}dz^j dz^l},$$

where

$$g_{jl} = g_{lj} = \frac{\partial x_k}{\partial z^j}\frac{\partial x_k}{\partial z^l} \tag{6.3}$$

are the *covariant*[2] *components of the metric tensor*[3] $\mathbf{g} \in Sym(\mathcal{V})$. We note that, as $\mathbf{g}$ defines a positive quadratic form (the length of a vector), it is a positive definite symmetric tensor, so

$$\det \mathbf{g} > 0. \tag{6.4}$$

Coming back to the vector notation, from Eq. (6.2), we get[4]

$$dx = dx_i \mathbf{e}_i = \frac{\partial x_i}{\partial z^k} dz^k \mathbf{e}_i;$$

introducing the vector $\mathbf{g}_k$,

$$\mathbf{g}_k := \frac{\partial x_i}{\partial z^k}\mathbf{e}_i, \tag{6.5}$$

we can write

$$dx = dz^k \mathbf{g}_k.$$

We see hence that a vector dx can be expressed as a linear combination of the vectors $\mathbf{g}_k$; these form therefore a basis, called the *local basis*. Generally, $\mathbf{g}_k \notin \mathcal{S}$, and it is clearly tangent to the lines $z^j = const.$ This can be seen in Fig. 6.2 for a two-dimensional case:

$$\begin{aligned} dx &= \lim_{\Delta\mathbf{x}\to 0} \Delta\mathbf{x} = \lim_{\Delta\mathbf{x}\to 0} \frac{x_i(z^1, z^2 + \Delta z^2, z^3) - x_i(z^1, z^2, z^3)}{\Delta z^2}\Delta z^2 \mathbf{e}_i \\ &= \frac{\partial x_i}{\partial z^2}\mathbf{e}_i dz^2 = \mathbf{g}_2 dz^2. \end{aligned}$$

Then,

$$\mathbf{g}_k \cdot \mathbf{g}_l = \frac{\partial x_i}{\partial z^k}\mathbf{e}_i \cdot \frac{\partial x_j}{\partial z^l}\mathbf{e}_j = \frac{\partial x_i}{\partial z^k}\frac{\partial x_j}{\partial z^l}\delta_{ij} = g_{kl}, \tag{6.6}$$

[2]The notion of co- and contravariant components is detailed in the next section.

[3]As usually done in the literature, we indicate the metric tensor by $\mathbf{g}$, i.e. a lowercase letter, though it is a second-rank tensor, not a vector.

[4]The differential dx is a vector because it is the difference of two infinitely close points; that is why it is not necessary to denote it in bold letters.

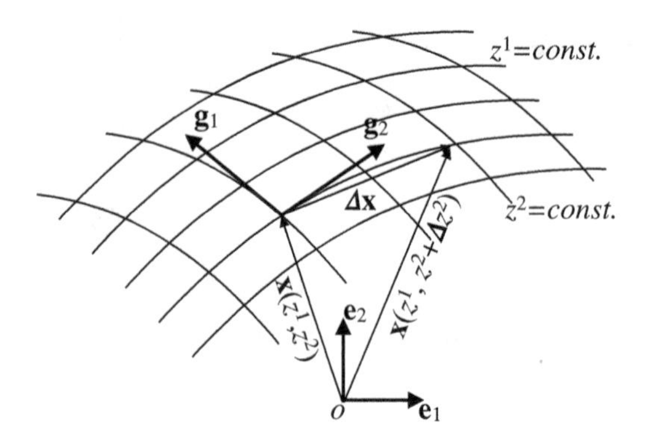

Figure 6.2: Tangent vectors to the curvilinear coordinates lines.

i.e. the components of the metric tensor $\mathbf{g}$ are the scalar products of the tangent vectors $\mathbf{g}_k$s. If the curvilinear coordinates are *orthogonal*, i.e. if $\mathbf{g}_h \cdot \mathbf{g}_k = 0 \ \forall h, k = 1, 2, 3, h \neq k$, then $\mathbf{g}$ is diagonal. If, in addition, $\mathbf{g}_k \in \mathcal{S} \ \forall k = 1, 2, 3$, then $\mathbf{g} = \mathbf{I}$: It is the case of Cartesian coordinates. As an example, let us consider the case of polar coordinates:

$$\begin{cases} x_1 = r \ \cos\theta, \\ x_2 = r \ \sin\theta, \end{cases} \qquad \begin{cases} z^1 = r = \sqrt{x_1^2 + x_2^2}, \\ z^2 = \theta = \arctan\dfrac{x_2}{x_1}. \end{cases}$$

Hence, from Fig. 6.3,

$$\begin{aligned} \mathbf{g}_1 &= \frac{\partial x_1}{\partial z^1}\mathbf{e}_1 + \frac{\partial x_2}{\partial z^1}\mathbf{e}_2 = \cos\theta\mathbf{e}_1 + \sin\theta\mathbf{e}_2 = \mathbf{e}_r, \\ \mathbf{g}_2 &= \frac{\partial x_1}{\partial z^2}\mathbf{e}_1 + \frac{\partial x_2}{\partial z^2}\mathbf{e}_2 = -r\sin\theta\mathbf{e}_1 + r\cos\theta\mathbf{e}_2 = r\mathbf{e}_\theta. \end{aligned}$$

We remark that $|\mathbf{g}_1| = 1$ but $|\mathbf{g}_2| \neq 1$, and it is variable with the position.

6.3 Co- and contravariant components

A geometrical way to introduce the concept of *covariant* and *contravariant* components is to consider how to represent a vector $\mathbf{v}$ in the $z-$system. There are basically two ways, cf. Fig. 6.4, referred, for the sake of simplicity, to a planar case:

(i) *Contravariant components*: $\mathbf{v}$ is projected parallel to z^1 and z^2; they are indicated by superscripts: $\mathbf{v} = (v^1, v^2, v^3)$.
(ii) *Covariant components*: $\mathbf{v}$ is projected perpendicularly to z^1 and z^2; they are indicated by subscripts: $\mathbf{v} = (v_1, v_2, v_3)$.

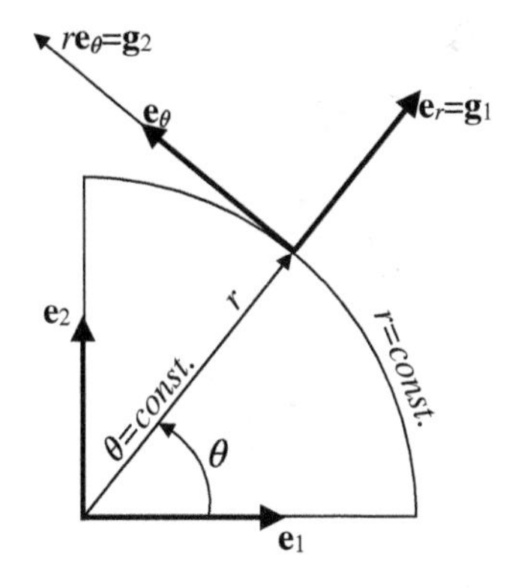

Figure 6.3: Tangent vectors to the polar coordinates lines.

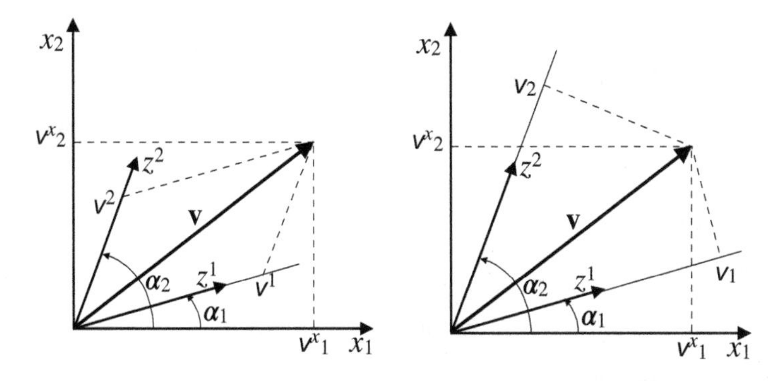

Figure 6.4: Contravariant (left) and covariant (right) components of a vector in a plane.

Still referring to the planar case in Fig. 6.4, if the Cartesian components[5] of $\mathbf{v}$ are $\mathbf{v} = (v_1^x, v_2^x)$, we get

$$\begin{cases} v^1 = h(v_1^x \sin\alpha_2 - v_2^x \cos\alpha_2), \\ v^2 = h(-v_1^x \sin\alpha_1 + v_2^x \cos\alpha_1), \end{cases} \quad \begin{cases} v_1 = v_1^x \cos\alpha_1 + v_2^x \sin\alpha_1, \\ v_2 = v_1^x \cos\alpha_2 + v_2^x \sin\alpha_2, \end{cases} \tag{6.7}$$

and conversely,

$$\begin{cases} v_1^x = v^1 \cos\alpha_1 + v^2 \cos\alpha_2, \\ v_2^x = v^1 \sin\alpha_1 + v^2 \sin\alpha_2, \end{cases} \quad \begin{cases} v_1^x = h(v_1 \sin\alpha_2 - v_2 \sin\alpha_1), \\ v_2^x = h(-v_1 \cos\alpha_2 + v_2 \cos\alpha_1), \end{cases} \tag{6.8}$$

with

$$h = \frac{1}{\sin(\alpha_2 - \alpha_1)}.$$

[5]In the following, we use the superscript x to indicate a Cartesian component: v_i^x is the ith Cartesian component of $\mathbf{v} \in \mathcal{V}$ and L_{ij}^x the ijth Cartesian component of $\mathbf{L} \in Lin(\mathcal{V})$.

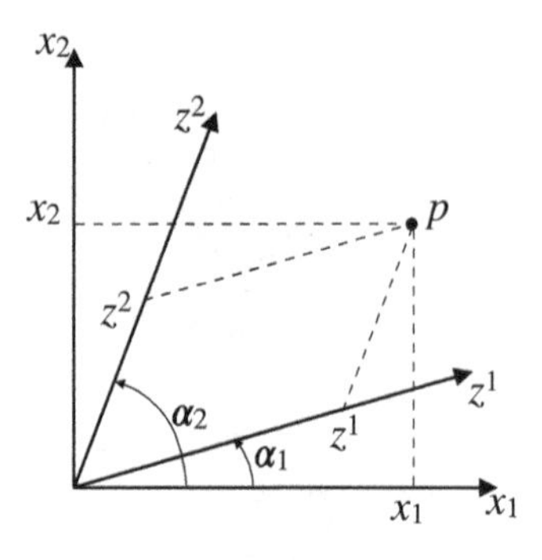

Figure 6.5: Relation between Cartesian and contravariant components.

It is apparent that the Cartesian coordinates are at the same time co- and contravariant. Still, on a planar scheme, we can see how to pass from a system of coordinates to another one, cf. Fig. 6.5. For a point p, the Cartesian coordinates (x_1, x_2) are related to the contravariant ones by

$$\begin{aligned} x_1 &= z^1 \cos\alpha_1 + z^2 \cos\alpha_2, \\ x_2 &= z^1 \sin\alpha_1 + z^2 \sin\alpha_2, \end{aligned}$$

and conversely,

$$\begin{aligned} z^1 &= h(x_1 \sin\alpha_2 - x_2 \cos\alpha_2), \\ z^2 &= h(-x_1 \sin\alpha_1 + x_2 \cos\alpha_1). \end{aligned}$$

So, differentiating, we get

$$\begin{aligned} \frac{\partial x_1}{\partial z^1} &= \cos\alpha_1, \quad \frac{\partial x_1}{\partial z^2} = \cos\alpha_2, \\ \frac{\partial x_2}{\partial z^1} &= \sin\alpha_1, \quad \frac{\partial x_2}{\partial z^2} = \sin\alpha_2 \end{aligned}$$

and

$$\begin{aligned} \frac{\partial z^1}{\partial x_1} &= h\sin\alpha_2, \quad \frac{\partial z^1}{\partial x_2} = -h\cos\alpha_2, \\ \frac{\partial z^2}{\partial x_1} &= -h\sin\alpha_1, \quad \frac{\partial z^2}{\partial x_2} = h\cos\alpha_1. \end{aligned}$$

Injecting these expressions into Eqs. (6.7) and (6.8) gives

$$\left.\begin{aligned} v_1 &= v_1^x \frac{\partial x_1}{\partial z^1} + v_2^x \frac{\partial x_2}{\partial z^1}, \\ v_2 &= v_1^x \frac{\partial x_1}{\partial z^2} + v_2^x \frac{\partial x_2}{\partial z^2}, \end{aligned}\right. \rightarrow v_i = \frac{\partial x_k}{\partial z^i} v_k^x \tag{6.9}$$

and

$$\begin{aligned} v^1 &= v_1^x \frac{\partial z^1}{\partial x_1} + v_2^x \frac{\partial z^1}{\partial x_2}, \\ v^2 &= v_1^x \frac{\partial z^2}{\partial x_1} + v_2^x \frac{\partial z^2}{\partial x_2}, \end{aligned} \quad \rightarrow \quad v^i = \frac{\partial z^i}{\partial x_k} v_k^x. \tag{6.10}$$

Now, if we calculate

$$g_{hi} v^i = g_{hi} \frac{\partial z^i}{\partial x_k} v_k^x$$

from Eq. (6.3) and by the chain rule,[6] we get

$$\begin{aligned} g_{hi} v^i &= \frac{\partial x_j}{\partial z^h} \frac{\partial x_j}{\partial z^i} \frac{\partial z^i}{\partial x_k} v_k^x \\ &= \frac{\partial x_j}{\partial z^h} \frac{\partial x_j}{\partial x_k} v_k^x = \frac{\partial x_j}{\partial z^h} \delta_{jk} v_k^x = \frac{\partial x_k}{\partial z^h} v_k^x = v_h, \end{aligned}$$

i.e. we obtain the rule of *lowering of the indices* for passing from contravariant to covariant components:

$$v_h = g_{hi} v^i.$$

Introducing the *inverse*[7] *to* g_{hi} as

$$g^{hi} = \frac{\partial z^h}{\partial x_k} \frac{\partial z^i}{\partial x_k}, \tag{6.11}$$

we get, again using the chain rule,

$$\begin{aligned} g^{hi} v_i = g^{hi} \frac{\partial x_k}{\partial z^i} v_k^x &= \frac{\partial z^h}{\partial x_j} \frac{\partial z^i}{\partial x_j} \frac{\partial x_k}{\partial z^i} v_k^x \\ &= \frac{\partial z^h}{\partial x_j} \frac{\partial x_k}{\partial x_j} v_k^x = \frac{\partial z^h}{\partial x_j} \delta_{jk} v_k^x = \frac{\partial z^h}{\partial x_k} v_k^x = v^h, \end{aligned}$$

which is the rule of *raising of the indices* for passing from covariant to contravariant components:

$$v^h = g^{hi} v_i.$$

---

[6]The reader can easily see that, in practice, the chain rule allows us to handle the derivatives as fractions.

[7]To prove that the contravariant components g^{pq} are the inverse of the covariant ones, g_{pq}, is direct:

$$g^{pq} g_{pq} = \frac{\partial z^p}{\partial x_k} \frac{\partial z^q}{\partial x_k} \frac{\partial x_j}{\partial z^p} \frac{\partial x_j}{\partial z^q} = \delta_{jk} \delta_{jk} = 1.$$

Again applying the chain rule, by Eq. (6.9), we get

$$\frac{\partial z^i}{\partial x^l}v_i = \frac{\partial z^i}{\partial x^l}\frac{\partial x_k}{\partial z^i}v_k^x = \frac{\partial x_k}{\partial x^l}v_k^x = \delta_{kl}v_k^x,$$

i.e.

$$v_k^x = \frac{\partial z^i}{\partial x^k}v_i, \tag{6.12}$$

which is the converse of Eq. (6.9). In a similar way, we get the converse of Eq. (6.10):

$$v_k^x = \frac{\partial x_k}{\partial z^i}v^i. \tag{6.13}$$

Let us now calculate the norm v of a vector $\mathbf{v}$; starting from the Cartesian components and using the last two results,

$$v = \sqrt{\mathbf{v}\cdot\mathbf{v}} = \sqrt{v_k^x v_k^x} = \sqrt{\frac{\partial z^i}{\partial x^k}v_i\frac{\partial z^j}{\partial x^k}v_j} = \sqrt{\frac{\partial z^i}{\partial x^k}\frac{\partial z^j}{\partial x^k}v_i v_j} = \sqrt{g^{ij}v_i v_j},$$

or also,

$$v = \sqrt{\mathbf{v}\cdot\mathbf{v}} = \sqrt{v_k^x v_k^x} = \sqrt{\frac{\partial x_k}{\partial z^i}v^i\frac{\partial x_k}{\partial z^j}v^j} = \sqrt{\frac{\partial x_k}{\partial z^i}\frac{\partial x_k}{\partial z^j}v^i v^j} = \sqrt{g_{ij}v^i v^j}$$

and even,

$$v = \sqrt{\mathbf{v}\cdot\mathbf{v}} = \sqrt{v_k^x v_k^x} = \sqrt{\frac{\partial z^i}{\partial x_k}v_i\frac{\partial x_k}{\partial z^j}v^j} = \sqrt{\frac{\partial z^i}{\partial x_k}\frac{\partial x_k}{\partial z^j}v_i v^j}$$
$$= \sqrt{\delta^i{}_j v_i v^j} = \sqrt{v_i v^i}.$$

Through Eq. (6.13) and by the definition of the tangent vectors to the lines of curvilinear coordinates, in Eq. (6.5), for a vector $\mathbf{v}$, we get

$$\mathbf{v} = v_i^x\mathbf{e}_i = v^k\frac{\partial x_i}{\partial z^k}\mathbf{e}_i = v^k\mathbf{g}_k.$$

We see hence that the contravariant components are actually the components of $\mathbf{v}$ in the basis composed of $\mathbf{g}_k$s, the tangents to the lines of curvilinear coordinates. In a similar manner, if we introduce the *dual basis* whose vectors $\mathbf{g}^k$ are defined as

$$\mathbf{g}^k := \frac{\partial z^k}{\partial x_i}\mathbf{e}_i \tag{6.14}$$

and proceeding in the same way, we obtain that

$$\mathbf{v} = v_i^x \mathbf{e}_i = v_k \frac{\partial z^k}{\partial x_i}\mathbf{e}_i = v_k \mathbf{g}^k,$$

i.e. the covariant components are actually the components of $\mathbf{v}$ in the dual basis. Finally, for a vector, we have, alternatively,

$$\mathbf{v} = v_i^x \mathbf{e}_i = v^k \mathbf{g}_k = v_k \mathbf{g}^k. \tag{6.15}$$

Just as for $\mathbf{g}_k$s, we have

$$\mathbf{g}^h \cdot \mathbf{g}^k = \left(\frac{\partial z^h}{\partial x_i}\mathbf{e}_i\right)\cdot\left(\frac{\partial z^k}{\partial x_j}\mathbf{e}_j\right) = \frac{\partial z^h}{\partial x_i}\frac{\partial z^k}{\partial x_j}\delta_{ij} = \frac{\partial z^h}{\partial x_i}\frac{\partial z^k}{\partial x_i} = g^{hk};$$

moreover,

$$\mathbf{g}^h \cdot \mathbf{g}_k = \left(\frac{\partial z^h}{\partial x_i}\mathbf{e}_i\right)\cdot\left(\frac{\partial x_j}{\partial z^k}\mathbf{e}_j\right) = \frac{\partial z^h}{\partial x_i}\frac{\partial x_j}{\partial z^k}\delta_{ij} = \frac{\partial z^h}{\partial x_i}\frac{\partial x_i}{\partial z^k} = \frac{\partial z^h}{\partial z^k} = \delta^h{}_k,$$

and by the symmetry of the scalar product,

$$\delta_h{}^k := \mathbf{g}_h \cdot \mathbf{g}^k = \mathbf{g}^k \cdot \mathbf{g}_h = \delta^k{}_h.$$

The last equations define the *orthogonality conditions* for the $\mathbf{g}$ vectors. Using these results and Eq. (6.15), we also have

$$v^k = \delta^k{}_h v^h = \mathbf{g}^k \cdot v^h \mathbf{g}_h = \mathbf{g}^k \cdot \mathbf{v} = \mathbf{g}^k \cdot v_h \mathbf{g}^h = g^{kh} v_h,$$
$$v_k = \delta_k{}^h v_h = \mathbf{g}_k \cdot v_h \mathbf{g}^h = \mathbf{g}_k \cdot \mathbf{v} = \mathbf{g}_k \cdot v^h \mathbf{g}_h = g_{kh} v^h,$$

thus finding again the rules of raising and lowering of the indices.

What was done for vectors can be transposed, using a similar approach, to tensors. In particular, for a second-rank tensor $\mathbf{L}$, we get

$$\begin{aligned} L^{ij} &= \frac{\partial z^i}{\partial x_h}\frac{\partial z^j}{\partial x_k}L^x_{hk}, \\ L_{ij} &= \frac{\partial x_h}{\partial z^i}\frac{\partial x_k}{\partial z^j}L^x_{hk} \end{aligned} \tag{6.16}$$

for the contravariant and covariant components, respectively, while we can also introduce the *mixed components*:

$$\begin{aligned} L^i{}_j &= \frac{\partial z^i}{\partial x_h}\frac{\partial x_k}{\partial z^j}L^x_{hk}, \\ L_i{}^j &= \frac{\partial x_h}{\partial z^i}\frac{\partial z^j}{\partial x_k}L^x_{hk}. \end{aligned} \tag{6.17}$$

Conversely,

$$\begin{aligned} L^x_{hk} &= \frac{\partial x_h}{\partial z^i}\frac{\partial x_k}{\partial z^j}L^{ij},\\ L^x_{hk} &= \frac{\partial z^i}{\partial x_h}\frac{\partial z^j}{\partial x_k}L_{ij},\\ L^x_{hk} &= \frac{\partial x_h}{\partial z^i}\frac{\partial z^j}{\partial x_k}L^i{}_j,\\ L^x_{hk} &= \frac{\partial z^i}{\partial x_h}\frac{\partial x_k}{\partial z^j}L_i{}^j. \end{aligned} \tag{6.18}$$

Also, for $\mathbf{L}$, the rule of lowering or raising the indices is valid:

$$L^{ij} = g^{ih}g^{jk}L_{hk}, \quad L_{ij} = g_{ih}g_{jk}L^{hk}. \tag{6.19}$$

From Eq. (6.18) and by the same definitions of g_{ij}, Eq. (6.3), and g^{ij}, Eq. (6.11), we get

$$\mathbf{L} = L^x_{ij}\mathbf{e}_i \otimes \mathbf{e}_j = \frac{\partial x_i}{\partial z^h}\frac{\partial x_j}{\partial z^k}L^{hk}\mathbf{e}_i \otimes \mathbf{e}_j = L^{hk}\mathbf{g}_h \otimes \mathbf{g}_k$$

and

$$\mathbf{L} = L^x_{ij}\mathbf{e}_i \otimes \mathbf{e}_j = \frac{\partial z^h}{\partial x_i}\frac{\partial z^k}{\partial x_j}L_{hk}\mathbf{e}_i \otimes \mathbf{e}_j = L_{hk}\mathbf{g}^h \otimes \mathbf{g}^k.$$

In a similar manner, the tensor mixed components are also found:

$$\mathbf{L} = L^x_{ij}\mathbf{e}_i \otimes \mathbf{e}_j = \frac{\partial x_i}{\partial z^h}\frac{\partial z^k}{\partial x_j}L^h{}_k\mathbf{e}_i \otimes \mathbf{e}_j = L^h{}_k\mathbf{g}_h \otimes \mathbf{g}^k$$

and

$$\mathbf{L} = L^x_{ij}\mathbf{e}_i \otimes \mathbf{e}_j = \frac{\partial z^k}{\partial x_j}\frac{\partial x_i}{\partial z^h}L_h{}^k\mathbf{e}_i \otimes \mathbf{e}_j = L_h{}^k\mathbf{g}^h \otimes \mathbf{g}_k.$$

We see hence that a second-rank tensor can be given with four different combinations of coordinates; even more complex is the case of higher-order tensors, which will not be treated here.

Still, by Eqs. (6.3) and (6.11) and applying the chain rule to $\delta^i_j = \dfrac{\partial z^i}{\partial z^j}$, we get

$$\begin{aligned}
g_{ij} &= \frac{\partial x_k}{\partial z^i}\frac{\partial x_k}{\partial z^j} = \frac{\partial x_h}{\partial z^i}\frac{\partial x_k}{\partial z^j}\delta_{hk},\\
g^{ij} &= \frac{\partial z^i}{\partial x_k}\frac{\partial z^j}{\partial x_k} = \frac{\partial z^i}{\partial x_h}\frac{\partial z^j}{\partial x_k}\delta_{hk},\\
\delta^i{}_j &= \frac{\partial z^i}{\partial x_h}\frac{\partial x_k}{\partial z^j}\delta_{hk},\\
\delta_i{}^j &= \frac{\partial x_h}{\partial z^i}\frac{\partial z^j}{\partial x_k}\delta_{hk}.
\end{aligned} \tag{6.20}$$

So, applying Eq. (6.18) to the identity tensor, we get

$$\mathbf{I} = \delta_{ij}\mathbf{e}_i \otimes \mathbf{e}_j = \frac{\partial x_i}{\partial z^h}\frac{\partial x_j}{\partial z^k} I^{hk}\mathbf{e}_i \otimes \mathbf{e}_j = I^{hk}\mathbf{g}_h \otimes \mathbf{g}_k,$$

but by Eqs. (6.16) and (6.20),

$$I^{hk} = \frac{\partial z^h}{\partial x_i}\frac{\partial z^k}{\partial x_j}\delta_{ij} = g^{hk},$$

so finally,

$$\mathbf{I} = g^{hk}\mathbf{g}_h \otimes \mathbf{g}_k.$$

Proceeding in a similar manner, we can also get

$$\mathbf{I} = g_{hk}\mathbf{g}^h \otimes \mathbf{g}^k = \delta^h_k \mathbf{g}_h \otimes \mathbf{g}^k = \delta^k_h \mathbf{g}^h \otimes \mathbf{g}_k.$$

We see hence that g_{hk}s represent $\mathbf{I}$ in covariant coordinates, g^{hk}s in the contravariant ones, and δ^k_hs and δ^h_ks in mixed coordinates.

6.4 Spatial derivatives of fields in curvilinear coordinates

Let φ be a spatial[8] scalar field, $\varphi : \mathcal{E} \to \mathbb{R}$. Generally,

$$\varphi = \varphi(z^j(x_i)),$$

or also,

$$\varphi = \varphi(x_j(z^k)),$$

[8]The term *spatial* here refers to differentiation with respect to spatial coordinates, which can be Cartesian or curvilinear.

where x_js, z^ks are, respectively, Cartesian and curvilinear coordinates related as in Eq. (6.1). By the chain rule,

$$\frac{\partial \varphi}{\partial x_j} = \frac{\partial \varphi}{\partial z^k}\frac{\partial z^k}{\partial x_j}, \tag{6.21}$$

and inversely,

$$\frac{\partial \varphi}{\partial z^k} = \frac{\partial \varphi}{\partial x_j}\frac{\partial x_j}{\partial z^k}.$$

We remark that the last quantity transforms like the components of a covariant vector, cf. Eq.(6.9).

The gradient of φ is the vector that in the Cartesian basis, cf. Eq. $(5.6)_1$, is given by

$$\nabla \varphi = \frac{\partial \varphi}{\partial x_j}\mathbf{e}_j,$$

so by Eqs. (6.14) and (6.21), we get that in the dual basis,

$$\nabla \varphi = \frac{\partial \varphi}{\partial z^k}\frac{\partial z^k}{\partial x_j}\mathbf{e}_j = \frac{\partial \varphi}{\partial z^k}\mathbf{g}^k.$$

We see hence that in curvilinear coordinates, the nabla operator, Eq. (5.7), is defined by

$$\nabla(\cdot) = \frac{\partial\ \cdot}{\partial z^k}\mathbf{g}^k. \tag{6.22}$$

The contravariant components of the gradient can be obtained by the covariant ones upon multiplication by the components of the inverse (contravariant) metric tensor, Eq. (6.11):

$$g^{hk}\frac{\partial \varphi}{\partial z^k} = \frac{\partial z^h}{\partial x_i}\frac{\partial z^k}{\partial x_i}\frac{\partial \varphi}{\partial x_j}\frac{\partial x_j}{\partial z^k} = \delta_{ij}\frac{\partial \varphi}{\partial x_j}\frac{\partial z^h}{\partial x_i} = \frac{\partial \varphi}{\partial x_j}\frac{\partial z^h}{\partial x_j} \rightarrow \nabla \varphi = \frac{\partial \varphi}{\partial x_j}\frac{\partial z^h}{\partial x_j}\mathbf{g}_h.$$

Let us now consider a vector field $\mathbf{v} : \mathcal{E} \rightarrow \mathcal{V}$; we want to calculate the spatial derivative of its Cartesian components. By the chain rule and Eq. (6.13), we get

$$\begin{aligned}\frac{\partial v_i^x}{\partial x_j} &= \frac{\partial v_i^x}{\partial z^k}\frac{\partial z^k}{\partial x_j} = \frac{\partial z^k}{\partial x_j}\frac{\partial}{\partial z^k}\left(\frac{\partial x_i}{\partial z^h}v^h\right) = \frac{\partial z^k}{\partial x_j}\left(\frac{\partial x_i}{\partial z^h}\frac{\partial v^h}{\partial z^k} + \frac{\partial^2 x_i}{\partial z^k \partial z^l}v^l\right)\\ &= \frac{\partial z^k}{\partial x_j}\frac{\partial x_i}{\partial z^h}\left(\frac{\partial v^h}{\partial z^k} + \frac{\partial z^h}{\partial x_m}\frac{\partial^2 x_m}{\partial z^k \partial z^l}v^l\right),\end{aligned}$$

whence

$$\frac{\partial z^h}{\partial x_i}\frac{\partial x_j}{\partial z^k}\frac{\partial v_i^x}{\partial x_j} = \frac{\partial v^h}{\partial z^k} + \frac{\partial z^h}{\partial x_m}\frac{\partial^2 x_m}{\partial z^k \partial z^l}v^l. \tag{6.23}$$

Comparing this result with Eq. $(6.17)_1$, we see that the first member actually corresponds to the components of a mixed tensor field, which is the *gradient of the vector field* $\mathbf{v}$, that we write as

$$v^h{}_{;k} = \frac{\partial v^h}{\partial z^k} + \Gamma^h_{kl} v^l, \tag{6.24}$$

where the functions

$$\Gamma^h_{kl} = \frac{\partial z^h}{\partial x_m} \frac{\partial^2 x_m}{\partial z^k \partial z^l} \tag{6.25}$$

are the *Christoffel symbols.* We immediately see that $\Gamma^h_{kl} = \Gamma^h_{lk}$. The quantity $v^h{}_{;k}$ is the *covariant derivative of the contravariant components* v^h. The proof that the Christoffel symbols can also be written as

$$\Gamma^h_{kl} = \frac{1}{2} g^{hm} \left(\frac{\partial g_{mk}}{\partial z^l} + \frac{\partial g_{ml}}{\partial z^k} - \frac{\partial g_{kl}}{\partial z^m} \right) \tag{6.26}$$

is left to the reader as an exercise.

Proceeding in a similar way for the covariant components of $\mathbf{v}$ but now using Eqs. (6.12) and $(6.17)_1$, we get

$$v_{h;k} = \frac{\partial v_h}{\partial z^k} - \Gamma^l_{kh} v_l,$$

which is the *covariant derivative of the covariant components* v_h.

Using Eqs. (6.23) and (6.24), we conclude that, cf. Eq. $(5.6)_3$,

$$\mathrm{div}\mathbf{v} = \frac{\partial v_i^x}{\partial x_i} = v^h{}_{;h}.$$

Then, applying the operator divergence so defined to the gradient of the scalar field φ, we obtain, in arbitrary coordinates z, the Laplacian $\Delta\varphi$ as

$$\Delta\varphi = \left(g^{hk} \frac{\partial \varphi}{\partial z^k} \right)_{;h} = \frac{\partial}{\partial z^h} \left(g^{hk} \frac{\partial \varphi}{\partial z^k} \right) + \Gamma^h_{hj} g^{jk} \frac{\partial \varphi}{\partial z^k}. \tag{6.27}$$

Using the definition of the nabla operator in curvilinear coordinates, Eq. (6.22), jointly to the fact that, cf. Section 5.5,

$$\Delta f := \mathrm{div}\nabla\varphi = \nabla \cdot \nabla\varphi,$$

we get the following representation of the Laplace operator in curvilinear coordinates:

$$\Delta(\cdot) = \nabla \cdot \nabla(\cdot) = \left(\frac{\partial}{\partial z^k} \left(\frac{\partial(\cdot)}{\partial z^h} \mathbf{g}^h \right) \right) \cdot \mathbf{g}^k = \frac{\partial^2(\cdot)}{\partial z^k \partial z^h} \mathbf{g}^h \cdot \mathbf{g}^k + \frac{\partial \mathbf{g}^h}{\partial z^k} \frac{\partial(\cdot)}{\partial z^h} \cdot \mathbf{g}^k$$
$$= \frac{\partial^2(\cdot)}{\partial z^k \partial z^h} g^{hk} + \frac{\partial \mathbf{g}^h}{\partial z^k} \cdot \mathbf{g}^k \frac{\partial(\cdot)}{\partial z^h}.$$

Let us now calculate the spatial derivatives of the components of a second-rank tensor **L**: By Eqs. $(6.18)_1$ and (6.25), we get

$$\frac{\partial L^x_{ij}}{\partial x_k} = \frac{\partial z^h}{\partial x_k}\frac{\partial}{\partial z^h}\left(\frac{\partial x_i}{\partial z^n}\frac{\partial x_j}{\partial z^p}L^{np}\right)$$
$$= \frac{\partial z^h}{\partial x_k}\frac{\partial x_i}{\partial z^n}\frac{\partial x_j}{\partial z^p}\left(\frac{\partial L^{np}}{\partial z^h} + \Gamma^n_{hr}L^{rp} + \Gamma^p_{hr}L^{nr}\right),$$

which implies that

$$\frac{\partial L^{np}}{\partial z^h} + \Gamma^n_{hr}L^{rp} + \Gamma^p_{hr}L^{nr} = \frac{\partial x_k}{\partial z^h}\frac{\partial z^n}{\partial x_i}\frac{\partial z^p}{\partial x_j}\frac{\partial L^x_{ij}}{\partial x_k}. \tag{6.28}$$

So, using Eq. (6.16), we can conclude that the expression

$$L^{np}{}_{;h} = \frac{\partial L^{np}}{\partial z^h} + \Gamma^n_{hr}L^{rp} + \Gamma^p_{hr}L^{nr} \tag{6.29}$$

represents the covariant derivative of the contravariant components of the second-rank tensor **L**. In a similar manner, this time by Eq. $(6.18)_2$, we obtain the covariant derivatives of the covariant components of **L**:

$$\frac{\partial L_{np}}{\partial z^h} - \Gamma^r_{nh}L_{rp} - \Gamma^r_{ph}L_{nr} = \frac{\partial x_k}{\partial z^h}\frac{\partial x_i}{\partial z^n}\frac{\partial x_j}{\partial z^p}\frac{\partial L^x_{ij}}{\partial x_k},$$

i.e.

$$L_{np;h} = \frac{\partial L_{np}}{\partial z^h} - \Gamma^r_{ph}L_{nr} - \Gamma^r_{nh}L_{pr}. \tag{6.30}$$

The same procedure with Eqs. $(6.18)_{3,4}$ gives the covariant derivatives of the mixed components[9] of **L**:

$$L^n_{p;h} = \frac{\partial L^n{}_p}{\partial z^h} + \Gamma^n_{hr}L^r{}_p - \Gamma^r_{ph}L^n_r,$$
$$L^n_{p;h} = \frac{\partial L^n_p}{\partial z^h} - \Gamma^r_{ph}L^n_r + \Gamma^n_{hr}L^r_p. \tag{6.31}$$

If in Eqs. (6.28) and (6.29) we set $p = h$, we get

$$L^{nh}_{;h} = \frac{\partial L^{nh}}{\partial z^h} + \Gamma^n_{hr}L^{rh} + \Gamma^h_{hr}L^{nr} = \frac{\partial x_k}{\partial z^h}\frac{\partial z^n}{\partial x_i}\frac{\partial z^h}{\partial x_j}\frac{\partial L^x_{ij}}{\partial x_k}$$
$$= \delta_{kj}\frac{\partial z^n}{\partial x_i}\frac{\partial L^x_{ij}}{\partial x_k} = \frac{\partial z^n}{\partial x_i}\frac{\partial L^x_{ij}}{\partial x_j},$$

which are the components of the contravariant vector field div**L**.

[9]Equations (6.29)–(6.31) represent the different forms of the components of an operator depending upon three indices, i.e. of a third-rank tensor: ∇**L**, the *gradient of* **L**.

6.5 Exercises

1. Write $\mathbf{g}$ and ds for cylindrical coordinates.
2. Write $\mathbf{g}$ and ds for spherical coordinates.
3. Find the length of a helix traced on a circular cylinder of radius R between the angles θ and $\theta + 2\pi$.
4. A curve is traced as a quarter circle of radius R, see Fig. 6.6, with ρ proportional to θ. When the quarter circle is rolled into a cone, the curve appears as indicated in the figure. Determine the length ℓ of the curve, first using the polar coordinates in the plane of the quarter circle, then the cylindrical ones for the case of the curve on the cone (exercise given in the book by Müller, see the suggested texts).
5. Calculate $\mathbf{g}$ for a planar system of coordinates composed of two axes z^1 and z^2 inclined, respectively, at α_1 and α_2 on the axis x_1. Then, find the vectors $\mathbf{g}_k$ and $\mathbf{g}^k$, $k = 1, 2$, check the orthogonality conditions $\mathbf{g}^h \cdot \mathbf{g}_k = \delta^h_k$, determine the norm of these vectors, and design them.
6. Calculate $\mathbf{g}_i$s for a system of spherical coordinates.
7. In the plane, elliptical coordinates are defined by the relations

   $$x_1 = c\ \cosh z^1 \cos z^2,\quad x_2 = c\ \sinh z^1 \sin z^2,\quad z^1 \in (0, \infty),\ z^2 \in [0, 2\pi);$$

   show that the lines $z^1 = const.$ and $z^2 = const.$ are confocal ellipses and hyperbole, determine the axes of the ellipses in terms of the parameter c, discuss the limit case of ellipses that degenerate into a crack, and determine its length. Finally, find $\mathbf{g}, \mathbf{g}_1$, and $\mathbf{g}_2$.
8. Determine the co- and contravariant components of a tensor $\mathbf{L}$ in cylindrical coordinates.

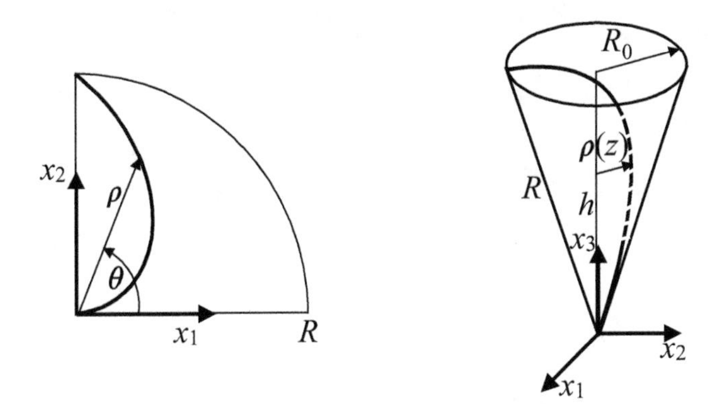

Figure 6.6: Curve in a plane and on a cone.

9. Determine the co- and contravariant components of a tensor $\mathbf{L}$ in spherical coordinates.
10. Show that

$$\mathrm{tr}\mathbf{L} = L_{ii}^{x} = g_{ij}L^{ij} = g^{ij}L_{ij} = L^{i}{}_{i} = L_{j}{}^{j}.$$

11. Prove Eq. (6.26).
12. Prove the *lemma of Ricci*:

$$\frac{\partial g_{jk}}{\partial z^h} = \Gamma^{i}_{jh} g_{ik} + \Gamma^{i}_{kh} g_{ji}.$$

13. Using Eq. (6.26), find the Christoffel symbols for the cylindrical, spherical, and elliptical (in the plane) coordinates.
14. Write the Laplacian Δf of a spatial scalar field f in cylindrical and spherical coordinates.
15. Prove that

$$g^{np}{}_{;h} = g_{np;h} = 0.$$

Chapter 7

Surfaces in $\mathcal{E}$

7.1 Surfaces in $\mathcal{E}$, coordinate lines and tangent planes

A function $\mathbf{f}(u, v) : \Omega \subset \mathbb{R}^2 \rightarrow \mathcal{E}$ of class $\geq$C^1 and such that its Jacobian

$$J = \begin{bmatrix} \dfrac{\partial f_1}{\partial u} & \dfrac{\partial f_1}{\partial v} \\ \dfrac{\partial f_2}{\partial u} & \dfrac{\partial f_2}{\partial v} \\ \dfrac{\partial f_3}{\partial u} & \dfrac{\partial f_3}{\partial v} \end{bmatrix}$$

has maximum rank (rank[J] = 2) defines a *surface in* $\mathcal{E}$, see Fig. 7.1. We say also that $\mathbf{f}$ is an *immersion* of Ω into $\mathcal{E}$ and that the subset $\Sigma \subset \mathcal{E}$ image of $\mathbf{f}$ is the *support* or *trace* of the surface $\mathbf{f}$.

As usual, we indicate the derivatives with respect to the variables u and v by, for example, $\dfrac{\partial \mathbf{f}}{\partial u} = \mathbf{f}_{,u}$, etc. The condition on the rank of J is equivalent to impose that

$$\mathbf{f}_{,u}(u, v) \times \mathbf{f}_{,v}(u, v) \neq \mathbf{o} \ \ \forall (u, v) \in \Omega. \tag{7.1}$$

This allows us to introduce the *normal to the surface* $\mathbf{f}$ is the vector $\mathbf{N} \in \mathcal{S}$ defined as

$$\mathbf{N} := \frac{\mathbf{f}_{,u} \times \mathbf{f}_{,v}}{|\mathbf{f}_{,u} \times \mathbf{f}_{,v}|}. \tag{7.2}$$

A *regular point of* Σ is a point where $\mathbf{N}$ is defined; if $\mathbf{N}$ is defined $\forall p \in \Sigma$, then the surface is said to be *regular*.

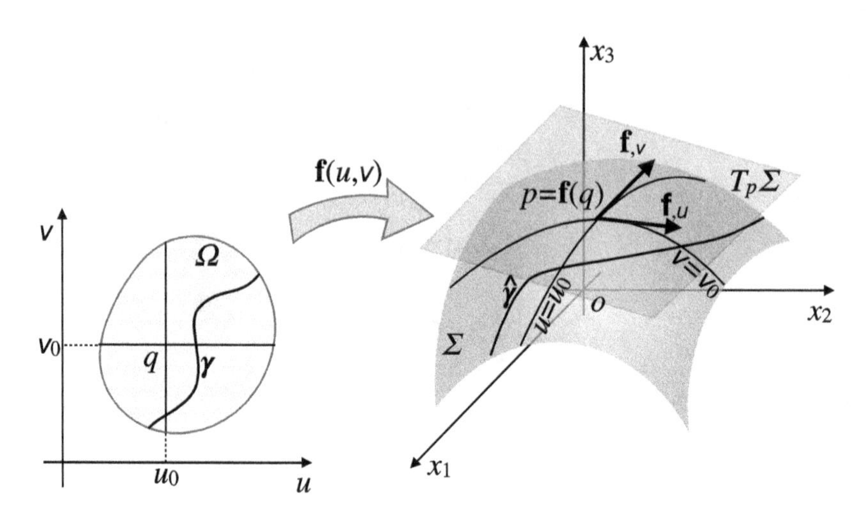

Figure 7.1: General scheme of a surface and of the tangent space at a point p.

A function $\gamma(t) : G \subset \mathbb{R} \to \Omega$ whose parametric equation is $\gamma(t) = (u(t), v(t))$ describes a curve in Ω whose image, through $\mathbf{f}$, is the curve, see Fig. 7.1,

$$\widehat{\gamma}(t) = \mathbf{f}(u(t), v(t)) : G \subset \mathbb{R} \to \Sigma \subset \mathcal{E}.$$

As a special case of curve in Ω, let us consider the curves of the type $v = v_0$ or $u = u_0$, with u_0, v_0 being some constants. Then, their image through $\mathbf{f}$ are two curves $\mathbf{f}(u, v_0), \mathbf{f}(u_0, v)$ on Σ called *coordinate lines*, see again Fig. 7.1. The *tangent vectors to the coordinate lines* are, respectively, the vectors $\mathbf{f}_{,u}(u, v_0)$ and $\mathbf{f}_{,v}(u_0, v)$, while the tangent to a curve $\widehat{\gamma}(t) = \mathbf{f}(u(t), v(t))$ is the vector

$$\widehat{\gamma}'(t) = \mathbf{f}_{,u}\frac{du}{dt} + \mathbf{f}_{,v}\frac{dv}{dt}, \tag{7.3}$$

i.e. the tangent vector to any curve on Σ is a linear combination of the tangent vectors to the coordinate lines. We remark that the tangent vectors $\mathbf{f}_{,u}(u, v_0)$ and $\mathbf{f}_{,v}(u_0, v)$ are necessarily non-null and linearly independent as a consequence of the assumption on the rank of J and hence of the existence of $\mathbf{N}$, i.e. of the regularity of Σ. They determine a plane that contains the tangents to all the curves on Σ passing by $p = \mathbf{f}(u_0, v_0)$ and form a basis on this plane, called the *natural basis.* Such a plane is the *tangent plane to* Σ *in* p and is indicated by $T_p\Sigma$; this plane is actually the space spanned by $\mathbf{f}_{,u}(u, v_0)$ and $\mathbf{f}_{,v}(u_0, v)$ and is also called the *tangent vector space.*

Let us consider two open subsets $\Omega_1, \Omega_2 \subset \mathbb{R}^2$; a *diffeomorphism*[1] *of class* C^k between Ω_1 and Ω_2 is a bijective map $\vartheta : \Omega_1 \to \Omega_2$ of class C^k with also its inverse of class C^k; the diffeomorphism is *smooth* if $k = \infty$.

Let Ω_1, Ω_2 be two open subsets of $\mathbb{R}^2$, $\mathbf{f} : \Omega_2 \to \mathcal{E}$ a surface, and $\vartheta : \Omega_1 \to \Omega_2$ a smooth diffeomorphism. Then, the surface $\mathbf{F} = \mathbf{f} \circ \vartheta : \Omega_1 \to \mathcal{E}$ is a *change in parameterization* for $\mathbf{f}$. In practice, the function defining the surface changes, but not Σ, its trace in $\mathcal{E}$. Let (U, V) be the coordinates in Ω_1 and (u, v) in Ω_2. Then, by the chain rule,

$$\mathbf{F}_{,U} = \mathbf{f}_{,u} \frac{\partial u}{\partial U} + \mathbf{f}_{,v} \frac{\partial v}{\partial U},$$
$$\mathbf{F}_{,V} = \mathbf{f}_{,u} \frac{\partial u}{\partial V} + \mathbf{f}_{,v} \frac{\partial v}{\partial V},$$

or denoting by J_ϑ the Jacobian of ϑ,

$$\left\{ \begin{matrix} \mathbf{F}_{,U} \\ \mathbf{F}_{,V} \end{matrix} \right\} = [J_\vartheta]^\top \left\{ \begin{matrix} \mathbf{f}_{,u} \\ \mathbf{f}_{,v} \end{matrix} \right\},$$

whence, making the cross product, one gets immediately

$$\mathbf{F}_{,U} \times \mathbf{F}_{,V} = \det[J_\vartheta]\ \mathbf{f}_{,u} \times \mathbf{f}_{,v}.$$

This result shows that the regularity of the surface, condition (7.1), the tangent plane, and the tangent space vector *do not depend upon the parameterization of* Σ. From the last equation, we also get

$$\mathbf{N}(U, V) = \operatorname{sgn}(\det[J_\vartheta])\ \mathbf{N}(u, v);$$

we say that the change in parameterization *preserves the orientation if* $\det[J_\vartheta] > 0$ and that it *inverses the parameterization* in the opposite case.

7.2 Surfaces of revolution

A *surface of revolution* is a surface whose trace is obtained by letting a plane curve, say γ, rotate around an axis, say x_3. To be more specific and without loss of generality, let $\gamma : G \subset \mathbb{R} \to \mathbb{R}^2$ be a regular curve of the

[1]The definition of diffeomorphism, of course, can be given for subsets of $\mathbb{R}^n$, $n \geq 1$; here, we bound the definition to the case of interest.

plane $x_2 = 0$, whose parametric equation is

$$\gamma(u) : \begin{cases} x_1 = \varphi(u), \\ x_3 = \psi(u), \end{cases} \quad \varphi(u) > 0 \ \forall u \in G. \tag{7.4}$$

Then, the subset $\Sigma_\gamma \subset \mathcal{E}$ defined by

$$\Sigma_\gamma := \left\{(x_1, x_2, x_3) \in \mathcal{E} | x_1^2 + x_2^2 = \varphi^2(u), x_3 = \psi(u), u \in G\right\}$$

is the trace of a surface of revolution of the curve $\boldsymbol{\gamma}(u)$ around the axis x_3. A general parameterization of such a surface is

$$\mathbf{f}(u,v) : G \times (-\pi, \pi] \to \mathcal{E} | \quad \begin{cases} x_1 = \varphi(u)\cos v, \\ x_2 = \varphi(u)\sin v, \\ x_3 = \psi(u). \end{cases} \tag{7.5}$$

It is readily checked that this parameterization actually defines a regular surface:

$$\mathbf{f}_{,u} = \begin{Bmatrix} \varphi'(u)\cos v \\ \varphi'(u)\sin v \\ \psi'(u) \end{Bmatrix}, \quad \mathbf{f}_{,v} = \begin{Bmatrix} -\varphi(u)\sin v \\ \varphi(u)\cos v \\ 0 \end{Bmatrix} \to \mathbf{f}_{,u} \times \mathbf{f}_{,v}$$

$$= \begin{Bmatrix} -\varphi(u)\psi'(u)\cos v \\ -\varphi(u)\psi'(u)\sin v \\ \varphi(u)\varphi'(u) \end{Bmatrix}$$

so that

$$|\mathbf{f}_{,u} \times \mathbf{f}_{,v}| = \varphi^2(u)(\varphi'^2(u) + \psi'^2(u)) \neq 0 \ \forall u \in G$$

for being $\boldsymbol{\gamma}(u)$ a regular curve, i.e. with $\boldsymbol{\gamma}'(u) \neq \mathbf{o} \ \forall u \in G$. A *meridian* is a curve in $\mathcal{E}$ intersection of the trace of $\mathbf{f}$, Σ_γ, with a plane containing the axis x_3; the equation of a meridian is obtained fixing the value of v, say $v = v_0$:

$$\begin{cases} x_1 = \varphi(u)\cos v_0, \\ x_2 = \varphi(u)\sin v_0, \\ x_3 = \psi(u). \end{cases}$$

A *parallel* is a curve in $\mathcal{E}$ intersection of Σ_γ with a plane orthogonal to x_3; the equation of a parallel, which is a circle with center on the axis x_3, is

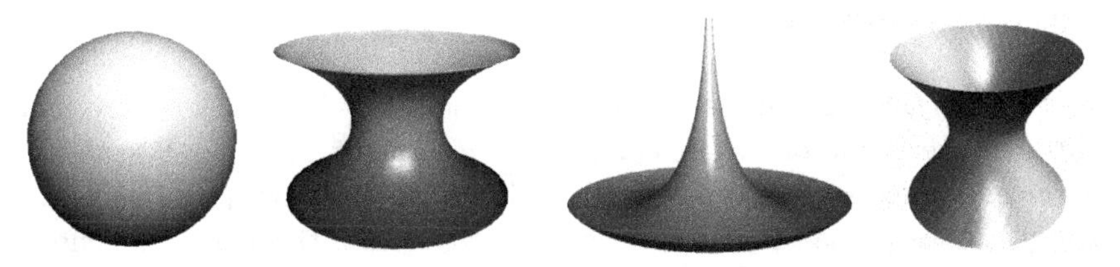

Figure 7.2: Surfaces of revolution: (from left) sphere, catenoid, pseudo-sphere, and hyperbolic hyperboloid.

obtained by fixing the value of u, say $u = u_0$:

$$\begin{cases} x_1 = \varphi(u_0)\cos v, \\ x_2 = \varphi(u_0)\sin v, \\ x_3 = \psi(u_0), \end{cases}$$

or also,

$$\begin{cases} x_1^2 + x_2^2 = \varphi(u_0)^2, \\ x_3 = \psi(u_0); \end{cases}$$

the radius of the circle is $\varphi(u_0)$. A *loxodrome* or *rhumb line* is a curve on Σ_γ crossing all the meridians at the same angle.[2] Some important examples of surfaces of revolution are:

- the *sphere*:

$$\mathbf{f}(u,v) : \left[-\frac{\pi}{2}, \frac{\pi}{2}\right] \times (-\pi, \pi] \to \mathcal{E} | \quad \begin{cases} x_1 = \cos u \cos v, \\ x_2 = \cos u \sin v, \\ x_3 = \sin v; \end{cases}$$

- the *catenoid*:

$$\mathbf{f}(u,v) : [-a, a] \times (-\pi, \pi] \to \mathcal{E} | \quad \begin{cases} x_1 = \cosh u \cos v, \\ x_2 = \cosh u \sin v, \\ x_3 = u; \end{cases}$$

- the *pseudo-sphere*:

$$\mathbf{f}(u,v) : [0, a] \times (-\pi, \pi] \to \mathcal{E} | \quad \begin{cases} x_1 = \sin u \cos v, \\ x_2 = \sin u \sin v, \\ x_3 = \cos u + \ln\left(\tan \dfrac{u}{2}\right); \end{cases} \tag{7.6}$$

- the *hyperbolic hyperboloid*:

$$\mathbf{f}(u,v) : [-a, a] \times (-\pi, \pi] \to \mathcal{E} | \quad \begin{cases} x_1 = \cos u - v \sin u, \\ x_2 = \sin u + v \cos u, \\ x_3 = v. \end{cases}$$

[2]This concept is important for marine and aerial navigation.

7.3 Ruled surfaces

A *ruled surface* (also named a *scroll*) is a surface with the property that through every one of its points, there is a straight line that lies on the surface. A ruled surface can be seen as the set of points swept by a moving straight line. We say that a surface is *doubly ruled* if through every one of its points, there are two distinct straight lines that lie on the surface.

Any ruled surface can be represented by a parameterization of the form

$$\mathbf{f}(u,v)=\boldsymbol{\gamma}(u)+v\boldsymbol{\lambda}(u), \tag{7.7}$$

where $\boldsymbol{\gamma}(u)$ is a regular smooth curve, the *directrix*, and $\boldsymbol{\lambda}(u)$ is a smooth curve. Fixing $u=u_0$ gives a *generator line* $\mathbf{f}(u_0,v)$ of the surface; the vectors $\boldsymbol{\lambda}(u)\neq\mathbf{o}$ describe the directions of the generators. Some important examples of ruled surfaces are:

- *Cones*: For these surfaces, all the straight lines pass through a point, the *apex* of the cone, choosing the apex as the origin, then it must be $\boldsymbol{\lambda}(u)=k\boldsymbol{\gamma}(u),\ k\in\mathbb{R}\rightarrow$

$$\mathbf{f}(u,v)=v\boldsymbol{\gamma}(u);$$

- *Cylinders*: A ruled surface is a cylinder $\iff \boldsymbol{\lambda}(u)=const.$ In this case, it is always possible to choose $\boldsymbol{\lambda}(u)\in\mathcal{S}$ and $\boldsymbol{\gamma}(u)$ a planar curve lying in a plane orthogonal to $\boldsymbol{\lambda}(u)$; in fact, it is sufficient to choose the curve $\boldsymbol{\gamma}^*(u)=(\mathbf{I}-\boldsymbol{\lambda}(u)\otimes\boldsymbol{\lambda}(u))\boldsymbol{\gamma}(u)$;
- *Helicoids*: A surface generated by rotating and simultaneously displacing a curve, the *profile curve*, along an axis is a helicoid. Any point of the profile curve is the starting point of a circular helix. Generally, we get a helicoid if

$$\boldsymbol{\gamma}(u)=(0,0,\varphi(u)),\quad \boldsymbol{\lambda}(u)=(\cos u,\sin u,0),\quad \varphi(u):\mathbb{R}\rightarrow\mathbb{R};$$

- *Möbius strip*: It is a ruled surface with

$$\boldsymbol{\gamma}(u)=(\cos 2u,\sin 2u,0),\quad \boldsymbol{\lambda}(u)=(\cos u\cos 2u,\cos u\sin 2u,\sin u).$$

7.4 First fundamental form of a surface

Let us consider two vectors of $T_p\Sigma$, say $\mathbf{w}_1,\mathbf{w}_2$; we want to calculate their scalar product in terms of their components in the natural basis $\{\mathbf{f}_{,u},\mathbf{f}_{,v}\}$

Figure 7.3: Ruled surfaces: (from left) elliptical cone, elliptical cylinder, helicoid, and Möbius strip.

of $T_p\Sigma$. If $\mathbf{w}_1 = a_1\mathbf{f}_{,u} + b_1\mathbf{f}_{,v}$ and $\mathbf{w}_2 = a_2\mathbf{f}_{,u} + b_2\mathbf{f}_{,v}$, then

$$\mathbf{w}_1 \cdot \mathbf{w}_2 = a_1 a_2 \mathbf{f}_{,u}^2 + (a_1 b_2 + a_2 b_1)\mathbf{f}_{,u} \cdot \mathbf{f}_{,v} + b_1 b_2 \mathbf{f}_{,v}^2,$$

which can be rewritten in the form

$$I(\mathbf{w}_1, \mathbf{w}_2) = \mathbf{w}_1 \cdot \mathbf{g}\mathbf{w}_2,$$

where[3]

$$\mathbf{g} = \begin{bmatrix} \mathbf{f}_{,u} \cdot \mathbf{f}_{,u} & \mathbf{f}_{,u} \cdot \mathbf{f}_{,v} \\ \mathbf{f}_{,v} \cdot \mathbf{f}_{,u} & \mathbf{f}_{,v} \cdot \mathbf{f}_{,v} \end{bmatrix}$$

is precisely the metric tensor $\mathbf{g}$ of Σ, cf. Eq. (6.6). In fact, $\mathbf{f}_{,u}$ and $\mathbf{f}_{,v}$ are the tangent vectors to the coordinate lines on Σ, i.e. they coincide with the vectors $\mathbf{g}_k$s.

$I(\mathbf{w}_1, \mathbf{w}_2)$ is the *first fundamental form* (or simply the *first form*) of $\mathbf{f}(u, v)$. If $\mathbf{w}_1 = \mathbf{w}_2 = \mathbf{w} = a\mathbf{f}_{,u} + b\mathbf{f}_{,v}$, then

$$I(\mathbf{w}) = \mathbf{w}^2 = a^2\mathbf{f}_{,u}^2 + 2ab\mathbf{f}_{,u} \cdot \mathbf{f}_{,v} + b^2\mathbf{f}_{,v}^2.$$

By the same definition of scalar product, $I(\mathbf{w}_1, \mathbf{w}_2)$ is a positive definite, bilinear, symmetric form $\forall \mathbf{w} \in T_p\Sigma$.

Through $I(\cdot, \cdot)$, we can calculate some important quantities regarding the geometry of Σ:

- Metric on Σ: $\forall ds \in \Sigma$,

$$ds^2 = ds \cdot ds = I(ds),$$

[3]Often, in texts on differential geometry, tensor $\mathbf{g}$ is indicated as

$$\mathbf{g} = \begin{bmatrix} E & F \\ F & G \end{bmatrix},$$

where $E := \mathbf{f}_{,u} \cdot \mathbf{f}_{,u}, F := \mathbf{f}_{,u} \cdot \mathbf{f}_{,v}, G := \mathbf{f}_{,v} \cdot \mathbf{f}_{,v}$.

so if

$$ds = \mathbf{f}_{,u}du + \mathbf{f}_{,v}dv,$$

then

$$ds^2 = \mathbf{f}_{,u}^2 du^2 + 2\mathbf{f}_{,u} \cdot \mathbf{f}_{,v} du\ dv + \mathbf{f}_{,v}^2 dv^2. \tag{7.8}$$

- Length ℓ of a curve $\gamma : [t_1, t_2] \subset \mathbb{R} \rightarrow \Sigma$: We know, from Eq. (4.4), that the length of a curve is the integral of the tangent vector:

$$\ell = \int_{t_1}^{t_2} |\gamma'(t)| dt = \int_{t_1}^{t_2} \sqrt{\gamma'(t) \cdot \gamma'(t)} dt$$

and hence, see Eq. (7.3), if we call $\mathbf{w} = (u', v')$ the tangent vector to γ, expressed by its components in the natural basis,

$$\begin{aligned} \ell &= \int_{t_1}^{t_2} \sqrt{u'^2 \mathbf{f}_{,u}^2 + 2u'v' \mathbf{f}_{,u} \cdot \mathbf{f}_{,v} + v'^2 \mathbf{f}_{,v}^2} dt \\ &= \int_{t_1}^{t_2} \sqrt{(u', v') \cdot \mathbf{g}\ (u', v')} dt \\ &= \int_{t_1}^{t_2} \sqrt{I(\mathbf{w})} dt. \end{aligned} \tag{7.9}$$

- Angle θ formed by two vectors $\mathbf{w}_1, \mathbf{w}_2 \in T_p\Sigma$:

$$\cos\theta = \frac{\mathbf{w}_1 \cdot \mathbf{w}_2}{|\mathbf{w}_1||\mathbf{w}_2|} = \frac{I(\mathbf{w}_1, \mathbf{w}_2)}{\sqrt{I(\mathbf{w}_1)}\sqrt{I(\mathbf{w}_2)}}.$$

- Area of a small surface on Σ: Let $\mathbf{f}_{,u}du$ and $\mathbf{f}_{,v}dv$ be two small vectors on Σ, forming together the angle θ that are the transformed through[4] $\mathbf{f} : \Omega \rightarrow \Sigma$, of two small orthogonal vectors $du, dv \in \Omega$; then, the area $d\mathcal{A}$ of the parallelogram determined by them is

$$\begin{aligned} d\mathcal{A} &= |\mathbf{f}_{,u}du \times \mathbf{f}_{,v}dv| = |\mathbf{f}_{,u} \times \mathbf{f}_{,v}| du\ dv = \sqrt{\mathbf{f}_{,u}^2 \mathbf{f}_{,v}^2 \sin^2\theta} du\ dv \\ &= \sqrt{\mathbf{f}_{,u}^2 \mathbf{f}_{,v}^2 (1 - \cos^2\theta)} du\ dv = \sqrt{\mathbf{f}_{,u}^2 \mathbf{f}_{,v}^2 - \mathbf{f}_{,u}^2 \mathbf{f}_{,v}^2 \cos^2\theta} du\ dv \\ &= \sqrt{\mathbf{f}_{,u}^2 \mathbf{f}_{,v}^2 - (\mathbf{f}_{,u} \cdot \mathbf{f}_{,v})^2} du\ dv = \sqrt{\det \mathbf{g}} du\ dv. \end{aligned}$$

[4]For the sake of conciseness, from now on, we indicate a surface as the function $\mathbf{f} : \Omega \rightarrow \Sigma$, with $\mathbf{f} = \mathbf{f}(u, v)$, $(u, v) \in \Omega \subset \mathbb{R}^2$ and $\Sigma \subset \mathcal{E}$.

The term $\sqrt{\det \mathbf{g}}$ is hence the *dilatation factor of the areas*; recalling Eq. (6.4), we see that the previous expression has a sense $\forall \mathbf{f}(u,v)$, i.e. for any parameterization of the surface.

7.5 Second fundamental form of a surface

Let $\mathbf{f}:\Omega\rightarrow\Sigma$ be a regular surface, $\{\mathbf{f}_{,u},\mathbf{f}_{,v}\}$ the natural basis for $T_p\Sigma$, and $\mathbf{N}\in\mathcal{S}$ the normal to Σ defined as in (7.2). We call the *map of Gauss* of Σ the map $\varphi_\Sigma:\Sigma\rightarrow\mathcal{S}$ that associates to each $p\in\Sigma$ its $\mathbf{N}:\ \varphi_\Sigma(p)=\mathbf{N}(p)$. To each subset $\sigma\subset\Sigma$, the map of Gauss associates hence a subset $\sigma_\mathcal{S}\subset\mathcal{S}$, Fig. 7.4 (e.g. the Gauss map of a plane is just a point on $\mathcal{S}$).

We want to study how $\mathbf{N}(p)$ varies at the varying of p on Σ. The idea is that the change of $\mathbf{N}(p)$ on Σ is related to the curvature of the surface.[5] For this purpose, we calculate the change in $\mathbf{N}$ per unit length of a curve $\gamma(s)\in\Sigma$, i.e. we study how $\mathbf{N}$ varies along any curve of Σ per unit of length of the curve itself; that is why we parameterize the curve with its arc length s.[6] Let $\mathbf{N}=N_i(u,v)\mathbf{e}_i$; then, if $\boldsymbol{\tau}\in\mathcal{S}$ is the tangent to the curve,

$$\begin{aligned}\frac{d\mathbf{N}}{ds}&=\frac{dN_i(u(s),v(s))}{ds}\mathbf{e}_i=\left(\frac{\partial N_i}{\partial u}\frac{du}{ds}+\frac{\partial N_i}{\partial v}\frac{dv}{ds}\right)\mathbf{e}_i\\&=\nabla N_i\cdot\boldsymbol{\tau}\mathbf{e}_i=(\mathbf{e}_i\otimes\nabla N_i)\boldsymbol{\tau}=(\nabla\mathbf{N})\,\boldsymbol{\tau}=\frac{d\mathbf{N}}{d\boldsymbol{\tau}}.\end{aligned}$$

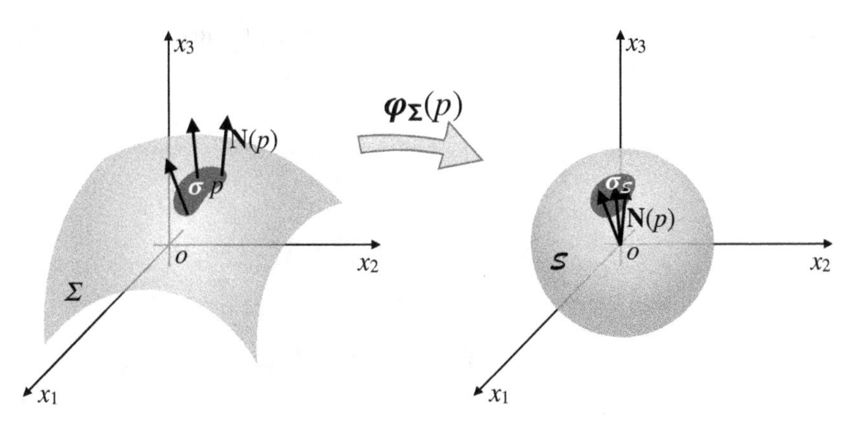

Figure 7.4: The map of Gauss.

[5] For curves, the curvature is linked to the change in $\boldsymbol{\tau}$, but for surfaces this should not be meaningful, as $\boldsymbol{\tau}$ is not unique $\forall p\in\Sigma$ while $\mathbf{N}$ is.

[6] Actually, it is also possible to introduce the following concepts more generally for any parameterization of the curve; anyway, for the sake of simplicity, we just use the parameter s in the following.

The change in $\mathbf{N}$ is hence related to the directional derivative of $\mathbf{N}$ along the tangent $\boldsymbol{\tau}$ to $\gamma(s)$, which is a linear operator on $T_p\Sigma$. Moreover, as $\mathbf{N} \in \mathcal{S}$, then, cf. Eq. (4.1),

$$\mathbf{N} \cdot \mathbf{N}_{,u} = \mathbf{N} \cdot \mathbf{N}_{,v} = 0 \Rightarrow \mathbf{N}_{,u}, \mathbf{N}_{,v} \in T_p\Sigma.$$

We then call the *Weingarten operator* $\mathcal{L}_W : T_p\Sigma \to T_p\Sigma$ the opposite of the directional derivative of $\mathbf{N}$:

$$\mathcal{L}_W(\boldsymbol{\tau}) := -\frac{d\mathbf{N}}{d\boldsymbol{\tau}}.$$

Hence,

$$\mathcal{L}_W(\mathbf{f}_{,u}) = -\mathbf{N}_{,u}, \quad \mathcal{L}_W(\mathbf{f}_{,v}) = -\mathbf{N}_{,v}. \tag{7.10}$$

Because $\mathcal{L}_W$ is linear, then there exists a tensor $\mathbf{X}$ on $T_p\Sigma$ such that

$$\mathcal{L}_W(\mathbf{v}) = \mathbf{X}\mathbf{v} \quad \forall \mathbf{v} \in T_p\Sigma. \tag{7.11}$$

For any two vectors $\mathbf{w}_1, \mathbf{w}_2 \in T_p\Sigma$, we define the *second fundamental form of a surface*, denoted by $II(\mathbf{w}_1, \mathbf{w}_2)$, the bilinear form

$$II(\mathbf{w}_1, \mathbf{w}_2) := I(\mathcal{L}_W(\mathbf{w}_1), \mathbf{w}_2).$$

Theorem 43 (Symmetry of the second fundamental form). $\forall \mathbf{w}_1, \mathbf{w}_2 \in T_p\Sigma,\ II(\mathbf{w}_1, \mathbf{w}_2) = II(\mathbf{w}_2, \mathbf{w}_1)$.

Proof. Because I and $\mathcal{L}_W$ are linear, it is sufficient to prove the thesis for the natural basis $\{\mathbf{f}_{,u}, \mathbf{f}_{,v}\}$ of $T_p\Sigma$, and by the symmetry of I, it is sufficient to prove that

$$I(\mathcal{L}_W(\mathbf{f}_{,u}), \mathbf{f}_{,v}) = I(\mathbf{f}_{,u}, \mathcal{L}_W(\mathbf{f}_{,v})),$$

i.e. that

$$I(-\mathbf{N}_{,u}, \mathbf{f}_{,v}) = I(\mathbf{f}_{,u}, -\mathbf{N}_{,v})$$

and in the end that

$$\mathbf{N}_{,u} \cdot \mathbf{f}_{,v} = \mathbf{f}_{,u} \cdot \mathbf{N}_{,v}.$$

For this purpose, we recall that

$$\mathbf{N} \cdot \mathbf{f}_{,u} = 0 = \mathbf{N} \cdot \mathbf{f}_{,v}.$$

So, differentiating the first equation by v and the second one by u, we get

$$\mathbf{N}_{,v} \cdot \mathbf{f}_{,u} = -\mathbf{N} \cdot \mathbf{f}_{,uv} = \mathbf{N}_{,u} \cdot \mathbf{f}_{,v}. \tag{7.12}$$

□

The second fundamental form defines a quadratic, bilinear symmetric form:

$$\begin{aligned} II(\mathbf{w}_1, \mathbf{w}_2) &= I(\mathcal{L}_W(\mathbf{w}_1), \mathbf{w}_2) = I(\mathbf{w}_1, \mathcal{L}_W(\mathbf{w}_2)) \\ &= I(\mathbf{w}_1, \mathbf{X}\mathbf{w}_2) = \mathbf{w}_1 \cdot \mathbf{g}\mathbf{X}\mathbf{w}_2 = \mathbf{w}_1 \cdot \mathbf{B}\mathbf{w}_2, \end{aligned}$$

where

$$\mathbf{B} := \mathbf{g}\mathbf{X}. \tag{7.13}$$

In the natural basis $\{\mathbf{f}_{,u}, \mathbf{f}_{,v}\}$ of $T_p\Sigma$, by Eq. (7.12), it is[7]

$$B_{ij} = II(\mathbf{f}_{,i}, \mathbf{f}_{,j}) = I(\mathcal{L}_W(\mathbf{f}_{,i}), \mathbf{f}_{,j}) = -\mathbf{N}_{,i} \cdot \mathbf{f}_{,j} = \mathbf{N} \cdot \mathbf{f}_{,ij}; \tag{7.14}$$

tensor $\mathbf{X}$ can then be calculated by Eq. (7.13):

$$\mathbf{X} = \mathbf{g}^{-1}\mathbf{B}. \tag{7.15}$$

By Eq. (7.14), because $\mathbf{f}_{,ij} = \mathbf{f}_{,ji}$ or simply because $II(\cdot, \cdot)$ is symmetric, we get that

$$\mathbf{B} = \mathbf{B}^\top.$$

7.6 Curvatures of a surface

Let $\mathbf{f} : \Omega \to \Sigma$ be a regular surface and $\boldsymbol{\gamma}(s) : G \subset \mathbb{R} \to \Sigma$ be a regular curve on Σ parameterized with the arc length s. We call the *curvature vector of* $\boldsymbol{\gamma}(s)$ the vector $\boldsymbol{\kappa}(s)$, defined as

$$\boldsymbol{\kappa}(s) := c(s)\boldsymbol{\nu}(s) = \boldsymbol{\gamma}''(s),$$

where $\boldsymbol{\nu}(s)$ is the principal normal to $\boldsymbol{\gamma}(s)$. By Eq. (4.11), it is also

$$\boldsymbol{\kappa}(s) = \boldsymbol{\gamma}''(s).$$

Then, we call the *normal curvature* $\kappa_N(s)$ *of* $\boldsymbol{\gamma}(s)$ the projection of $\boldsymbol{\kappa}(s)$ onto $\mathbf{N}(s)$, the normal to Σ:

$$\kappa_N(s) := \boldsymbol{\kappa}(s) \cdot \mathbf{N}(s) = c(s)\ \boldsymbol{\nu}(s) \cdot \mathbf{N}(s) = \boldsymbol{\gamma}''(s) \cdot \mathbf{N}(s).$$

Theorem 44. *The normal curvature $\kappa_N(s)$ of $\boldsymbol{\gamma}(s) \in \Sigma$ depends uniquely on $\boldsymbol{\tau}(s)$:*

$$\kappa_N(s) = \boldsymbol{\tau}(s) \cdot \mathbf{B}\boldsymbol{\tau}(s) = II(\boldsymbol{\tau}(s), \boldsymbol{\tau}(s)). \tag{7.16}$$

---

[7] In many texts on differential geometry, the following symbols are used:

$$\begin{aligned} L &= \mathbf{f}_{,uu} \cdot \mathbf{N} = -\mathbf{f}_{,u} \cdot \mathbf{N}_{,u}, \\ M &= \mathbf{f}_{,uv} \cdot \mathbf{N} = -\mathbf{f}_{,u} \cdot \mathbf{N}_{,v}, \\ N &= \mathbf{f}_{,vv} \cdot \mathbf{N} = -\mathbf{f}_{,v} \cdot \mathbf{N}_{,v}. \end{aligned}$$

Proof.

$$\gamma(s) = \gamma(u(s), v(s)) \rightarrow \boldsymbol{\tau}(s) = \gamma'(s) = \mathbf{f}_{,u}u' + \mathbf{f}_{,v}v';$$

therefore, $\boldsymbol{\tau} = (u', v')$ in the natural basis and

$$\boldsymbol{\kappa}(s) = \gamma''(s) = \mathbf{f}_{,u}u'' + \mathbf{f}_{,v}v'' + \mathbf{f}_{,uu}u'^2 + 2\mathbf{f}_{,uv}u'v' + \mathbf{f}_{,vv}v'^2,$$

and finally, by Eqs. (7.2) and (7.14),

$$\begin{aligned}\kappa_N(s) &= \gamma''(s) \cdot \mathbf{N}(s) = B_{11}u'^2 + 2B_{12}u'v' + B_{22}v'^2 = \boldsymbol{\tau} \cdot \mathbf{B}\boldsymbol{\tau} \\ &= II(\boldsymbol{\tau}, \boldsymbol{\tau}).\end{aligned}$$

□

Now, if $s = s(t)$ is a change in parameter for γ, then

$$\gamma'(t) = |\gamma'(t)|\boldsymbol{\tau}(t),$$

so by the linearity of $II(\cdot, \cdot)$, we get

$$II(\gamma'(t), \gamma'(t)) = |\gamma'(t)|^2 II(\boldsymbol{\tau}(t), \boldsymbol{\tau}(t)) = |\gamma'(t)|^2 \kappa_N(t)$$

and finally,

$$\kappa_N(t) = \frac{II(\gamma'(t), \gamma'(t))}{I(\gamma'(t), \gamma'(t))}.$$

To each point $p \in \Sigma$, it corresponds uniquely (in the assumption of regularity of the surface $\mathbf{f} : \Omega \rightarrow \Sigma$) to a tangent plane and a tangent space vector $T_p\Sigma$. In p, there are infinite tangent vectors to Σ, all of them belonging to $T_p\Sigma$. We can associate a curvature to each direction $\mathbf{t} \in T_p\Sigma$, i.e. to each tangent direction, in the following way: Let us consider the bundle $\mathcal{H}$ of planes whose support is the straight line through p and parallel to $\mathbf{N}$. Then, any plane $H \in \mathcal{H}$ is a *normal plane to Σ in p*; each normal plane is uniquely determined by a tangent direction $\mathbf{t}$, and the (planar) curve $\gamma_{N\mathbf{t}} := H \cap \Sigma$ is called a *normal section of* Σ. If $\boldsymbol{\nu}$ and $\mathbf{N}$ are, respectively, the principal normal to $\gamma_{N\mathbf{t}}$ and the normal to Σ in p, then

$$\boldsymbol{\nu} = \pm\mathbf{N}$$

for each normal section. We have, in this way, defined a function that to each tangent direction, $\mathbf{t} \in T_p\Sigma$ associates the normal curvature κ_N of the normal section $\gamma_{N\mathbf{t}}$:

$$\kappa_N : \mathcal{S} \cap T_p\Sigma \rightarrow \mathbb{R}| \quad \kappa_N(\mathbf{t}) = \frac{II(\mathbf{t}, \mathbf{t})}{I(\mathbf{t}, \mathbf{t})}.$$

By the bilinearity of the second fundamental form, $\kappa_N(\mathbf{t}) = \kappa_N(-\mathbf{t})$.

A point $p \in \Sigma$ is said to be a *umbilical point* if $\kappa_N(\mathbf{t}) = const.\ \forall \mathbf{t}$, it is a *planar point* if $\kappa_N(\mathbf{t}) = 0\ \forall \mathbf{t}$. In all the other points, κ_N takes a minimum and a maximum value on distinct directions $\mathbf{t} \in T_p\Sigma$.

Because $\mathbf{B} = \mathbf{B}^\top$, by the spectral theorem, there exists an orthonormal basis $\{\mathbf{u}_1, \mathbf{u}_2\}$ of $T_p\Sigma$ such that

$$\mathbf{B} = \beta_j \mathbf{u}_j \otimes \mathbf{u}_j,$$

with β_j the eigenvalues of $\mathbf{B}$. In such a basis, by Eq. (7.13), we get

$$\kappa_N(\mathbf{u}_i) = \frac{II(\mathbf{u}_{\underline{i}}, \mathbf{u}_{\underline{i}})}{I(\mathbf{u}_{\underline{i}}, \mathbf{u}_{\underline{i}})} = \frac{\mathbf{u}_{\underline{i}} \cdot \mathbf{B}\mathbf{u}_{\underline{i}}}{\mathbf{u}_{\underline{i}} \cdot \mathbf{g}\mathbf{u}_{\underline{i}}} = \frac{\mathbf{u}_{\underline{i}} \cdot \mathbf{g}\mathbf{X}\mathbf{u}_{\underline{i}}}{\mathbf{u}_{\underline{i}} \cdot \mathbf{g}\mathbf{u}_{\underline{i}}}, \quad i = 1, 2.$$

Then, because $\{\mathbf{u}_1, \mathbf{u}_2\}$ is an orthonormal basis, $\mathbf{g} = \mathbf{I}$ and

$$\kappa_N(\mathbf{u}_i) = \mathbf{u}_{\underline{i}} \cdot \mathbf{X}\mathbf{u}_{\underline{i}}, \quad i = 1, 2,$$

i.e. $\mathbf{X}$ and $\mathbf{B}$ share the same eigenvectors. Moreover, cf. Section 2.8, we know that the two directions $\mathbf{u}_1$ and $\mathbf{u}_2$ are the directions whereupon the quadratic form in the previous equation gets its maximum, κ_1, and minimum, κ_2, values, and in such a basis,

$$\mathbf{X} = \kappa_i \mathbf{u}_i \otimes \mathbf{u}_i.$$

We call κ_1 and κ_2 the *principal curvatures of* Σ *in* p and $\mathbf{u}_1, \mathbf{u}_2$ the *principal directions of* Σ *in* p, see Fig. 7.5.

We call the *Gaussian curvature* K the product of the principal curvatures:

$$K := \kappa_1 \kappa_2 = \det \mathbf{X}.$$

By Eq. (7.15) and the theorem of Binet, it is also

$$K = \frac{\det \mathbf{B}}{\det \mathbf{g}}. \tag{7.17}$$

We define the *mean curvature* H *of a surface*[8] $\mathbf{f} : \Omega \to \Sigma$ at a point $p \in \Sigma$ the mean of the principal curvatures at p:

$$H := \frac{\kappa_1 + \kappa_2}{2} = \frac{1}{2}\mathrm{tr}\mathbf{X}.$$

Of course, a change in parameterization of a surface can change the orientation, cf. Section 7.1, which induces a change in $\mathbf{N}$ into its opposite one and, by consequence, in the sign of the second fundamental form and hence in the normal and principal curvatures. These last are hence defined to less the sign, and the mean curvature too, while the principal directions, umbilicality, flatness, and Gaussian curvature are intrinsic to Σ, i.e. they do not depend on its parameterization.

[8]The concept of mean curvature of a surface was introduced for the first time by Sophie Germain in her celebrated work on the elasticity of plates (1815).

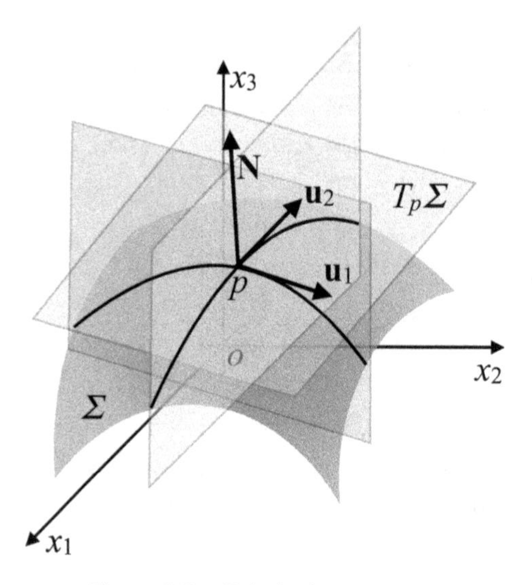

Figure 7.5: Principal curvatures.

7.7 The theorem of Rodrigues

The principal directions of curvature have a property that is specified by the following.

Theorem 45 (Theorem of Rodrigues). *Let* $\mathbf{f}(u,v)$ *be a surface of class at least* C^2 *and* $\boldsymbol{\lambda}=(\lambda_u,\lambda_v)\in T_p\Sigma$, *then*

$$\frac{d\mathbf{N}(p)}{d\boldsymbol{\lambda}}=-\kappa_\lambda\boldsymbol{\lambda}\tag{7.18}$$

if and only if $\boldsymbol{\lambda}$ *is a principal direction;* κ_λ *is the principal curvature relative to* $\boldsymbol{\lambda}$.

Proof. Let $\boldsymbol{\lambda}$ be a principal direction of $T_p\Sigma$. Because $\mathbf{N}\in\mathcal{S}$, then

$$\frac{d\mathbf{N}}{d\boldsymbol{\lambda}}\cdot\mathbf{N}=0;\tag{7.19}$$

moreover,

$$\frac{d\mathbf{N}}{d\boldsymbol{\lambda}}=\nabla\mathbf{N}\,\boldsymbol{\lambda}=\begin{bmatrix}0&0&0\\0&0&0\\\mathbf{N}_{,u}&\mathbf{N}_{,v}&1\end{bmatrix}\begin{Bmatrix}\lambda_u\\\lambda_v\\0\end{Bmatrix}=\mathbf{N}_{,u}\lambda_u+\mathbf{N}_{,v}\lambda_v.\tag{7.20}$$

Let $\boldsymbol{\mu} = (\mu_u, \mu_v)$ be the other principal direction of $T_p\Sigma$, then

$$\boldsymbol{\lambda} \cdot \boldsymbol{\mu} = 0 \;\rightarrow\; I(\boldsymbol{\lambda}, \boldsymbol{\mu}) = II(\boldsymbol{\lambda}, \boldsymbol{\mu}) = 0.$$

Moreover,

$$\frac{d\mathbf{N}}{d\boldsymbol{\lambda}} \cdot \boldsymbol{\mu} = -II(\boldsymbol{\lambda}, \boldsymbol{\mu}) = 0,$$

which implies, together with Eq. (7.19),

$$\frac{d\mathbf{N}}{d\boldsymbol{\lambda}} = \alpha\boldsymbol{\lambda}. \tag{7.21}$$

Therefore,

$$\frac{d\mathbf{N}}{d\boldsymbol{\lambda}} \cdot \boldsymbol{\lambda} = -II(\boldsymbol{\lambda}) = \alpha\boldsymbol{\lambda} \cdot \boldsymbol{\lambda} = \alpha I(\boldsymbol{\lambda})$$

and finally,

$$\alpha = -\frac{II(\boldsymbol{\lambda})}{I(\boldsymbol{\lambda})} = -\kappa_\lambda.$$

Contrarily, if we assume Eq. (7.21), as before, we get $\alpha = -\kappa_\lambda$, and to end, we just need to prove that $\boldsymbol{\lambda}$ is a principal direction. From Eqs. (7.20) and (7.21), we get

$$\lambda_u \mathbf{N}_{,u} + \lambda_v \mathbf{N}_{,v} = -\kappa_\lambda(\lambda_u \mathbf{f}_{,u} + \lambda_v \mathbf{f}_{,v}).$$

Projecting this equation onto $\mathbf{f}_{,u}$ and $\mathbf{f}_{,v}$ gives the two equations

$$\begin{aligned} L\lambda_u + M\lambda_v &= \kappa_\lambda(E\lambda_u + F\lambda_v), \\ M\lambda_u + N\lambda_v &= \kappa_\lambda(E\lambda_u + G\lambda_v), \end{aligned} \tag{7.22}$$

with the symbols E, F, G, L, M, and N defined in Notes 3 and 7 and used here for the sake of conciseness. Let $\mathbf{w} = (w_u, w_v) \in T_p\Sigma$ and consider the function

$$\zeta(\mathbf{w}, \kappa_\lambda) = II(\mathbf{w}) - \kappa_\lambda I(\mathbf{w});$$

it is easy to check that ζ, $\frac{\partial \zeta}{\partial w_u}$ and $\frac{\partial \zeta}{\partial w_v}$ take zero value for $\mathbf{w} = \boldsymbol{\lambda}_0$, with $\boldsymbol{\lambda}_0$ the eigenvector of the principal direction relative to κ_λ, which gives the system of equations

$$\begin{cases} II(\boldsymbol{\lambda}_0) - \kappa_\lambda I(\boldsymbol{\lambda}_0) = 0, \\ \dfrac{\partial II(\boldsymbol{\lambda}_0)}{\partial w_u} - \kappa_\lambda \dfrac{\partial II(\boldsymbol{\lambda}_0)}{\partial w_u} = 0, \\ \dfrac{\partial II(\boldsymbol{\lambda}_0)}{\partial w_v} - \kappa_\lambda \dfrac{\partial II(\boldsymbol{\lambda}_0)}{\partial w_v} = 0. \end{cases}$$

Developing the derivatives and making some standard operations, Eq. (7.22) is found again, which proves that $\boldsymbol{\lambda}$ is necessarily the principal direction relative to κ_λ. □

This theorems hence states that the derivative of $\mathbf{N}$ along a given direction is a vector parallel to such a direction only when this is a principal direction of curvature.

7.8 Classification of the points of a surface

Let $\mathbf{f}:\Omega\to\Sigma$ be a regular surface and $p\in\Sigma$ a non-planar point. Then, we say that

- p is an *elliptic point* if $K(p)>0$;
- p is a *hyperbolic point* if $K(p)<0$;
- p is a *parabolic point* if $K(p)=0$.

We remark that, by Eq. (7.17), because $\det\mathbf{g}>0$, Eq. (6.4), the value of $\det\mathbf{B}$ is sufficient to determine the type of a point on Σ.

Theorem 46. *If p is an elliptical point of σ, then there exists a neighborhood $U\in\Sigma$ of p such that all the points $q\in U$ belong to the same half-space into which $\mathcal{E}$ is divided by the tangent plane $T_p\Sigma$.*

Proof. For the sake of simplicity and without loss of generality, we can always chose a parameterization $\mathbf{f}(u,v)$ of the surface such that $p=\mathbf{f}(0,0)$. Expanding $\mathbf{f}(u,v)$ into a Taylor's series around $(0,0)$, we get the position of a point $q=\mathbf{f}(u,v)\in\Sigma$ in the neighborhood of p (though not indicated for the sake of brevity, all the derivatives are intended to be calculated at $(0,0)$):

$$\mathbf{f}(u,v)=\mathbf{f}_{,u}u+\mathbf{f}_{,v}v+\frac{1}{2}(\mathbf{f}_{,uu}u^2+2\mathbf{f}_{,uv}uv+\mathbf{f}_{,vv}v^2)+o(u^2+v^2).$$

The distance with sign $d(q)$ of $q\in\Sigma$ from the tangent plane $T_p\Sigma$ is the projection onto $\mathbf{N}$, i.e.:

$$\begin{aligned}d(q)&=\frac{1}{2}(\mathbf{f}_{,uu}u^2+2\mathbf{f}_{,uv}uv+\mathbf{f}_{,vv}v^2)\cdot\mathbf{N}+o(u^2+v^2)\\&=\frac{1}{2}(B_{11}u^2+2B_{12}uv+B_{22}v^2)+o(u^2+v^2),\end{aligned}$$

or, equivalently, once we set $\mathbf{w}=u\mathbf{f}_{,u}+v\mathbf{f}_{,v}$,

$$d(q)=II(\mathbf{w},\mathbf{w})+o(u^2+v^2).\tag{7.23}$$

If p is an elliptic point, the principal curvatures have the same sign because $K = \kappa_1\kappa_2 > 0 \Rightarrow$ the sign of $II(\mathbf{w}, \mathbf{w})$ does not depend upon $\mathbf{w}$, i.e. upon the tangent vector. As a consequence, the sign of $d(q)$ does not change with $\mathbf{w} \Rightarrow \forall q \in U,\ \Sigma$ is on the same side of the tangent plane $T_p\Sigma$. □

Theorem 47. *If p is a hyperbolic point of Σ, then for each neighborhood $U \in \Sigma$ of p, there are points $q \in U$ that are in half-spaces on the opposite sides with respect to the tangent plane $T_p\Sigma$.*

Proof. The proof is identical to that of the previous theorem until Eq. (7.23); now, if p is a hyperbolic point, the principal curvatures have opposite signs, and by consequence $d(q)$ changes of sign at least two times in any neighborhood U of $p \Rightarrow$, there are points $q \in U$ lying in half-spaces on the opposite sides with respect to the tangent plane $T_p\Sigma$. □

In a parabolic point, there are different possibilities: Σ is on one side of the space with respect to $T_p\Sigma$, as for the case of a cylinder, or not, like, for example, for the points $(0, v)$ of the surface, see Fig. 7.6,

$$\begin{cases} x = (u^3 + 2)\cos v, \\ y = (u^3 + 2)\sin v, \\ z = -u. \end{cases}$$

This is also the case for planar points: e.g., the point $(0, 0, 0)$ is a planar point for both the surfaces

$$z = x^4 + y^4, \quad z = x^3 - 3xy^2,$$

but in the first case, all of the surface is on one side of the tangent plane, while it is on both sides for the second case (the so-called *monkey's saddle*), see Fig. 7.7.

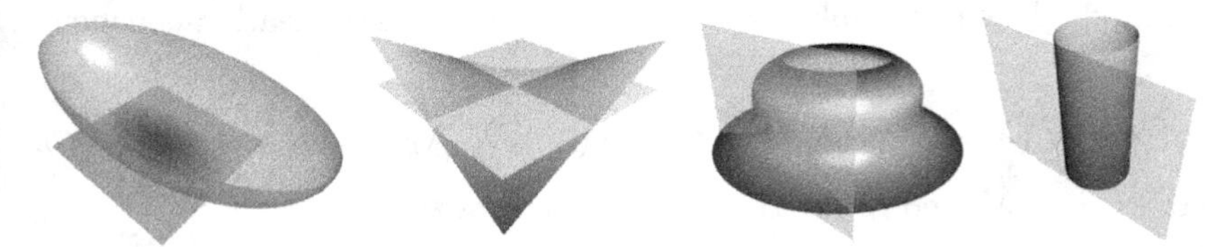

Figure 7.6: Elliptic (left), hyperbolic (center), and parabolic (last two on the right) points.

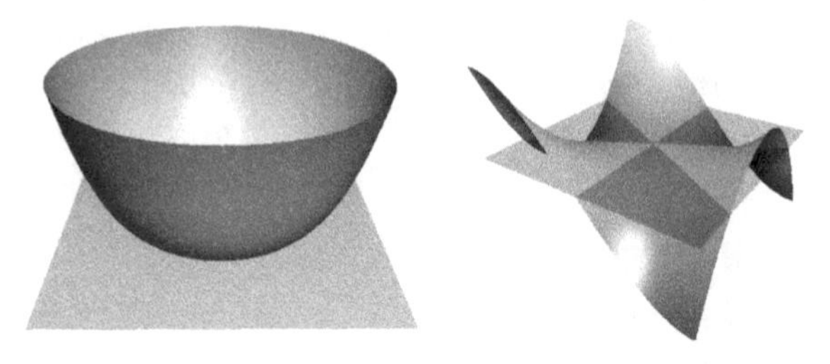

Figure 7.7: Two different planar points.

7.9 Developable surfaces

Let us now consider a ruled surface $\mathbf{f}:\Omega\to\Sigma$ as in Eq. (7.7); then,

$$\mathbf{f}_{,u}=\boldsymbol{\gamma}'+v\boldsymbol{\lambda}',\quad \mathbf{f}_{,v}=\boldsymbol{\lambda},\quad \mathbf{f}_{,u}\times\mathbf{f}_{,v}=\boldsymbol{\gamma}'\times\boldsymbol{\lambda}+v\boldsymbol{\lambda}'\times\boldsymbol{\lambda},\quad \mathbf{f}_{,uv}=\boldsymbol{\lambda}',\quad \mathbf{f}_{,vv}=\mathbf{o}.$$

Consequently, $B_{22}=\mathbf{N}\cdot\mathbf{f}_{,vv}=0\ \Rightarrow\ \det\mathbf{B}=-B_{12}^2$: The points of Σ are hyperbolic or parabolic. Namely, the parabolic points are those with

$$B_{12}=\mathbf{N}\cdot\mathbf{f}_{,uv}=\frac{\mathbf{f}_{,u}\times\mathbf{f}_{,v}}{|\mathbf{f}_{,u}\times\mathbf{f}_{,v}|}\cdot\mathbf{f}_{,uv}=0$$
$$\iff\quad(\boldsymbol{\gamma}'\times\boldsymbol{\lambda}+v\boldsymbol{\lambda}'\times\boldsymbol{\lambda})\cdot\boldsymbol{\lambda}'=\boldsymbol{\gamma}'\times\boldsymbol{\lambda}\cdot\boldsymbol{\lambda}'=0.$$

We remark that the ruled surfaces made of parabolic points have *null Gaussian curvature everywhere:* $K=0$.

Let us consider ruled surfaces having only parabolic points; then, we have the following.

Theorem 48. *For a ruled surface* $\mathbf{f}(u,v)=\boldsymbol{\gamma}(u)+v\boldsymbol{\lambda}(u)$, *the following are equivalents:*

(i) $\boldsymbol{\gamma}',\boldsymbol{\lambda},\boldsymbol{\lambda}'$ *are linearly dependent;*
(ii) $\mathbf{N}_{,v}=\mathbf{o}$.

Proof. Condition (ii) implies that $\mathbf{N}$ does not change along a straight line lying on the ruled surface $\Rightarrow\mathbf{f}_{,u}\times\mathbf{f}_{,v}=\boldsymbol{\gamma}'\times\boldsymbol{\lambda}+v\boldsymbol{\lambda}'\times\boldsymbol{\lambda}$ does not depend on v as well. This is possible $\iff\ \boldsymbol{\gamma}'\times\boldsymbol{\lambda}$ and $\boldsymbol{\lambda}'\times\boldsymbol{\lambda}$ are linearly dependent, i.e. $\iff$

$$(\boldsymbol{\gamma}'\times\boldsymbol{\lambda})\times(\boldsymbol{\lambda}'\times\boldsymbol{\lambda})=(\boldsymbol{\lambda}'\times\boldsymbol{\lambda}\cdot\boldsymbol{\gamma}')\boldsymbol{\lambda}-(\boldsymbol{\lambda}'\times\boldsymbol{\lambda}\cdot\boldsymbol{\lambda})\boldsymbol{\gamma}'=(\boldsymbol{\lambda}'\times\boldsymbol{\lambda}\cdot\boldsymbol{\gamma}')\boldsymbol{\lambda}=\mathbf{o},$$

i.e. when $\boldsymbol{\lambda},\boldsymbol{\lambda}'$, and $\boldsymbol{\gamma}'$ are coplanar, which proves the thesis. □

We say that a ruled surface is *developable* if one of the conditions of Theorem 48 is satisfied. A developable surface is a surface that can be

Figure 7.8: Ruled surface of the tangents to a cylindrical helix.

flattened without distortion onto a plane, i.e. it can be bent without stretching or shearing or vice versa and it can be obtained by transforming a plane. We remark that only ruled surfaces are developable (but not all the ruled surfaces are developable).

It is immediate to check that a cylinder or a cone are developable surfaces, while the helicoid, hyperbolic hyperboloid, or hyperbolic paraboloid are not. Another classical example of developable surface is the *ruled surface of the tangents to a curve*: Let $\boldsymbol{\gamma}(t): G \subset \mathbb{R} \rightarrow \mathcal{E}$ be a regular smooth curve; then, the ruled surface of the tangents to $\boldsymbol{\gamma}$ is the surface $\mathbf{f}(u,v): G \times \mathbb{R} \rightarrow \Sigma$ defined by

$$\mathbf{f}(u,v) = \boldsymbol{\gamma}(u) + v\boldsymbol{\gamma}'(u).$$

Figure 7.8 shows the ruled surface of the tangents to a cylindrical helix.

7.10 Points of a surface of revolution

Let us now consider a surface of revolution $\mathbf{f}: \Omega \rightarrow \Sigma_\gamma$ as in Eq. (7.5) and, for the sake of simplicity, let u be the natural parameter of the curve in Eq. (7.4) generating the surface. Then,

$$\varphi'^2(u) + \psi'^2(u) = 1, \quad \psi''(u)\varphi'(u) - \psi'(u)\varphi''(u) = c(u).$$

We can then calculate:

- the vectors of the natural basis:

$$\mathbf{f}_{,u} = \left\{ \begin{array}{c} \varphi'(u)\cos v \\ \varphi'(u)\sin v \\ \psi'(u) \end{array} \right\}, \quad \mathbf{f}_{,v} = \left\{ \begin{array}{c} -\varphi(u)\sin v \\ \varphi(u)\cos v \\ 0 \end{array} \right\};$$

- the normal to the surface

$$\mathbf{N} = \left\{ \begin{array}{c} -\psi'(u)\cos v \\ -\psi'(u)\sin v \\ \varphi'(u) \end{array} \right\};$$

- the metric tensor (i.e. the first fundamental form):

$$\mathbf{g} = \begin{bmatrix} 1 & 0 \\ 0 & \varphi^2(u) \end{bmatrix};$$

- the second derivatives of $\mathbf{f}$:

$$\mathbf{f}_{,uu} = \begin{Bmatrix} \varphi''(u)\cos v \\ \varphi''(u)\sin v \\ \psi''(u) \end{Bmatrix}, \quad \mathbf{f}_{,uv} = \begin{Bmatrix} -\varphi'(u)\sin v \\ \varphi'(u)\cos v \\ 0 \end{Bmatrix},$$

$$\mathbf{f}_{,vv} = \begin{Bmatrix} -\varphi(u)\cos v \\ -\varphi(u)\sin v \\ 0 \end{Bmatrix};$$

- tensor $\mathbf{B}$ (i.e. the second fundamental form):

$$\mathbf{B} = \begin{bmatrix} c(u) & 0 \\ 0 & \varphi(u)\psi'(u) \end{bmatrix};$$

- the Gaussian curvature K:

$$K = \det \mathbf{X} = \frac{\det \mathbf{B}}{\det \mathbf{g}} = \frac{c(u)\psi'(u)}{\varphi(u)}.$$

Therefore, the points of Σ_γ where $c(u)$ and $\psi'(u)$ have the same sign are elliptic, but hyperbolic otherwise.[9] Parabolic points correspond to inflexion points of $\gamma(u)$ if $c(u) = 0$ or to points with horizontal tangent to $\gamma(u)$ if $\psi'(u) = 0$.

As an example, let us consider the case of the pseudo-sphere, Eq. (7.6). Then,

$$\varphi(u) = \sin u, \quad \psi(u) = \cos u + \ln \tan \frac{u}{2}.$$

Some simple calculations give

$$\psi'(u) = -\sin u + \frac{1}{\sin u}, \quad c(u) = -\frac{|\tan u|}{|\cot u|};$$

as a consequence,

$$K = \frac{c(u)\psi'(u)}{\varphi(u)} = -\frac{(-\sin u + \frac{1}{\sin u})|\tan u|}{\sin u|\cot u|} = -1.$$

Finally, $K = const. = -1$, which is the reason for the name of this surface.

[9]Recall that in a revolution surface, $\varphi(u) > 0 \ \forall u$.

7.11 Lines of curvature, conjugated directions, asymptotic directions

A *line of curvature* is a curve on a surface with the property of being tangent, at each point, to a principal direction.

Theorem 49. *The lines of curvature of a surface are the solutions to the differential equation*

$$X_{21}u'^2 + (X_{22} - X_{11})u'v' - X_{12}v'^2 = 0.$$

Proof. A curve $\gamma(t) : G \subset \mathbb{R} \to \Sigma \subset \mathcal{E}$ is a line of curvature $\iff$

$$\gamma'(t) = \mathbf{f}_{,u}u' + \mathbf{f}_{,v}v'$$

is an eigenvector of $\mathbf{X}(t)$ $\forall t$, i.e. $\iff$ there exists a function $\mu(t)$ such that

$$\mathbf{X}(t)\gamma'(t) = \mu(t)\gamma'(t) \quad \forall t.$$

In the natural basis of $T_p\Sigma$, this condition reads as (we omit the dependence upon t for the sake of conciseness)

$$\begin{bmatrix} X_{11} & X_{12} \\ X_{21} & X_{22} \end{bmatrix} \begin{Bmatrix} u' \\ v' \end{Bmatrix} = \mu \begin{Bmatrix} u' \\ v' \end{Bmatrix},$$

which is satisfied $\iff$ the two vectors at the left- and right-hand sides are proportional, i.e. $\iff$

$$\det \begin{bmatrix} X_{11}u' + X_{12}v' & u' \\ X_{21}u' + X_{22}v' & v' \end{bmatrix} = 0 \to X_{21}u'^2 + (X_{22} - X_{11})u'v' - X_{12}v'^2 = 0.$$

□

As a corollary, if $\mathbf{X}$ is diagonal, then the coordinate lines are, at the same time, principal directions and lines of curvature.

Theorem 50. *A curve $\gamma(u) : G \subset \mathbb{R} \to \Sigma$ is a line of curvature $\iff$ the surface*

$$\mathbf{f}(u, v) = \gamma(u) + v\mathbf{N}(\gamma(u)), \tag{7.24}$$

is developable.

Proof. From Theorem 48, $\mathbf{f}(u, v)$ is developable $\iff$ $\gamma' \cdot \mathbf{N} \times \mathbf{N}' = 0$. Because γ' and $\mathbf{N}' \in T_p\Sigma$, which is orthogonal to $\mathbf{N}$, the surface will be developable $\iff$ $\gamma' \times \mathbf{N}' = \mathbf{o}$. Moreover, writing

$$\gamma' = \mathbf{f}_{,u}u' + \mathbf{f}_{,v}v',$$

it is

$$\mathbf{N}' = \mathbf{N}_{,u}u' + \mathbf{N}_{,v}v' = -\mathcal{L}_W(\boldsymbol{\gamma}'),$$

hence $\mathbf{f}(u,v)$ is developable $\iff$ $\mathcal{L}_W(\boldsymbol{\gamma}') \times \boldsymbol{\gamma}' = \mathbf{o}$, i.e. when $\boldsymbol{\gamma}'$ is a principal direction. □

The curve in Eq. (7.24) is called the *ruled surface of the normals.*

Let p be a non-planar point of a surface $\mathbf{f} : \Omega \to \Sigma$ and $\mathbf{v}_1, \mathbf{v}_2$ two vectors of $T_p\Sigma$. We say that $\mathbf{v}_1$ and $\mathbf{v}_2$ are *conjugated* if $II(\mathbf{v}_1, \mathbf{v}_2) = 0$. The directions corresponding to $\mathbf{v}_1$ and $\mathbf{v}_2$ are called *conjugated directions.* Hence, the principal directions at a point p are conjugated; if p is an umbilical point, any two orthogonal directions are conjugated.

The direction of a vector $\mathbf{v} \in T_p\Sigma$ is said to be *asymptotic* if it is *autoconjugated*, i.e. if $II(\mathbf{v}, \mathbf{v}) = 0$. An asymptotic direction is hence a direction where the normal curvature is null. In a hyperbolic point, there are two asymptotic directions; in a parabolic point only one; and in an elliptic point, there are no asymptotic directions. An *asymptotic line* is a curve on a surface with the property of being tangent at every point to an asymptotic direction. The asymptotic lines are the solution to the differential equation

$$II(\boldsymbol{\gamma}', \boldsymbol{\gamma}') = 0 \ \to \ B_{11}u'^2 + 2B_{12}u'v' + B_{22}v'^2 = 0;$$

in particular, if $B_{11} = B_{22} = 0$ and $\mathbf{B} \neq \mathbf{O}$, then the coordinate lines are asymptotic lines. Asymptotic lines exist only in the regions where $K \leq 0$.

7.12 Dupin's conical curves

The *conical curves of Dupin* are the real curves in $T_p\Sigma$ whose equations are

$$II(\mathbf{v}, \mathbf{v}) = \pm 1, \quad \mathbf{v} \in \mathcal{S}.$$

Let $\{\mathbf{u}_1, \mathbf{u}_2\}$ be the basis of the principal directions. Using polar coordinates, we can write

$$\mathbf{v} = \rho\mathbf{e}_\rho, \quad \mathbf{e}_\rho = \cos\theta\mathbf{u}_1 + \sin\theta\mathbf{u}_2.$$

Therefore,

$$II(\mathbf{v}, \mathbf{v}) = \rho^2 II(\mathbf{e}_\rho, \mathbf{e}_\rho) = \rho^2\kappa_N(\mathbf{e}_\rho),$$

and the conicals' equations are

$$\rho^2(\kappa_1\cos^2\theta + \kappa_2\sin^2\theta) = \pm 1.$$

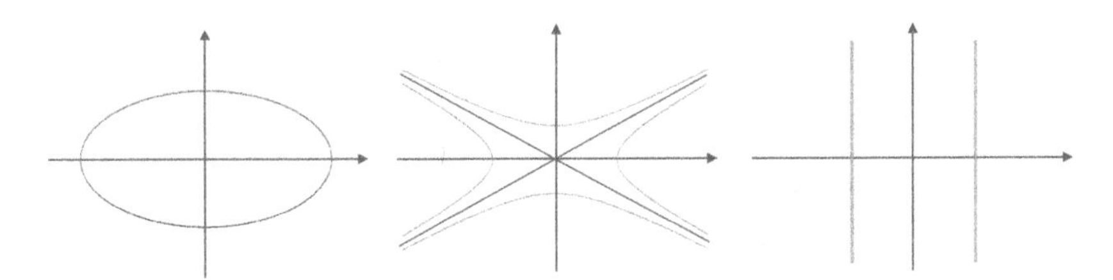

Figure 7.9: The conical curves of Dupin (from left): elliptic, hyperbolic, and parabolic points.

With the Cartesian coordinates $\xi = \rho\cos\theta, \eta = \rho\sin\theta$, we get

$$\kappa_1\xi^2 + \kappa_2\eta^2 = \pm 1.$$

The type of conical curves depend upon the kind of point on Σ:

- *Elliptical points*: The principal curvatures have the same sign $\rightarrow$ one of the conical curves is an ellipse, the other one the null set (actually, it is not a real curve).
- *Hyperbolic points*: The principal curves have opposite signs $\rightarrow$ the conical curves are conjugated hyperbolae whose asymptotes coincide with the asymptotic directions.
- *Parabolic points*: At least one of the principal curvatures is null $\rightarrow$ one of the conical curves degenerates into couple of parallel straight lines, corresponding to the asymptotic direction, the other one is the null set.

The three possible cases are depicted in Fig. 7.9.

7.13 The Gauss–Weingarten equations

Let $\mathbf{f} : \Omega \rightarrow \Sigma$ be a surface; for any point $p \in \Sigma$, consider the basis $\{\mathbf{f}_{,u}, \mathbf{f}_{,v}, \mathbf{N}\}$, also called the *Gauss' basis*. It is the equivalent of the Frenet–Serret basis for the surfaces. We want to calculate the derivatives of the vectors of this basis, i.e. we want to obtain, for the surfaces, something equivalent to the Frenet–Serret equations.

Generally, $\mathbf{N} \in \mathcal{S}$ and $\mathbf{N}\cdot\mathbf{f}_{,u} = \mathbf{N}\cdot\mathbf{f}_{,v} = 0$, but $\mathbf{f}_{,u}, \mathbf{f}_{,v} \notin \mathcal{S}$ and $\mathbf{f}_{,u}\cdot\mathbf{f}_{,v} \neq 0$. In other words, we are dealing with a case of non-orthogonal (curvilinear) coordinates. So, if w is the coordinate along the normal $\mathbf{N}$, let us call, for the sake of convenience,

$$u = z^1, \ v = z^2$$

while, for the vectors,

$$\mathbf{f}_{,u} = \mathbf{f}_{,1} = \mathbf{g}_1, \ \mathbf{f}_{,v} = \mathbf{f}_{,2} = \mathbf{g}_2,$$

with $\mathbf{g}_1, \mathbf{g}_2$ exactly the $\mathbf{g}$ vectors of the coordinate lines on Σ. Then,

$$\begin{aligned} \frac{\partial \mathbf{g}_i}{\partial z^j} \cdot \mathbf{g}_i &= \frac{1}{2} \frac{\partial (\mathbf{g}_i \cdot \mathbf{g}_i)}{\partial z^j} = \frac{1}{2} \frac{\partial g_{ii}}{\partial z^j}, \\ \frac{\partial \mathbf{g}_i}{\partial z^i} \cdot \mathbf{g}_j &= \frac{\partial (\mathbf{g}_i \cdot \mathbf{g}_j)}{\partial z^i} - \frac{\partial \mathbf{g}_i}{\partial z^j} \cdot \mathbf{g}_i = \frac{\partial g_{ij}}{\partial z^i} - \frac{1}{2} \frac{\partial g_{ii}}{\partial z^j}, \end{aligned} \qquad i, j = 1, 2,$$

where for the last equation, we have used the identity

$$\frac{\partial \mathbf{g}_j}{\partial z^i} = \mathbf{f}_{,ji} = \mathbf{f}_{,ij} = \frac{\partial \mathbf{g}_i}{\partial z^j}, \quad i, j = 1, 2.$$

Using Eq. (6.26), it can be proved that it is also[10]

$$\frac{\partial \mathbf{g}_i}{\partial z^j} \cdot \mathbf{g}_h = \Gamma^h_{ij} \quad i, j, h = 1, 2.$$

Moreover, by Eq. (7.14),

$$\frac{\partial \mathbf{g}_i}{\partial z^j} \cdot \mathbf{N} = \mathbf{f}_{,ij} \cdot \mathbf{N} = B_{ij} \quad i, j = 1, 2,$$

and by Eqs. (7.10) and (7.11),

$$\frac{\partial \mathbf{N}}{\partial z^i} \cdot \mathbf{g}_j = -\mathcal{L}_W(\mathbf{g}_i) \cdot \mathbf{g}_j = -\mathbf{X}\mathbf{g}_i \cdot \mathbf{g}_j = -X_{ji}, \quad i, j = 1, 2,$$

while, because $\mathbf{N} \in \mathcal{S}$, then from Eq. (4.1),

$$\frac{\partial \mathbf{N}}{\partial z^i} \cdot \mathbf{N} = 0 \quad \forall i = 1, 2.$$

[10]The proof is rather cumbersome and it is omitted here; in many texts on differential geometry, the Christoffel symbols are just introduced in this way, as the projection of the derivatives of vectors $\mathbf{g}_i$s onto the same vectors, i.e. as the coefficients of the Gauss equations.

Finally, the decomposition of the derivatives of the vectors of the basis $\{\mathbf{f}_{,u}, \mathbf{f}_{,v}, \mathbf{N}\}$ onto these same vectors gives the equations

$$\begin{aligned} \frac{\partial \mathbf{g}_i}{\partial z^j} &= \Gamma^h_{ij}\mathbf{g}_h + B_{ij}\mathbf{N}, \\ \frac{\partial \mathbf{N}}{\partial z^j} &= -X_{ij}\mathbf{g}_i, \end{aligned} \qquad i,j = 1,2; \tag{7.25}$$

these are the *Gauss–Weingarten equations* (the first one is due to Gauss and the second to Weingarten).

Now, if we make the scalar product of the Gauss equations by $\mathbf{g}_1$ and $\mathbf{g}_2$, i.e.

$$\mathbf{g}_k \cdot \frac{\partial \mathbf{g}_i}{\partial z^j} = \mathbf{g}_k \cdot (\Gamma^h_{ij}\mathbf{g}_h + B_{ij}\mathbf{N}), \quad i,j,k = 1,2,$$

we get the following three systems of equations:

$$\begin{cases} \Gamma^1_{11}g_{11} + \Gamma^2_{11}g_{21} = \dfrac{1}{2}\dfrac{\partial g_{11}}{\partial z^1}, \\[2ex] \Gamma^1_{11}g_{12} + \Gamma^2_{11}g_{22} = \dfrac{\partial g_{12}}{\partial z^1} - \dfrac{1}{2}\dfrac{\partial g_{11}}{\partial z^2}; \end{cases} \tag{7.26}$$

$$\begin{cases} \Gamma^1_{12}g_{11} + \Gamma^2_{12}g_{21} = \dfrac{1}{2}\dfrac{\partial g_{11}}{\partial z^2}, \\[2ex] \Gamma^1_{12}g_{12} + \Gamma^2_{12}g_{22} = \dfrac{1}{2}\dfrac{\partial g_{22}}{\partial z^1}; \end{cases} \tag{7.27}$$

$$\begin{cases} \Gamma^1_{22}g_{11} + \Gamma^2_{22}g_{21} = \dfrac{\partial g_{12}}{\partial z^2} - \dfrac{1}{2}\dfrac{\partial g_{22}}{\partial z^1}, \\[2ex] \Gamma^1_{22}g_{12} + \Gamma^2_{22}g_{22} = \dfrac{1}{2}\dfrac{\partial g_{22}}{\partial z^2}. \end{cases} \tag{7.28}$$

The determinant of each one of these systems is simply $\det \mathbf{g} \neq 0 \rightarrow$ it is possible to express the Christoffel symbols as functions of the g_{ij}s and of their derivatives, i.e. as functions of the first fundamental form (the metric tensor).

7.14 The *theorema egregium*

The following theorem is a fundamental result due to Gauss:

Theorem 51 (Theorema egregium). *The Gaussian curvature K of a surface $\mathbf{f}(u,v) : \Omega \rightarrow \Sigma$ depends only upon the first fundamental form of* $\mathbf{f}$.

Proof. Let us write the identity

$$\frac{\partial^2 \mathbf{g}_1}{\partial z^1 \partial z^2} = \frac{\partial^2 \mathbf{g}_1}{\partial z^2 \partial z^1}$$

using the Gauss equations $(7.25)_1$:

$$\begin{aligned} &\Gamma^1_{11}\mathbf{g}_{1,2} + \Gamma^2_{11}\mathbf{g}_{2,2} + B_{11}\mathbf{N}_{,2} + \Gamma^1_{11,2}\mathbf{g}_1 + \Gamma^2_{11,2}\mathbf{g}_2 + B_{11,2}\mathbf{N} \\ &\quad = \Gamma^1_{12}\mathbf{g}_{1,1} + \Gamma^2_{12}\mathbf{g}_{1,2} + B_{12}\mathbf{N}_{,1} + \Gamma^1_{12,1}\mathbf{g}_1 + \Gamma^2_{12,1}\mathbf{g}_2 + B_{12,1}\mathbf{N}, \end{aligned}$$

where, for the sake of shortness, we have abridged $\dfrac{\partial(\cdot)}{\partial z^j}$ by $(\cdot)_{,j}$. Then, we use again Eqs. (7.25) to express $\mathbf{g}_{1,1}, \mathbf{g}_{1,2}, \mathbf{g}_{2,2}, \mathbf{N}_{,1}$ and $\mathbf{N}_{,2}$; after doing that and equating to 0, the coefficient of $\mathbf{g}_2$, we get

$$B_{11}X_{22} - B_{12}X_{21} = \Gamma^1_{11}\Gamma^2_{12} + \Gamma^2_{11}\Gamma^2_{22} + \Gamma^2_{11,2} - \Gamma^1_{12}\Gamma^2_{11} - \Gamma^2_{12}\Gamma^2_{12} - \Gamma^2_{12,1};$$

by Eq. (7.13), we get that

$$B_{11} = g_{11}X_{11} + g_{12}X_{21}, \quad B_{12} = g_{11}X_{12} + g_{12}X_{22},$$

which on injecting into the previous equation gives

$$g_{11} \det \mathbf{X} = \Gamma^1_{11}\Gamma^2_{12} + \Gamma^2_{11}\Gamma^2_{22} + \Gamma^2_{11,2} - \Gamma^1_{12}\Gamma^2_{11} - \Gamma^2_{12}\Gamma^2_{12} - \Gamma^2_{12,1}. \quad (7.29)$$

Setting equal to zero the coefficient of $\mathbf{g}_1$, a similar expression can also be obtained for g_{12}. Because $\mathbf{g}$ is positive definite, it is not possible that $g_{11} = g_{12} = 0$. So, remembering that $K = \det \mathbf{X}$ and the result of the previous section, we see that it is possible to express K through the coefficients of the first fundamental form and of its derivatives. □

7.15 Minimal surfaces

A *minimal surface* is a surface $\mathbf{f} : \Omega \rightarrow \Sigma$ having the mean curvature $H = 0 \, \forall p \in \Sigma$. Typical minimal surfaces are the catenoid and the helicoid.[11] Other minimal surfaces are the *Enneper's surface*:

$$\begin{cases} x_1 u - \dfrac{u^3}{3} + uv^2, \\ x_2 = v - \dfrac{v^3}{3} + u^2 v, \\ x_3 = u^2 - v^2, \end{cases}$$

and the *Costa's* and *Schwarz's* surfaces, see Fig. 7.10.

[11] Minimal surfaces have some interesting applications in the mechanics of tensile structures composed of prestressed membranes. Also, it can be shown that a soap film, when not bounding a closed region, takes the form of a minimal surface.

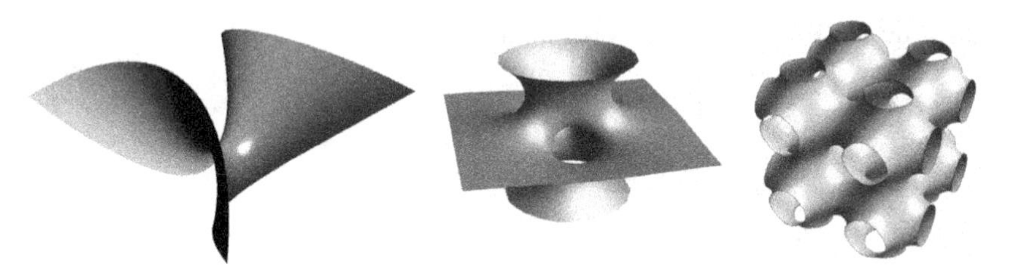

Figure 7.10: The minimal surfaces of (from left) Enneper, Costa, and Schwarz.

Theorem 52. *The non-planar points of a minimal surface are hyperbolic.*

Proof. This is a direct consequence of the definition of mean curvature H and of hyperbolic points: $H = 0 \iff \kappa_1\kappa_2 < 0$. □

Let $\mathbf{f} : \Omega \to \Sigma$ be a regular surface and Q a subset of Ω with its boundary ∂Q a closed regular curve in Ω; then, $R = \mathbf{f}(Q) \subset \Sigma$ is a *simple region* of Σ. Let $h : Q \to \mathbb{R}$ be a smooth function. Then, we call the *normal variation of* R the map $\varphi : Q \times (-\epsilon, \epsilon) \to \mathcal{E}$ defined by

$$\varphi(u, v, t) = \mathbf{f}(u, v) + t\ h(u, v)\mathbf{N}(u, v).$$

For each fixed t, $\varphi(u, v, t)$ is a surface with

$$\begin{aligned}\varphi_{,u}(u, v, t) &= \mathbf{f}_{,u}(u, v) + t\ h(u, v)\mathbf{N}_{,u}(u, v) + t\ h_{,u}(u, v)\mathbf{N}(u, v),\\ \varphi_{,v}(u, v, t) &= \mathbf{f}_{,v}(u, v) + t\ h(u, v)\mathbf{N}_{,v}(u, v) + t\ h_{,v}(u, v)\mathbf{N}(u, v).\end{aligned}$$

If the first fundamental form of $\mathbf{f}$ is represented by the metric tensor $\mathbf{g}$, we look for the metric tensor $\mathbf{g}^t$ representing the first fundamental form of $\varphi(u, v, t)$ $\forall t$:

$$\begin{aligned}g_{11}^t &= \varphi_{,u} \cdot \varphi_{,u} = g_{11} + 2t\ h\ \mathbf{f}_{,u} \cdot \mathbf{N}_{,u} + t^2(h^2\mathbf{N}_{,u}^2 + h_{,u}^2),\\ g_{12}^t &= \varphi_{,u} \cdot \varphi_{,v} = g_{12} + t\ h(\mathbf{f}_{,u} \cdot \mathbf{N}_{,v} + \mathbf{f}_{,v} \cdot \mathbf{N}_{,u}) + t^2(h^2\mathbf{N}_{,u} \cdot \mathbf{N}_{,v} + h_{,u}h_{,v}),\\ g_{22}^t &= \varphi_{,v} \cdot \varphi_{,v} = g_{22} + 2t\ h\ \mathbf{f}_{,v} \cdot \mathbf{N}_{,v} + t^2(h^2\mathbf{N}_{,v}^2 + h_{,v}^2),\end{aligned}$$

and by Eq. (7.14),

$$\begin{aligned}g_{11}^t &= g_{11} - 2t\ h\ B_{11} + t^2(h^2\mathbf{N}_{,u}^2 + h_{,u}^2),\\ g_{12}^t &= g_{12} - 2t\ h\ B_{12} + t^2(h^2\mathbf{N}_{,u} \cdot \mathbf{N}_{,v} + h_u h_{,v}),\\ g_{22}^t &= g_{22} - 2t\ h\ B_{22} + t^2(h^2\mathbf{N}_{,v}^2 + h_{,v}^2),\end{aligned}$$

whence

$$\det \mathbf{g}^t = \det \mathbf{g} - 2th(g_{11}B_{22} - 2g_{12}B_{12} + g_{22}B_{11}) + o(t^2).$$

Then, by Eq. (7.15), we get easily that

$$g_{11}B_{22} - 2g_{12}B_{12} + g_{22}B_{11} = 2H \det \mathbf{g}$$

so that

$$\det \mathbf{g}^t = \det \mathbf{g}(1 - 4thH) + o(t^2).$$

We can now calculate the area $\mathcal{A}(t)$ of the simple region $R^t = \varphi(u, v, t)$ corresponding to the subset Q:

$$\mathcal{A}^t = \int_Q \sqrt{\det \mathbf{g}(1 - 4thH) + o(t^2)}dudv.$$

For $\epsilon \ll 1$, $\mathcal{A}^t$ is differentiable, and its derivative at $t = 0$ is

$$\left[\frac{d\mathcal{A}^t}{dt}\right]_{t=0} = -\int_Q 2hH\sqrt{\det \mathbf{g}}dudv.$$

Theorem 53. *A surface* $\mathbf{f} : \Omega \to \Sigma$ *is minimal* $\iff \left[\frac{d\mathcal{A}^t}{dt}\right]_{t=0} = 0 \ \forall R \subset \Sigma$ *and for each normal variation.*

Proof. If $\mathbf{f}$ is minimal, the condition is clearly satisfied ($H = 0$). Conversely, let us suppose that $\exists p = \mathbf{f}(\overline{u}, \overline{v}) \in \Sigma | H(p) \neq 0$. Consider $r_1, r_2 \in \mathbb{R}$ such that $|H| \neq 0$ in the circle D_2 with center p and radius r_2 and $|H| > \frac{1}{2}|H(p)|$ in the circle D_1 with center p and radius r_1. Then, we chose a smooth function $h(u, v)$ such that (i) $h = H$ inside D_1, (ii) $hH > 0$ inside D_2, and (iii) $h = 0$ outside D_2. For the normal variation defined by such $h(u, v)$, we have

$$\begin{aligned}
-\left[\frac{d\mathcal{A}^t}{dt}\right]_{t=0} &= \int_{D_2} 2hH\sqrt{\det \mathbf{g}}du\ dv \geq \int_{D_1} 2H^2\sqrt{\det \mathbf{g}}du\ dv \\
&\geq \int_{D_1} \frac{H(p)^2}{2}\sqrt{\det \mathbf{g}}du\ dv = \frac{H(p)^2}{2}\mathcal{A}(\mathbf{f}(D_1)) \\
&\Rightarrow \left[\frac{d\mathcal{A}^t}{dt}\right]_{t=0} < 0,
\end{aligned}$$

which contradicts the hypothesis. □

The meaning of this theorem justifies the name of minimal surfaces: These are the surfaces that have the minimal area among all the surfaces that share the same boundary.

7.16 Geodesics

Let $\mathbf{f}:\Omega\to\Sigma$ be a surface and $\gamma(t):G\subset\mathbb{R}\to\Sigma$ a curve on Σ. A vector function $\mathbf{w}(t):G\to T_{\gamma(t)}\Sigma$ is called a *vector field*[12] *along* $\gamma(t)$. We call the *covariant derivative of* $\mathbf{w}(t)$ *along* $\gamma(t)$ the vector field $D_\gamma\mathbf{w}(t):G\to\mathcal{V}$ defined as[13]

$$D_\gamma\mathbf{w}:=(\mathbf{I}-\mathbf{N}\otimes\mathbf{N})\frac{d\mathbf{w}}{dt},$$

i.e. the projection of the derivative of $\mathbf{w}$ onto $T_{\gamma(t)}\Sigma$. It is always possible to decompose $\mathbf{w}(t)$ into its components in the natural basis $\{\mathbf{f}_{,u},\mathbf{f}_{,v}\}$:

$$\mathbf{w}(t)=w_1(t)\mathbf{f}_{,u}(\gamma(t))+w_2(t)\mathbf{f}_{,v}(\gamma(t)).$$

Differentiating, we get (a prime here denotes the derivative with respect to t)

$$\mathbf{w}'=w_1'\mathbf{f}_{,u}+w_1(\mathbf{f}_{,uu}u'+\mathbf{f}_{,uv}v')+w_2'\mathbf{f}_{,v}+w_2(\mathbf{f}_{,uv}u'+\mathbf{f}_{,vv}v'),$$

and using the Gauss equations, Eq. $(7.25)_1$, we obtain (summation of the dummy indexes, where u_1 stands for u and u_2 for v)

$$\mathbf{w}'=\mathbf{f}_{,k}w_k'+(\Gamma_{ij}^k\mathbf{f}_{,k}+B_{ij}\mathbf{N})w_iu_j',\quad i,j,k=1,2$$

so that the projection onto $T_{\gamma(t)}\Sigma$, i.e. $D_\gamma\mathbf{w}(t)$, is

$$D_\gamma\mathbf{w}=(w_k'+\Gamma_{ij}^k w_iu_j')\mathbf{f}_{,k}. \tag{7.30}$$

A *parallel vector field* $\mathbf{w}$ *along* γ is a vector field having $\mathbf{D}_\gamma\mathbf{w}=\mathbf{o}\ \forall t$. A regular curve γ is a *geodesic of* Σ if the vector field γ' of the vectors tangent to γ is parallel along γ.

Theorem 54. *A curve γ is a geodesic of $\Sigma \iff \boldsymbol{\nu}\times\mathbf{N}=\mathbf{o}$.*

Proof. If γ is a geodesic, then the derivative of its tangent γ' has a component only along $\mathbf{N}$, i.e. $\gamma''\times\mathbf{N}=\mathbf{o}\Rightarrow\gamma'\cdot\gamma''=0$. The principal normal to γ, $\boldsymbol{\nu}$, is orthogonal to $\gamma'\Rightarrow\boldsymbol{\nu}\times\mathbf{N}=\mathbf{o}$. Vice versa, if $\boldsymbol{\nu}\times\mathbf{N}=\mathbf{o}$, then γ'' is orthogonal to $\gamma'\Rightarrow\mathbf{D}_\gamma\gamma'=\mathbf{o}$, i.e. γ is a geodesic. □

Theorem 55. *If γ is a geodesic, then $|\gamma'|=const.$*

Proof. In a geodesic, $\gamma'\cdot\gamma''=0\Rightarrow\dfrac{d(\gamma'\cdot\gamma')}{dt}=0\Rightarrow|\gamma'|=const.$ □

---

[12] More precisely, $\mathbf{w}(t)$ is a curve of vectors; however, it is normally called a vector field along a curve.

[13] The operator that gives the projection of $\mathbf{w}$ onto a vector orthogonal to $\mathbf{N}\in\mathcal{S}$, i.e. onto $T_{\gamma(t)}\Sigma$, is $\mathbf{I}-\mathbf{N}\otimes\mathbf{N}$, cf. Exercise 2, Chapter 2.

This result shows that in a geodesic, the parameter is always the natural parameter s. Let $\gamma(s)$ be a curve on Σ parameterized by the arc length s. We call the *geodesic curvature of* $\gamma(s)$ the function

$$\kappa_g := D_\gamma \boldsymbol{\tau} \cdot (\mathbf{N} \times \boldsymbol{\tau}),$$

where $\boldsymbol{\tau} = \gamma' \in \mathcal{S}$ is the tangent vector to γ. Because $\mathbf{N} \times \boldsymbol{\tau} \in \mathcal{S}$ lies in $T_\gamma\Sigma$, the component of $\boldsymbol{\tau}'$ orthogonal to $T_\gamma\Sigma$ gives a null contribution to κ_g, so we can also write

$$\kappa_g = \boldsymbol{\tau}' \cdot (\mathbf{N} \times \boldsymbol{\tau}).$$

Theorem 56. *A regular curve* $\gamma(s)$ *is a geodesic* $\iff \kappa_g = 0\ \forall s$.

Proof. If γ is a geodesic, clearly $\kappa_g = 0$. Vice versa, if $\kappa_g = 0$, then $\boldsymbol{\tau}, \boldsymbol{\tau}'$ and $\mathbf{N}$ are linearly dependent, i.e. coplanar. Because $\boldsymbol{\tau}' \cdot \boldsymbol{\tau} = \mathbf{N} \cdot \boldsymbol{\tau} = 0 \Rightarrow \boldsymbol{\tau}' \times \mathbf{N} = \mathbf{o} \Rightarrow$ by Theorem 54, γ is a geodesic. □

Let us now write Eq. (7.30) in the particular case of $\mathbf{w} = \gamma'$, i.e. $w_1 = u', w_2 = v'$:

$$D_\gamma \mathbf{w} = (u_k'' + \Gamma_{ij}^k u_i' u_j')\mathbf{f}_{,k};$$

therefore, the geodesics are the solutions to the system of differential equations

$$\begin{cases} u'' + \Gamma_{11}^1 u'^2 + 2\Gamma_{12}^1 u'v' + \Gamma_{22}^1 v'^2 = 0, \\ v'' + \Gamma_{11}^2 u'^2 + 2\Gamma_{12}^2 u'v' + \Gamma_{22}^2 v'^2 = 0. \end{cases} \tag{7.31}$$

It can be shown that $\forall p \in \Sigma$ and $\forall \mathbf{w}(p) \in T_p\Sigma$, the geodesic is unique.

Let p be a point of a regular surface $\mathbf{f} : \Omega \to \Sigma$ and $\boldsymbol{\alpha}(v) : G \subset \mathbb{R} \to \Sigma$ a smooth regular curve on Σ, with v being the natural parameter and such that $p = \boldsymbol{\alpha}(0)$. Consider the geodesic γ_v passing through $q = \boldsymbol{\alpha}(v)$ and such that $\gamma_v'(0) = \mathbf{N}(\boldsymbol{\alpha}(v)) \times \boldsymbol{\tau}(v)$, with $\boldsymbol{\tau}(v)$ the (unit) tangent vector to $\boldsymbol{\alpha}(v)$. Consider the map $\mathbf{f}(u,v) : \Omega \to \Sigma$ defined by posing $\mathbf{f}(u,v) = \gamma_v(u)$; this is a surface whose coordinates (u,v) are called *semigeodesic coordinates.*

Let us see the form that the first fundamental form (i.e. the metric tensor $\mathbf{g}$), the Christoffel symbols, and the Gaussian curvature take in semigeodesic coordinates. Curves $\mathbf{f}(u, v_0) = \gamma_{v_0}(u)$ are geodesics, and u is hence their natural parameter. Therefore, $\mathbf{f}_{,u} \in \mathcal{S} \Rightarrow g_{11} = 1$. Then, $\mathbf{f}_{uu}(u, v_0)$ is the derivative of the tangent vector to a geodesic $\mathbf{f}(u, v_0) = \gamma_{v_0}(u) \Rightarrow \mathbf{f}_{uu}(u, v_0)$ does not have a component along the tangent, hence Eq. $\Rightarrow \Gamma_{11}^1 = \Gamma_{11}^2 = 0$. Then, by Eq. $(7.26)_1$, we get $g_{12,u} = 0 \Rightarrow g_{12}$ does not depend upon

$u \Rightarrow g_{12}(u,v) = g_{12}(0,v)\ \forall u$. Moreover, let θ be the angle between the curve $\boldsymbol{\alpha}$, i.e. between the coordinate line $\mathbf{f}(0,v)$, whose tangent vector is $\mathbf{f}_{,v}(0,v)$, and the geodesic $\gamma_v(u)$, whose tangent vector at $(0,v)$ is $\gamma_v'(u)$. Then, $\theta = \dfrac{\pi}{2}$ because $\gamma_v'(0) = \mathbf{N}(\boldsymbol{\alpha}(v)) \times \boldsymbol{\tau}(v)$. As a consequence, $g_{12}(0,v) = 0 \Rightarrow g_{12}(u,v) = 0\ \forall (u,v) \in \Omega$. Finally, setting $\mathbf{g}_{22} = g$,

$$\mathbf{g} = \begin{bmatrix} 1 & 0 \\ 0 & g \end{bmatrix},$$

with $g > 0$ because $\mathbf{g}$ is positive definite. Through systems (7.26)–(7.28), we obtain

$$\Gamma^1_{12} = 0, \quad \Gamma^2_{12} = \frac{g_{,u}}{2g}, \quad \Gamma^1_{22} = -\frac{g_{,u}}{2}, \quad \Gamma^2_{22} = \frac{g_{,v}}{2g},$$

and using Eq. (7.29), we obtain

$$K = \det \mathbf{X} = -\frac{g_{,uu}}{2g} + \frac{g^2_{,u}}{4g^2}.$$

Given two points $p_1, p_2 \in \Sigma$, we define the *distance* $d(p_1, p_2)$ as the infimum of the lengths of the curves on Σ joining the two points. We end with an important characterization of geodesics.

Theorem 57. *Geodesics are the curves of minimal distance between two points of a surface.*

Proof. Let $\gamma : G \subset \mathbb{R} \to \Sigma$ be a geodesic on Σ, parameterized with the arc length, and $\boldsymbol{\alpha}$ a smooth regular curve through p and orthogonal to γ. Through $\boldsymbol{\alpha}$, we set up a system of semigeodesic coordinates in a neighborhood U of p. With an opportune parameterization $\boldsymbol{\alpha}(t)$ in such coordinates, we can get $p = \mathbf{f}(0,0)$ and γ described by the equation $v = 0$. Let $q \in U$ be a point in γ, and consider a regular curve connecting p with q. The length $\ell(p,q)$ of such a curve is

$$\ell(p,q) = \int_p^q \sqrt{u'^2 + g\, v'^2}dt \geq \left| \int_p^q u'dt \right| = |u_q - u_p|.$$

Observing that $p = (u_p, 0), q = (u_q, 0)$, we remark that $|u_q - u_p|$ is exactly the length of γ between p and q because γ is parameterized with its arc length. □

There is another, direct and beautiful, way to show that geodesics are the shortest path lines: the use of the method of calculus of variations.[14] The length $\ell(p,q)$ of a curve $\gamma(t)\in\Sigma$ between two points p and q is given by the functional (7.9); it depends upon the first fundamental form, i.e. upon the metric tensor $\mathbf{g}$ on σ. For the sake of conciseness, let $\mathbf{w}=(w^1_{,t},w^2_{,t})$ be the tangent vector to the curve $\gamma(w_1,w_2)\in\Sigma$. Then,

$$\ell(p,q)=\int_p^q\sqrt{I(\mathbf{w})}dt=\int_p^q\sqrt{\mathbf{w}\cdot\mathbf{g}\mathbf{w}}dt.$$

The curve $\gamma(t)$ that minimizes $\ell(p,q)$ is the solution to the *Euler–Lagrange equations*

$$\frac{d}{dt}\frac{\partial F}{\partial \mathbf{w}_{,t}}-\frac{\partial F}{\partial \mathbf{w}}=\mathbf{o}\ \rightarrow\ \frac{d}{dt}\frac{\partial F}{\partial w^k_{,t}}-\frac{\partial F}{\partial w^k}=0,\quad k=1,2,$$

where

$$F(\mathbf{w},\mathbf{w}_{,t},t)=\sqrt{\mathbf{w}\cdot\mathbf{g}\mathbf{w}}=\sqrt{g_{ij}w^i_{,t}w^j_{,t}}.$$

It is more direct, and equivalent, to minimize $J^2(t)$, i.e. to write the Euler–Lagrange equations for

$$\Phi(\mathbf{w},\mathbf{w}_{,t},t):=F^2(\mathbf{w},\mathbf{w}_{,t},t)=g_{ij}w^i_{,t}w^j_{,t}.$$

Therefore,

$$\frac{\partial\Phi}{\partial w^k_{,t}}=2g_{jk}w^j_{,t},$$

$$\frac{\partial\Phi}{\partial w^k}=\frac{\partial g_{hj}}{\partial w^k}w^h_{,t}w^j_{,t},\qquad j,h,k=1,2.$$

$$\frac{d}{dt}\frac{\partial\Phi}{\partial w^k_{,t}}=2\left(g_{jk}w^j_{,tt}+\frac{dg_{jk}}{dt}w^j_{,t}\right)=2\left(g_{jk}w^j_{,tt}+\frac{\partial g_{jk}}{\partial w^l}w^l_{,t}w^j_{,t}\right),$$

[14]The reader is referred to texts on the calculus of variations for an insight into this matter, cf. the suggested texts. Here, we just recall the fundamental fact to be used in the proof concerning geodesics: Let

$$J(t)=\int_a^b F(\mathbf{x},\mathbf{x}',t)dt$$

be a functional to be minimized by a proper choice of the function $\mathbf{x}(t)$ (in the case of the geodesics, $J=\ell(p,q)$); then, such a minimizing function can be found as a solution to the *Euler–Lagrange equations*

$$\frac{d}{dt}\frac{\partial F}{\partial \mathbf{x}'}-\frac{\partial F}{\partial \mathbf{x}}=\mathbf{o}.$$

The Euler–Lagrange equations are hence

$$g_{jk}w^j_{,tt} + \frac{\partial g_{jk}}{\partial w^h} w^h_{,t} w^j_{,t} - \frac{1}{2}\frac{\partial g_{hj}}{\partial w^k} w^h_{,t} w^j_{,t} = 0, \quad j,h,k = 1,2,$$

which can be rewritten as

$$g_{jk}w^j_{,tt} + \frac{1}{2}\left(\frac{\partial g_{jk}}{\partial w^h} + \frac{\partial g_{hk}}{\partial w^j} - \frac{\partial g_{hj}}{\partial w^k}\right) w^h_{,t} w^j_{,t} = 0, \quad j,h,k = 1,2.$$

Multiplying by g^{lk}, we get

$$g^{lk}g_{jk}w^j_{,tt} + \frac{1}{2}g^{lk}\left(\frac{\partial g_{jk}}{\partial w^h} + \frac{\partial g_{hk}}{\partial w^j} - \frac{\partial g_{hj}}{\partial w^k}\right) w^h_{,t} w^j_{,t} = 0, \quad j,h,k,l = 1,2,$$

and finally, because

$$g^{lk}g_{jk} = \delta_{lj}$$

and by Eq. (6.26), we get

$$w^l_{,tt} + \Gamma^l_{jh} w^j_{,t} w^h_{,t}, \quad j,h,l = 1,2.$$

These are the differential equations whose solution is the curve of minimal length between two points of Σ; comparing these equations with those of a geodesic of Σ, Eq. (7.31), we see that they are the same: The geodesics of a surface are hence the curves of minimal distance on the surface.

The Christoffel symbols of a plane are all null; as a consequence, the geodesic lines of a plane are straight lines. In fact, only such lines have a constant derivative.

Through systems (7.26)—(7.28), we can calculate the Christoffel symbols for a revolution surface, Eq. (7.5), which are all null excepted

$$\Gamma^2_{12} = \frac{\varphi'}{\varphi}, \quad \Gamma^1_{22} = -\varphi\,\varphi',$$

so the system of differential equations (7.31) becomes

$$\begin{cases} u'' - \varphi\,\varphi' v'^2 = 0, \\ v'' + 2\dfrac{\varphi'}{\varphi}u'v' = 0. \end{cases} \tag{7.32}$$

It is direct to check that the meridians $(u = t, v = v_0)$ are geodesic lines, while the parallels $(u = u_0, v = t)$ are geodesics $\iff \varphi'(u_0) = 0$.

7.17 The Gauss–Codazzi compatibility conditions

Let us consider a surface Σ whose points are determined by the vector function $\mathbf{r} : \Omega \subset \mathbb{R}^2 \to \Sigma \subset \mathcal{E}, \mathbf{r}(\alpha_1, \alpha_2) = x_i(\alpha_1, \alpha_2)\boldsymbol{\epsilon}_i$, with $\boldsymbol{\epsilon}_i, i = 1, 2, 3$, the vectors of the orthonormal basis of the reference frame $\mathcal{R} = \{o; \boldsymbol{\epsilon}_1, \boldsymbol{\epsilon}_2, \boldsymbol{\epsilon}_3\}$ and the parameters α_1, α_2 chosen in such a way that the lines $\alpha_1 = const., \alpha_2 = const.$ are the lines of curvature, i.e. the tangent at each point to the principal directions of curvature and hence mutually orthogonal.[15] With such a choice, cf. Eq. (7.8),

$$ds^2 = A_1^2 d\alpha_1^2 + A_2^2 d\alpha_2^2,$$

with

$$A_1 = \sqrt{\mathbf{r}_{,\alpha_1}^2} = \sqrt{\frac{dx_i}{d\alpha_1}\frac{dx_i}{d\alpha_1}},$$
$$A_2 = \sqrt{\mathbf{r}_{,\alpha_2}^2} = \sqrt{\frac{dx_i}{d\alpha_2}\frac{dx_i}{d\alpha_2}},$$

which are the so-called *Lamé's parameters*. We remark that along the lines of curvatures, i.e. the lines $\alpha_i = const., i = 1, 2$, which in short from now on, we call the *lines* α_i, it is

$$ds_1 = A_1 d\alpha_1,$$
$$ds_2 = A_2 d\alpha_2,$$

and hence,

$$\begin{aligned} \boldsymbol{\lambda}_1 &= \frac{ds_1}{d\alpha_1} = A_1\mathbf{e}_1, \\ \boldsymbol{\lambda}_2 &= \frac{ds_2}{d\alpha_2} = A_2\mathbf{e}_2 \end{aligned} \tag{7.33}$$

are the vectors tangent to the lines of curvature. Let

$$\mathbf{e}_1 = \frac{1}{A_1}\mathbf{r}_{,\alpha_1}, \quad \mathbf{e}_2 = \frac{1}{A_2}\mathbf{r}_{,\alpha_2}, \quad \mathbf{e}_3 = \mathbf{e}_1 \times \mathbf{e}_2 (= \mathbf{N}); \tag{7.34}$$

these three vectors form the orthonormal (local) natural basis $\mathsf{e} = \{\mathbf{e}_1, \mathbf{e}_2, \mathbf{e}_3\}$. We always make the choice of α_1, α_2 such that $\mathbf{e}_3$ is always directed toward the convex side of Σ if the point is elliptic or parabolic or toward the side of the centers of negative curvature if the point is hyperbolic.

[15]Here, the symbol $\mathbf{r}$ is preferred to $\mathbf{f}$, like α_1 to u and α_2 to v, to recall that we have made the particular choice of coordinate lines that are lines of curvature. All the developments could be done in a more general case, but this choice is made to obtain simpler relations, which preserves anyway the generality.

We consider a vector $\mathbf{v} = \mathbf{v}(p),\ p \in \Sigma$,

$$\mathbf{v} = v_1\mathbf{e}_1 + v_2\mathbf{e}_2 + v_3\mathbf{e}_3,$$

and we want to calculate how it transforms when p changes. To this end, we need to calculate how $\mathbf{e}_1, \mathbf{e}_2, \mathbf{e}_3$ change with α_1, α_2. Let $q \in \Sigma$ be a point in the neighborhood of p on the line α_i, and let us first consider the change in $\mathbf{e}_3$ in passing from p to q. Because p and q belong to the same line α_i, by the theorem of Rodrigues, we get (no summation on i in the following equations)

$$\frac{\partial \mathbf{e}_3}{\partial \boldsymbol{\lambda}_i} = -\kappa_i \boldsymbol{\lambda}_i,\ i = 1, 2,$$

i.e. by Eq. (7.33),

$$\frac{\partial \mathbf{e}_3}{\partial \alpha_i} = \frac{A_i}{R_i}\mathbf{e}_i,$$

with

$$R_i = -\frac{1}{\kappa_i}$$

the (principal) radius of curvature along the line α_i. The minus sign in the previous equation is due to the choice made above for orienting $\mathbf{e}_3 = \mathbf{N}$, which gives always $\mathbf{N} = -\boldsymbol{\nu}$, with $\boldsymbol{\nu}$ the principal normal to the line α_i. This result can also be obtained directly, see Fig. 7.11:

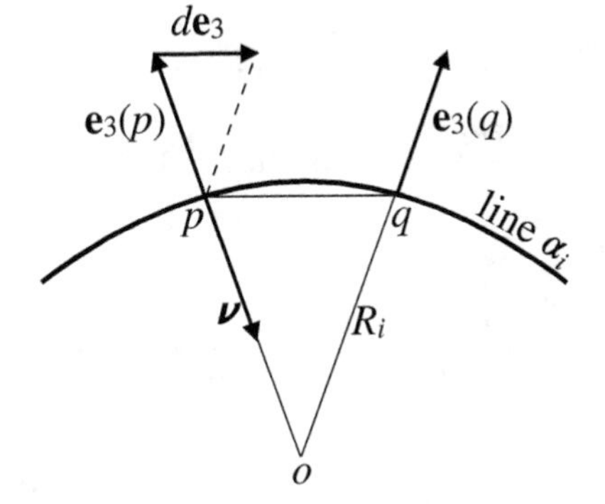

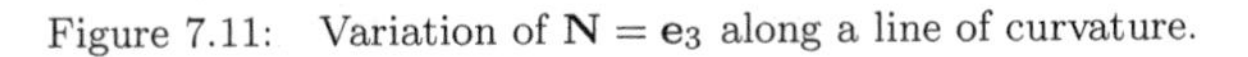
Figure 7.11: Variation of $\mathbf{N} = \mathbf{e}_3$ along a line of curvature.

$$\mathbf{e}_3(q) = \mathbf{e}_3(p) + d\mathbf{e}_3$$

and in the limit of $q \to p$, $d\mathbf{e}_3$ tends to be parallel to $q - p$ and

$$\lim_{q \to p}(q - p) = \boldsymbol{\lambda}_i = A_i\mathbf{e}_i.$$

By the similitude of the triangles, it is evident that

$$\frac{|d\mathbf{e}_3|}{|\mathbf{e}_3|} = \frac{|q - p|}{R_i};$$

moreover,

$$d\mathbf{e}_3 = \frac{\partial \mathbf{e}_3}{\partial \alpha_i} d\alpha_i \mathbf{e}_i.$$

Finally, as $|\mathbf{e}_3| = 1$, we get again

$$\frac{\partial \mathbf{e}_3}{\partial \alpha_i} = \frac{A_i}{R_i}\mathbf{e}_i. \tag{7.35}$$

Implicitly, in this last proof, we have used the theorem of Rodrigues because we have assumed that $d\mathbf{e}_3$ is parallel to $\boldsymbol{\lambda}_i$, as it is, because line α_i is a line of curvature.

We now move on to determine the changes in $\mathbf{e}_1$ and $\mathbf{e}_2$; for this purpose, we remark that

$$\frac{\partial \mathbf{r}_{,\alpha_1}}{\partial \alpha_2} = \frac{\partial^2 \mathbf{r}}{\partial \alpha_2 \partial \alpha_1} = \frac{\partial^2 \mathbf{r}}{\partial \alpha_1 \partial \alpha_2} = \frac{\partial \mathbf{r}_{,\alpha_2}}{\partial \alpha_1},$$

so by Eq. (7.34), we get

$$\frac{\partial (A_1 \mathbf{e}_1)}{\partial \alpha_2} = \frac{\partial (A_2 \mathbf{e}_2)}{\partial \alpha_1}. \tag{7.36}$$

Let us study now $\frac{\partial \mathbf{e}_j}{\partial \alpha_i}$; as $|\mathbf{e}_j| = 1, j = 1, 2,$

$$\frac{\partial \mathbf{e}_j}{\partial \alpha_i} \cdot \mathbf{e}_j = 0 \quad \forall i, j = 1, 2. \tag{7.37}$$

Because $\mathbf{e}_1 \cdot \mathbf{e}_2 = 0$,

$$\frac{\partial \mathbf{e}_1}{\partial \alpha_1} \cdot \mathbf{e}_2 = \frac{\partial (\mathbf{e}_1 \cdot \mathbf{e}_2)}{\partial \alpha_1} - \mathbf{e}_1 \cdot \frac{\partial \mathbf{e}_2}{\partial \alpha_1} = -\mathbf{e}_1 \cdot \frac{\partial \mathbf{e}_2}{\partial \alpha_1}.$$

By Eq. (7.36), we get

$$\frac{\partial \mathbf{e}_2}{\partial \alpha_1} = \frac{1}{A_2} \frac{\partial (A_1 \mathbf{e}_1)}{\partial \alpha_2} - \frac{1}{A_2} \frac{\partial A_2}{\partial \alpha_1} \mathbf{e}_2,$$

which when inserted into the previous equation gives, by Eq. (7.37),

$$\begin{aligned}\frac{\partial \mathbf{e}_1}{\partial \alpha_1}\cdot\mathbf{e}_2 &= -\frac{1}{A_2}\frac{\partial(A_1\mathbf{e}_1)}{\partial\alpha_2}\cdot\mathbf{e}_1+\frac{1}{A_2}\frac{\partial A_2}{\partial\alpha_1}\mathbf{e}_2\cdot\mathbf{e}_1\\ &= -\frac{A_1}{A_2}\frac{\partial\mathbf{e}_1}{\partial\alpha_2}\cdot\mathbf{e}_1-\frac{1}{A_2}\frac{\partial A_1}{\partial\alpha_2}\mathbf{e}_1\cdot\mathbf{e}_1=-\frac{1}{A_2}\frac{\partial A_1}{\partial\alpha_2}.\end{aligned}$$

Then, because $\mathbf{e}_1\cdot\mathbf{e}_3=0$,

$$\frac{\partial\mathbf{e}_1}{\partial\alpha_1}\cdot\mathbf{e}_3=\frac{\partial(\mathbf{e}_1\cdot\mathbf{e}_3)}{\partial\alpha_1}-\mathbf{e}_1\cdot\frac{\partial\mathbf{e}_3}{\partial\alpha_1}=-\mathbf{e}_1\cdot\frac{\partial\mathbf{e}_3}{\partial\alpha_1},$$

and by Eq. (7.35),

$$\frac{\partial\mathbf{e}_3}{\partial\alpha_1}=\frac{A_1}{R_1}\mathbf{e}_1,$$

so finally,

$$\frac{\partial\mathbf{e}_1}{\partial\alpha_1}\cdot\mathbf{e}_3=-\frac{A_1}{R_1}.$$

Again, through Eqs. (7.36) and (7.37), we get

$$\begin{aligned}\frac{\partial\mathbf{e}_1}{\partial\alpha_2}\cdot\mathbf{e}_2 &= \frac{1}{A_1}\frac{\partial(A_2\mathbf{e}_2)}{\partial\alpha_1}\cdot\mathbf{e}_2-\frac{1}{A_1}\frac{\partial A_1}{\partial\alpha_2}\mathbf{e}_1\cdot\mathbf{e}_2=\frac{A_2}{A_1}\frac{\partial\mathbf{e}_2}{\partial\alpha_1}\cdot\mathbf{e}_2\\ &\quad+\frac{1}{A_1}\frac{\partial A_2}{\partial\alpha_1}\mathbf{e}_2\cdot\mathbf{e}_2=\frac{1}{A_1}\frac{\partial A_2}{\partial\alpha_1}\end{aligned}$$

and also, by Eq. (7.35),

$$\frac{\partial\mathbf{e}_1}{\partial\alpha_2}\cdot\mathbf{e}_3=\frac{\partial(\mathbf{e}_1\cdot\mathbf{e}_3)}{\partial\alpha_2}-\mathbf{e}_1\cdot\frac{\partial\mathbf{e}_3}{\partial\alpha_2}=-\mathbf{e}_1\cdot\frac{\partial\mathbf{e}_3}{\partial\alpha_2}=-\frac{A_2}{R_2}\mathbf{e}_1\cdot\mathbf{e}_2=0.$$

The derivatives of $\mathbf{e}_2$ can be found in a similar way, and resuming, we have

$$\begin{aligned}\frac{\partial\mathbf{e}_1}{\partial\alpha_1}&=-\frac{1}{A_2}\frac{\partial A_1}{\partial\alpha_2}\mathbf{e}_2-\frac{A_1}{R_1}\mathbf{e}_3,\\ \frac{\partial\mathbf{e}_1}{\partial\alpha_2}&=\frac{1}{A_1}\frac{\partial A_2}{\partial\alpha_1}\mathbf{e}_2,\\ \frac{\partial\mathbf{e}_2}{\partial\alpha_1}&=\frac{1}{A_2}\frac{\partial A_1}{\partial\alpha_2}\mathbf{e}_1,\\ \frac{\partial\mathbf{e}_2}{\partial\alpha_2}&=-\frac{1}{A_1}\frac{\partial A_2}{\partial\alpha_1}\mathbf{e}_1-\frac{A_2}{R_2}\mathbf{e}_3,\\ \frac{\partial\mathbf{e}_3}{\partial\alpha_1}&=\frac{A_1}{R_1}\mathbf{e}_1,\\ \frac{\partial\mathbf{e}_3}{\partial\alpha_2}&=\frac{A_2}{R_2}\mathbf{e}_2.\end{aligned}\tag{7.38}$$

Passing now to the second-order derivatives, imposing the equality of mixed derivatives, gives some important differential relations between the Lamé's parameters A_i and the radii of curvatures R_i. In fact, from the identity

$$\frac{\partial^2 \mathbf{e}_3}{\partial \alpha_1 \partial \alpha_2} = \frac{\partial^2 \mathbf{e}_3}{\partial \alpha_2 \partial \alpha_1}$$

and Eqs. $(7.38)_{5,6}$, we get

$$\frac{\partial}{\partial \alpha_2}\left(\frac{A_1}{R_1}\mathbf{e}_1\right) = \frac{\partial}{\partial \alpha_1}\left(\frac{A_2}{R_2}\mathbf{e}_2\right),$$

whence

$$\frac{\partial}{\partial \alpha_2}\left(\frac{A_1}{R_1}\right)\mathbf{e}_1 + \frac{A_1}{R_1}\frac{\partial \mathbf{e}_1}{\partial \alpha_2} = \frac{\partial}{\partial \alpha_1}\left(\frac{A_2}{R_2}\right)\mathbf{e}_2 + \frac{A_2}{R_2}\frac{\partial \mathbf{e}_2}{\partial \alpha_1}.$$

Inserting now Eqs. $(7.38)_{2,3}$ into the last result and rearranging the terms gives

$$\left[\frac{\partial}{\partial \alpha_2}\left(\frac{A_1}{R_1}\right) - \frac{1}{R_2}\frac{\partial A_1}{\partial \alpha_2}\right]\mathbf{e}_1 - \left[\frac{\partial}{\partial \alpha_1}\left(\frac{A_2}{R_2}\right) - \frac{1}{R_1}\frac{\partial A_2}{\partial \alpha_1}\right]\mathbf{e}_2 = 0,$$

which to be true needs that the two following conditions be identically satisfied:

$$\begin{aligned}\frac{\partial}{\partial \alpha_2}\left(\frac{A_1}{R_1}\right) - \frac{1}{R_2}\frac{\partial A_1}{\partial \alpha_2} &= 0,\\ \frac{\partial}{\partial \alpha_1}\left(\frac{A_2}{R_2}\right) - \frac{1}{R_1}\frac{\partial A_2}{\partial \alpha_1} &= 0.\end{aligned} \tag{7.39}$$

The above equations are known as the *Codazzi conditions*. Let us now consider the other identity

$$\frac{\partial^2 \mathbf{e}_1}{\partial \alpha_1 \partial \alpha_2} = \frac{\partial^2 \mathbf{e}_1}{\partial \alpha_2 \partial \alpha_1};$$

again using Eq. (7.38), with some standard operations, this identity can be transformed to

$$\begin{aligned}&\left[\frac{\partial}{\partial \alpha_1}\left(\frac{1}{A_1}\frac{\partial A_2}{\partial \alpha_1}\right) + \frac{\partial}{\partial \alpha_2}\left(\frac{1}{A_2}\frac{\partial A_1}{\partial \alpha_2}\right) + \frac{A_1}{R_1}\frac{A_2}{R_2}\right]\mathbf{e}_2\\ &\quad + \left[\frac{\partial}{\partial \alpha_2}\left(\frac{A_1}{R_1}\right) - \frac{1}{R_2}\frac{\partial A_1}{\partial \alpha_2}\right]\mathbf{e}_3 = 0.\end{aligned}$$

Again, for this equation to be identically satisfied, each of the expressions in square brackets must vanish, which gives two more differential conditions, but of which only the first one is new, as the second one corresponds to Eq. $(7.39)_1$. The new condition is hence

$$\frac{\partial}{\partial\alpha_1}\left(\frac{1}{A_1}\frac{\partial A_2}{\partial\alpha_1}\right)+\frac{\partial}{\partial\alpha_2}\left(\frac{1}{A_2}\frac{\partial A_1}{\partial\alpha_2}\right)+\frac{A_1}{R_1}\frac{A_2}{R_2}=0, \tag{7.40}$$

which is known as the *Gauss condition*. The last identity

$$\frac{\partial^2\mathbf{e}_2}{\partial\alpha_1\partial\alpha_2}=\frac{\partial^2\mathbf{e}_2}{\partial\alpha_2\partial\alpha_1}$$

does not add any independent condition, which can be easily checked. The meaning of the *Gauss–Codazzi conditions*, Eqs. (7.39) and (7.40), is that of *compatibility conditions*: Only when these conditions are satisfied by functions A_1, A_2, R_1, and R_2, then such functions represent the Lamé's parameters and the principal radii of curvature of a surface, i.e. only in this case, they define a surface, except for its position in space. The Gauss–Codazzi conditions are important in establishing the equations of the classical theory of shells.

7.18 Exercises

1. Prove that a function of the type $x_3 = f(x_1, x_2)$, with $f : \Omega \subset \mathbb{R}^2 \to \mathbb{R}$ smooth, defines a surface.
2. Show that the catenoid is the rotation surface of a catenary, then find its Gaussian curvature.
3. Show that the pseudo-sphere is the rotation surface of a tractrice and explain why the surface has this name (hint: look for its Gaussian curvature).
4. Prove that the regularity of a cone $\mathbf{f}(u, v) = v\gamma(u)$ is satisfied at each point except at the apex and at the points on the straight lines tangent to $\gamma(u)$.
5. Prove that the hyperbolic hyperboloid is a doubly ruled surface, and determine the angle θ formed by two straight lines belonging to the two sets of lines on the surface, see the left panel of Fig. 7.12.
6. Prove that the hyperbolic paraboloid whose Cartesian equation is $x_3 = x_1x_2$, right panel of Fig. 7.12, is a doubly ruled surface, and determine the angle θ formed by two straight lines belonging to the two sets of lines on the surface. Where does $\theta = \dfrac{\pi}{2}$?

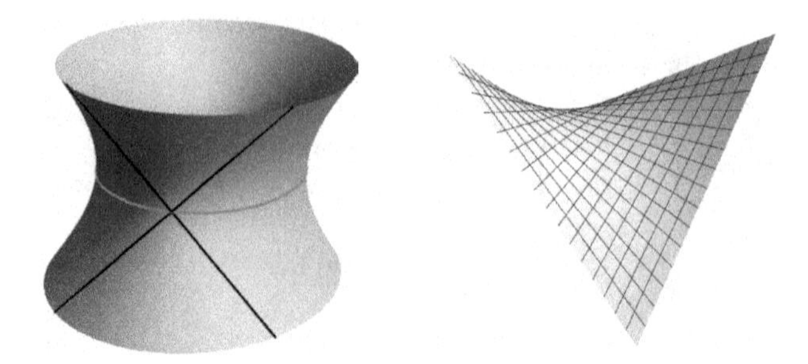

Figure 7.12: A hyperbolic hyperboloid (left) and a hyperbolic paraboloid (right).

7. Consider the parameterization

$$\mathbf{f}(u,v) = (1-v)\boldsymbol{\gamma}(u) + v\boldsymbol{\lambda}(u),$$

with

$$\boldsymbol{\gamma}(u) = (\cos(u-\alpha), \sin(u-\alpha), -1), \quad \boldsymbol{\lambda}(u) = (\cos(u+\alpha), \sin(u+\alpha), 1).$$

Show that:

- for $\alpha = 0$, one gets a cylinder with equation $x_1^2 + x_2^2 = 1$;
- for $\alpha = \dfrac{\pi}{2}$, one gets a cone with equation $x_1^2 + x_2^2 = x_3^2$;
- for $0 < \alpha < \dfrac{\pi}{2}$, one gets a hyperbolic hyperboloid with equation

$$\frac{x_1^2 + x_2^2}{\cos^2\alpha} - \frac{x_3^2}{\cot^2\alpha} = 1.$$

8. Calculate the metric tensor of a sphere of radius R, write its first fundamental form, and determine the area of the sector of a surface between the longitudes θ_1 and θ_2 and the length of the parallel at the latitude $\pi/4$ between these two longitudes.
9. Prove that the surface defined by

$$\mathbf{f}(u,v) : \Omega = \mathbb{R} \times (-\pi, \pi] \to \mathcal{E} |\ \mathbf{f}(u,v) = \left(\frac{\cos v}{\cosh u}, \frac{\sin v}{\cosh u}, \frac{\sinh u}{\cosh u}\right)$$

is a sphere. Then, show that the image of any straight line on Ω is a loxodromic line on the sphere.
10. Calculate the vectors of the natural basis, the tensors $\mathbf{g}, \mathbf{B}, \mathbf{X}$, and the first and second fundamental forms for the catenoid.

11. Calculate the same for the helicoid of parametric equation

$$\mathbf{f}(u,v) = \gamma(u) + v\boldsymbol{\lambda}(u),$$

with

$$\gamma(u) = (0,0,u), \quad \boldsymbol{\lambda}(u) = (\cos u, \sin u, 0).$$

12. Show that the catenoid and the helicoid are made of hyperbolic points.
13. Determine the geodesic lines of a circular cylinder.

Suggested Texts

There are many textbooks on tensors, differential geometry, and calculus of variations. The style, content, and language of such books greatly depend upon the scientific community the authors belong to: pure or applied mathematicians, theoretical mechanicians or engineers. It is hence difficult to suggest some readings in the domain, and ultimately, it is mostly a matter of personal taste.

This textbook is greatly inspired by some classical methods, style, and language that are typical in the community of theoretical mechanics; the following few suggested readings, among several possible others, belong to such a kind of scientific literature. They are classical textbooks, and though the list is far from being exhaustive, they constitute a solid basis for the topics briefly developed in this manuscript, where the objective is to present mathematics for mechanics.

A good introduction to tensor algebra and analysis, which greatly inspired the content of this manuscript, are the two introductory chapters of the classical textbook,

- M. E. Gurtin: *An Introduction to Continuum Mechanics*. Academic Press, 1981,

and also, in a similar style, the long article,

- P. Podio-Guidugli: A primer in elasticity. *Journal of Elasticity*, 58: 1–104, 2000.

A short, effective introduction to tensor algebra and differential geometry of curves can be found in the following text of exercises on analytical mechanics:

- P. Biscari, C. Poggi, E. G. Virga: *Mechanics Notebook*. Liguori Editore, 1999.

A classical textbook on linear algebra that is recommended is

- P. R. Halmos: *Finite-Dimensional Vector Spaces*. Van Nostrand Reynold, 1958.

In the previous textbooks, tensor algebra in curvilinear coordinates is not developed; an introduction to this topic, especially intended for physicists and engineers, can be found in

- W. H. Müller: *An Expedition to Continuum Theory*. Springer, 2014,

which has largely influenced Chapter 6.

Two modern and application-oriented textbooks on differential geometry of curves and surfaces are

- V. A. Toponogov: *Differential Geometry of Curves and Surfaces – A Concise Guide*. Birkhäuser, 2006 and
- A. Pressley: *Elementary Differential Geometry*. Springer, 2010.

A short introduction to the differential geometry of surfaces, oriented toward the mechanics of shells, can be found in the classical book

- V. V. Novozhilov: *Thin Shell Theory*. Noordhoff LTD., 1964.

For what concerns the calculus of variations, a valid textbook on the matter (but not only) is

- R. Courant, D. Hilbert: *Methods of Mathematical Physics*. Interscience Publishers, 1953.

Two very good and classical textbooks with an introduction to the calculus of variations for engineers are

- C. Lanczos: *The Variational Principles of Mechanics*. University of Toronto Press, 1949 and
- H. L. Langhaar: *Energy Methods in Applied Mechanics*. Wiley, 1962.

Solutions to the Exercises

Chapter 1

1. Suppose $\mathbf{o}_1 \neq \mathbf{o}_2$; then, apply to a point p the definition of vector null for both of them.
2. Use $\mathbf{v} + \mathbf{o} = \mathbf{v}$ and obtain the scalar product with a vector $\mathbf{w}$.
3. Obtain the norm of $\mathbf{v} + \mathbf{o} = \mathbf{v}$ and use the above result.
4. $|\mathbf{u} - \mathbf{v}| = |\mathbf{u} + \mathbf{v}| \iff |\mathbf{u} - \mathbf{v}|^2 = |\mathbf{u} + \mathbf{v}|^2 \Rightarrow (\mathbf{u} - \mathbf{v}) \cdot (\mathbf{u} - \mathbf{v}) = (\mathbf{u} + \mathbf{v}) \cdot (\mathbf{u} + \mathbf{v}) \Rightarrow \mathbf{u} \cdot \mathbf{v} = 0$.
5. By linearity, $\forall \mathbf{v} = v_i \mathbf{e}_i, \psi(\mathbf{v}) = v_i \psi(\mathbf{e}_i)$; moreover, $\mathbf{u} \cdot \mathbf{v} = u_i v_i$, so setting $\mathbf{u} = \psi(\mathbf{e}_i)\mathbf{e}_i, \psi(\mathbf{v}) = \mathbf{u} \cdot \mathbf{v}$. Uniqueness: Suppose $\exists \mathbf{u}_1 \neq \mathbf{u}_2 | \psi(\mathbf{v}) = \mathbf{u}_1 \cdot \mathbf{v} = \mathbf{u}_2 \cdot \mathbf{v} \Rightarrow (\mathbf{u}_1 - \mathbf{u}_2) \cdot \mathbf{v} = 0 \ \forall \mathbf{v} \iff \mathbf{u}_1 - \mathbf{u}_2 = \mathbf{o} \Rightarrow \mathbf{u}_1 = \mathbf{u}_2$.
6. Let θ_u and θ_v be the angles formed by $\mathbf{w}$ with $\mathbf{u}$ and $\mathbf{v}$, respectively; then, $\mathbf{w} \cdot \mathbf{u} = \mathbf{w} \cdot \mathbf{v} \Rightarrow uw \cos\theta_u = vw \cos\theta_v \Rightarrow \cos\theta_u = \cos\theta_v$, as $\mathbf{u}, \mathbf{v} \in \mathcal{S}$;
   $\theta_u = \theta_v \Rightarrow \cos\theta_u = \cos\theta_v \Rightarrow uw\cos\theta_u = vw\cos\theta_v \Rightarrow \mathbf{u} \cdot \mathbf{w} = \mathbf{v} \cdot \mathbf{w}$.
7. (i) Coplanar vectors: Let p be a point of the plane of the vectors $\Rightarrow \mathbf{M}_p^r \cdot \mathbf{R} = 0$ because $\mathbf{R} \in$ to the plane, while $\mathbf{M}_p^r = (p_i - p) \times \mathbf{v}^{p_i}$ is of course orthogonal to it. Then, let q be a point $\notin$ to the plane of the vectors $\rightarrow \mathbf{M}_q^r = \mathbf{M}_p^r + (p - q) \times \mathbf{R} \Rightarrow \mathbf{M}_q^r \cdot \mathbf{R} = \mathbf{M}_p^r \cdot \mathbf{R} + (p - q) \times \mathbf{R} \cdot \mathbf{R} = \mathbf{R} \times \mathbf{R} \cdot (p - q) = 0$.
   (ii) Parallel vectors: Let $\mathbf{e} \in \mathcal{S}| \ \mathbf{v}^{p_i} = \alpha_i \mathbf{e} \ \forall i = 1, \dots, n \Rightarrow \mathbf{R} = \sum_{i=1}^n \alpha_i \mathbf{e} \Rightarrow$
   $\forall o \in \mathcal{E}$, $\mathbf{M}_o^r = \sum_{i=1}^n (p_i - o) \times \alpha_i \mathbf{e} \Rightarrow \mathbf{M}_o^r \cdot \mathbf{R} = (\sum_{i=1}^n \alpha_i (p_i - o)) \times \mathbf{e} \cdot (\sum_{i=1}^n \alpha_i)\mathbf{e} = 0$.

8. It is a direct consequence of the theorem of reduction of systems of applied vectors for the case $\mathbf{R} = \mathbf{o}$, with o being any point.
9. If $\mathbf{R}$ is applied to p, then the system can be reduced to $\mathbf{R}^p$ plus $\mathbf{M}_p^r = \sum_{i=1}^n (p - p) \times \mathbf{v}_i^p = \mathbf{o}$.
10. (i) By the theorem of reduction, if $\mathbf{R}$ is applied to o, the system is reduced to $\mathbf{R}^o$ and $\mathbf{M}_o^r \Rightarrow$ to only $\mathbf{R}$ if $\mathbf{M}_o^r = \mathbf{o}$.
    (ii) Because for coplanar or parallel vectors $\forall o \in \mathcal{A},\ \mathbf{M}_o^r = \mathbf{o}$, then the system is equivalent to $\mathbf{R}$ applied to any point of $\mathcal{A}$.
11. If a system is equilibrated, then by definition, any equivalent system is equilibrated. Conversely, if there exists another system equilibrated and equivalent, then also the system in object is equilibrated by the relation of equivalence, and this is true for any other equivalent system, which hence must be equilibrated.
12. Let $\mathbf{v}^p, \mathbf{v}^q = -(\mathbf{v}^p)^q$ be two opposite vectors applied to p and q, respectively $\Rightarrow \mathbf{R} = \mathbf{v}^p + \mathbf{v}^q = \mathbf{o}$; moreover, $\mathbf{M}_p^r = (q - p) \otimes \mathbf{v}^q \Rightarrow \forall o \neq p,\ \mathbf{M}_o^r = \mathbf{M}_p^r + (p - o) \times \mathbf{R} = \mathbf{M}_p^r \Rightarrow \mathbf{M}_o^r = \mathbf{o}\ \forall o \in \mathcal{E} \iff q - p = \mathbf{o}$.
13. If all the vectors pass through a point p, then the system is equivalent to $\mathbf{R}^p$ (Exercise 9), and if $\mathbf{R} = \mathbf{o}$, then the system is equilibrated.

Chapter 2

1. $\forall \mathbf{u}, \mathbf{L}, \mathbf{L}\mathbf{u} = \mathbf{L}(\mathbf{u} + \mathbf{o}) = \mathbf{L}\mathbf{u} + \mathbf{L}\mathbf{o} \iff \mathbf{L}\mathbf{o} = \mathbf{o}$.
2. $\mathbf{u} \in \mathcal{S} \rightarrow (\mathbf{u} \otimes \mathbf{u})\mathbf{v} = v \cos\theta \mathbf{u}$; $(\mathbf{I} - \mathbf{u} \otimes \mathbf{u})\mathbf{v} = \mathbf{v} - v\cos\theta\mathbf{u}$, which is orthogonal to $\mathbf{u} : (\mathbf{I} - \mathbf{u} \otimes \mathbf{u})\mathbf{v} \cdot \mathbf{u} = \mathbf{v} \cdot \mathbf{u} - \mathbf{v} \cdot \mathbf{u}\ \mathbf{u} \cdot \mathbf{u} = \mathbf{o}$, as $\mathbf{u} \in \mathcal{S}$.
3. (i) $\forall \mathbf{a}, \mathbf{b} \in \mathcal{V}, \mathbf{a} \cdot (\alpha\mathbf{A})\mathbf{b} = (\alpha\mathbf{A})^\top \mathbf{a} \cdot \mathbf{b}$, and by the linearity of the scalar product, $\mathbf{a} \cdot (\alpha\mathbf{A})\mathbf{b} = \alpha\mathbf{a} \cdot \mathbf{A}\mathbf{b} = \alpha\mathbf{A}^\top\mathbf{a} \cdot \mathbf{b} \Rightarrow (\alpha\mathbf{A})^\top = \alpha\mathbf{A}^\top$.
   (ii) $\forall \mathbf{a}, \mathbf{b} \in \mathcal{V}, \mathbf{a} \cdot (\mathbf{A} + \mathbf{B})\mathbf{b} = (\mathbf{A} + \mathbf{B})^\top \mathbf{a} \cdot \mathbf{b}$, and by linearity of the scalar product and of tensors, $\mathbf{a} \cdot (\mathbf{A} + \mathbf{B})\mathbf{b} = \mathbf{a} \cdot \mathbf{A}\mathbf{b} + \mathbf{a} \cdot \mathbf{B}\mathbf{b} = \mathbf{A}^\top\mathbf{a} \cdot \mathbf{b} + \mathbf{B}^\top\mathbf{a} \cdot \mathbf{b} = (\mathbf{A}^\top + \mathbf{B}^\top)\mathbf{a} \cdot \mathbf{b} \Rightarrow (\mathbf{A} + \mathbf{B})^\top = \mathbf{A}^\top + \mathbf{B}^\top$.
   (iii) $\forall \mathbf{u}, \mathbf{v} \in \mathcal{V}\ \ \mathbf{u} \cdot (\mathbf{a} \otimes \mathbf{b})\mathbf{A}\mathbf{v} = \mathbf{u} \cdot (\mathbf{a} \otimes \mathbf{b})(\mathbf{A}\mathbf{v}) = (\mathbf{a} \otimes \mathbf{b})^\top \mathbf{u} \cdot (\mathbf{A}\mathbf{v}) = (\mathbf{b} \otimes \mathbf{a})\mathbf{u} \cdot (\mathbf{A}\mathbf{v}) = \mathbf{A}^\top(\mathbf{b} \otimes \mathbf{a})\mathbf{u} \cdot \mathbf{v} = \mathbf{A}^\top(\mathbf{a} \cdot \mathbf{u}\,\mathbf{b}) \cdot \mathbf{v} = \mathbf{a} \cdot \mathbf{u}\ \mathbf{A}^\top\mathbf{b} \cdot \mathbf{v} = ((\mathbf{A}^\top\mathbf{b}) \otimes \mathbf{a})\mathbf{u} \cdot \mathbf{v} = \mathbf{u} \cdot ((\mathbf{A}^\top\mathbf{b}) \otimes \mathbf{a})^\top \mathbf{v} = \mathbf{u} \cdot (\mathbf{a} \otimes \mathbf{A}^\top\mathbf{b})\mathbf{v}$.
4. By linearity and the definition of $\mathbf{O} :\ \forall \mathbf{u} \in \mathcal{V},\ (\mathbf{L} + \mathbf{O})\mathbf{u} = \mathbf{L}\mathbf{u} + \mathbf{O}\mathbf{u} = \mathbf{L}\mathbf{u} \iff \mathbf{L} + \mathbf{O} = \mathbf{L}$.
5. (i) $\mathrm{tr}\mathbf{I} = \mathrm{tr}(\delta_{ij}\mathbf{e}_i \otimes \mathbf{e}_j) = \delta_{ij}\mathrm{tr}(\mathbf{e}_i \otimes \mathbf{e}_j) = \delta_{ij}\mathbf{e}_i \cdot \mathbf{e}_j = \delta_{ij}\delta_{ij} = \delta_{ii} = 3$.
   (ii) $\mathrm{tr}\mathbf{L} = \mathrm{tr}(\mathbf{L} + \mathbf{O}) = \mathrm{tr}\mathbf{L} + \mathrm{tr}\mathbf{O} \iff \mathrm{tr}\mathbf{O} = 0$.
6. $\mathrm{tr}(\mathbf{A}\mathbf{B}) = \mathrm{tr}((A_{ij}\mathbf{e}_i \otimes \mathbf{e}_j)(B_{hk}\mathbf{e}_h \otimes \mathbf{e}_k)) = A_{ij}B_{hk}\mathrm{tr}((\mathbf{e}_i \otimes \mathbf{e}_j)(\mathbf{e}_h \otimes \mathbf{e}_k))$ $= A_{ij}B_{hk}\mathbf{e}_j \cdot \mathbf{e}_h \mathrm{tr}(\mathbf{e}_i \otimes \mathbf{e}_k) = A_{ij}B_{hk}\mathbf{e}_j \cdot \mathbf{e}_h\ \mathbf{e}_i \cot \mathbf{e}_k = A_{ij}B_{hk}\delta_{jh}\delta_{ik} =$

$A_{ij}B_{ji}$; in a similar way, we prove that $\mathrm{tr}(\mathbf{BA}) = B_{ij}A_{ji}$; because i, j are dummy indexes, $A_{ij}B_{ji} = B_{ij}A_{ji} \Rightarrow \mathrm{tr}(\mathbf{AB}) = \mathrm{tr}(\mathbf{BA})$.

7. (i) $\mathbf{L}^\top \cdot \mathbf{M}^\top = \mathrm{tr}((\mathbf{L}^\top)^\top \mathbf{M}^\top) = \mathrm{tr}(\mathbf{LM}^\top) = \mathrm{tr}(\mathbf{M}^\top \mathbf{L}) = \mathbf{M} \cdot \mathbf{L} = \mathbf{L} \cdot \mathbf{M}$.
   (ii) $\mathbf{LM} \cdot \mathbf{N} = \mathrm{tr}((\mathbf{LM})^\top \mathbf{N}) = \mathrm{tr}(\mathbf{M}^\top \mathbf{L}^\top \mathbf{N}) = \mathbf{M} \cdot \mathbf{L}^\top \mathbf{N}$;
   $$= \mathrm{tr}(\mathbf{N}(\mathbf{LM})^\top) = \mathrm{tr}((\mathbf{NM}^\top)\mathbf{L}^\top) = \mathrm{tr}(\mathbf{L}^\top(\mathbf{NM}^\top)) = \mathbf{L} \cdot \mathbf{NM}^\top.$$
8. (i) $(\mathbf{a} \otimes \mathbf{b})(\mathbf{c} \otimes \mathbf{d}) = ((\mathbf{a} \otimes \mathbf{b})(\mathbf{c} \otimes \mathbf{d}))_{ij} \mathbf{e}_i \otimes \mathbf{e}_j = (\mathbf{a} \otimes \mathbf{b})_{ik} (\mathbf{c} \otimes \mathbf{d})_{kj} \mathbf{e}_i \otimes \mathbf{e}_j$ $= a_i b_k c_k d_j \mathbf{e}_i \otimes \mathbf{e}_j = \mathbf{b} \cdot \mathbf{c}\ a_i d_j \mathbf{e}_i \otimes \mathbf{e}_j = \mathbf{b} \cdot \mathbf{c}\ \mathbf{a} \otimes \mathbf{d}$.
   (ii) $\mathbf{A}(\mathbf{a} \otimes \mathbf{b}) = \mathbf{A}(\mathbf{a} \otimes \mathbf{b})_{ij} \mathbf{e}_i \otimes \mathbf{e}_j = A_{ik}(\mathbf{a} \otimes \mathbf{b})_{kj} \mathbf{e}_i \otimes \mathbf{e}_j = A_{ik} a_k b_j \mathbf{e}_i \otimes \mathbf{e}_j$ $= (\mathbf{Aa})_i b_j \mathbf{e}_i \otimes \mathbf{e}_j = (\mathbf{Aa}) \otimes \mathbf{b}$.
9. $\mathbf{L} \cdot \mathbf{u} \otimes \mathbf{u} = \mathrm{tr}(\mathbf{L}^\top(\mathbf{u} \otimes \mathbf{u})) = \mathrm{tr}((\mathbf{L}^\top \mathbf{u}) \otimes \mathbf{u}) = \mathbf{L}^\top \mathbf{u} \cdot \mathbf{u} = \mathbf{u} \cdot \mathbf{Lu}$.
10. $\mathbf{A} = \mathbf{A}^\top, \mathbf{B} = -\mathbf{B}^\top \Rightarrow \mathbf{A} \cdot \mathbf{B} = \mathbf{A}^\top \cdot \mathbf{B}^\top = \mathbf{A} \cdot (-\mathbf{B}) = -\mathbf{A} \cdot \mathbf{B} \iff \mathbf{A} \cdot \mathbf{B} = 0$.
11. (i) $\mathbf{A} = \mathbf{A}^\top \Rightarrow \mathbf{A} \cdot \mathbf{L} = \mathbf{A} \cdot (\mathbf{L}^s + \mathbf{L}^a) = \mathbf{A} \cdot \mathbf{L}^s + \mathbf{A} \cdot \mathbf{L}^a = \mathbf{A} \cdot \mathbf{L}^s$.
    (ii) $\mathbf{B} = -\mathbf{B}^\top \Rightarrow \mathbf{B} \cdot \mathbf{L} = \mathbf{B} \cdot (\mathbf{L}^s + \mathbf{L}^a) = \mathbf{B} \cdot \mathbf{L}^s + \mathbf{B} \cdot \mathbf{L}^a = \mathbf{B} \cdot \mathbf{L}^a$.
12. (i) $\mathbf{A} \cdot (\mathbf{BCD}) = \mathrm{tr}(\mathbf{A}^\top \mathbf{BCD}) = \mathrm{tr}((\mathbf{B}^\top \mathbf{A})^\top \mathbf{CD}) = (\mathbf{B}^\top \mathbf{A}) \cdot (\mathbf{CD})$.
    (ii) $\mathbf{A} \cdot (\mathbf{BCD}) = (\mathbf{BCD}) \cdot \mathbf{A} = \mathrm{tr}((\mathbf{BCD})^\top \mathbf{A}) = \mathrm{tr}(\mathbf{A}(\mathbf{D}^\top \mathbf{C}^\top \mathbf{B}^\top)) =$ $\mathrm{tr}((\mathbf{AD}^\top)(\mathbf{C}^\top \mathbf{B}^\top)) = \mathrm{tr}((\mathbf{C}^\top \mathbf{B}^\top)(\mathbf{AD}^\top)) = \mathrm{tr}((\mathbf{BC})^\top (\mathbf{AD}^\top)) =$ $\mathbf{BC} \cdot \mathbf{AD}^\top = \mathbf{AD}^\top \cdot \mathbf{BC}$.
13. $\mathbf{L} \in Sym(\mathcal{V}) \Rightarrow \mathbf{L} \cdot \mathbf{W} = 0$, as already proved. Now, if $\mathbf{L} \cdot \mathbf{W} = 0\ \forall \mathbf{W} \in Skw(\mathcal{V})$, suppose $\mathbf{L} \notin Sym(\mathcal{V}) \Rightarrow \mathbf{L} = \mathbf{L}^s + \mathbf{L}^a \Rightarrow \mathbf{L} \cdot \mathbf{W} = \mathbf{L}^s \cdot \mathbf{W} + \mathbf{L}^a \cdot \mathbf{W} = \mathbf{L}^a \cdot \mathbf{W} = 0$; if in $Skw(\mathcal{V})$, we chose $\mathbf{W} = \mathbf{L}^a$, we get $\mathbf{W} \cdot \mathbf{L}^a = \mathbf{L}^a \cdot \mathbf{L}^a = 0 \iff \mathbf{L}^a = \mathbf{O} \Rightarrow \mathbf{L} \in Sym(\mathcal{V})$.
14. $I_2 = \frac{1}{2}(\mathrm{tr}^2\mathbf{L} - \mathrm{tr}\mathbf{L}^2) = \frac{1}{2}(L_{ii}L_{jj} - L_{ij}L_{ji})$ $= L_{11}L_{22} + L_{11}L_{33} + L_{22}L_{33} - L_{12}L_{21} - L_{13}L_{31} - L_{23}L_{32}$.
15. $\mathbf{a} \times \mathbf{b} \cdot \mathbf{c} = \begin{bmatrix} 0 & -a_3 & a_2 \\ a_3 & 0 & -a_1 \\ -a_2 & a_1 & 0 \end{bmatrix} \begin{Bmatrix} b_1 \\ b_2 \\ b_3 \end{Bmatrix} \cdot \begin{Bmatrix} c_1 \\ c_2 \\ c_3 \end{Bmatrix} = a_2 b_3 c_1 - a_3 b_2 c_1 + a_3 b_1 c_2 -$ $a_1 b_3 c_2 + a_1 b_2 c_3 - a_2 b_1 c_3 = \det \begin{bmatrix} 0 & -a_3 & a_2 \\ a_3 & 0 & -a_1 \\ -a_2 & a_1 & 0 \end{bmatrix}$.
16. Let $\mathbf{L}_1 \neq \mathbf{L}_2$ be two distinct inverse tensors of $\mathbf{L}$; then, $\mathbf{L}_1\mathbf{L} = \mathbf{I} = \mathbf{L}_2\mathbf{L} \Rightarrow \mathbf{L}_1\mathbf{L} - \mathbf{L}_2\mathbf{L} = \mathbf{O} \Rightarrow (\mathbf{L}_1 - \mathbf{L}_2)\mathbf{L} = \mathbf{O}\ \forall \mathbf{L} \iff \mathbf{L}_1 - \mathbf{L}_2 = \mathbf{O} \Rightarrow \mathbf{L}_1 = \mathbf{L}_2$.
17. $(\mathbf{a} \otimes \mathbf{b})_{ij} = a_i b_j$; it is then sufficient to write the matrix representing $(\mathbf{a} \otimes \mathbf{b})$ and to compute its determinant.

18. $(\alpha\mathbf{L})^{-1}(\alpha\mathbf{L}) = \alpha(\alpha\mathbf{L})^{-1}\mathbf{L} = \mathbf{I} \Rightarrow (\alpha\mathbf{L})^{-1}\mathbf{L} = \frac{1}{\alpha}\mathbf{I} \Rightarrow (\alpha\mathbf{L})^{-1}\mathbf{L}\mathbf{L}^{-1} = \frac{1}{\alpha}\mathbf{I}\mathbf{L}^{-1} \Rightarrow (\alpha\mathbf{L})^{-1} = \frac{1}{\alpha}\mathbf{L}^{-1}$.
19. $|\mathbf{W}^2| = \mathbf{W}\cdot\mathbf{W} = \mathrm{tr}(\mathbf{W}^\top\mathbf{W}) = -\mathrm{tr}(\mathbf{W}\mathbf{W}) = \mathrm{tr}(\mathbf{I}-\mathbf{w}\otimes\mathbf{w}) = 3-1 = 2 \Rightarrow \mathbf{W}\mathbf{W} = -\frac{1}{2}|\mathbf{W}^2|(\mathbf{I}-\mathbf{w}\otimes\mathbf{w})$.
20. Let $\mathbf{w}_1 = (a_1, b_1, c_1), \mathbf{w}_2 = (a_2, b_2 c_2)$; then, form $\mathbf{W}_1, \mathbf{W}_2$ and compute the two scalar products.
21. $\mathbf{u}\times\mathbf{v} = \mathbf{o} \iff \mathbf{v} = k\mathbf{u}, k\in\mathbb{R}$. So, $\mathbf{u}\times\mathbf{v} = \mathbf{o} \Rightarrow \mathbf{u}\otimes\mathbf{v} = k\mathbf{u}\otimes\mathbf{u} \in Sym(\mathcal{V})$. Conversely, if $\mathbf{u}\otimes\mathbf{v}\in Sym(\mathcal{V})$, then $\forall\mathbf{w},\ \mathbf{w}\cdot\mathbf{v}\,\mathbf{u} = (\mathbf{u}\otimes\mathbf{v})\mathbf{w} = (\mathbf{v}\otimes\mathbf{u})\mathbf{w} = \mathbf{w}\cdot\mathbf{u}\,\mathbf{v} \Rightarrow \mathbf{v} = \frac{\mathbf{w}\cdot\mathbf{v}}{\mathbf{w}\cdot\mathbf{u}}\mathbf{u} \Rightarrow \mathbf{u}\times\mathbf{v} = \mathbf{o}$.
22. $\mathbf{L} = \mathbf{L}^\top \Rightarrow (\mathbf{R}\mathbf{L}\mathbf{R}^\top)^\top = \mathbf{R}\mathbf{L}^\top\mathbf{R}^\top = \mathbf{R}\mathbf{L}\mathbf{R}^\top$; moreover, $\mathbf{u}\cdot\mathbf{L}\mathbf{u} > 0\ \forall\mathbf{u} \Rightarrow \mathbf{u}\cdot(\mathbf{R}\mathbf{L}\mathbf{R}^\top)\mathbf{u} = (\mathbf{R}^\top\mathbf{u})\cdot\mathbf{L}(\mathbf{R}^\top\mathbf{u}) > 0$.
23. (i) $\det(\mathbf{L}^{sph}-\lambda\mathbf{I}) = \det\left(\frac{1}{3}\mathrm{tr}\mathbf{L}\mathbf{I} - \lambda^{sph}\mathbf{I}\right) = \left(\frac{1}{3}\mathrm{tr}\mathbf{L}\mathbf{I} - \lambda^{sph}\mathbf{I}\right)^3\det\mathbf{I}$
$= \left(\frac{1}{3}\mathrm{tr}\mathbf{L}\mathbf{I} - \lambda^{sph}\mathbf{I}\right)^3 = 0 \Rightarrow \lambda_i^{sph} = \frac{1}{3}\mathrm{tr}\mathbf{L},\ i = 1,2,3.$
(ii) $(\mathbf{L}^{sph}-\lambda_i^{sph}\mathbf{I})\mathbf{v} = \mathbf{o}\ \forall i = 1,2,3 \Rightarrow \left(\mathbf{L}^{sph} - \frac{1}{3}\mathrm{tr}\mathbf{L}\mathbf{I}\right)\mathbf{v} = \mathbf{o} \Rightarrow$
$(\mathbf{L}^{sph}-\mathbf{L}^{sph})\mathbf{v} = \mathbf{o} \Rightarrow \mathbf{O}\mathbf{v} = \mathbf{o}$, which is true $\forall\mathbf{v}$.
24. $\det(\mathbf{L}^{dev}-\lambda^{dev}\mathbf{I}) = \det(\mathbf{L}-\mathbf{L}^{dev}-\lambda^{dev}\mathbf{I}) = \det\left(\mathbf{L}-\left(\frac{1}{3}\mathrm{tr}\mathbf{L}+\lambda^{dev}\right)\mathbf{I}\right) = \det\left(\mathbf{L}-\left(\lambda^{sph}+\lambda^{dev}\right)\mathbf{I}\right) = 0 \Rightarrow \lambda = \lambda^{sph}+\lambda^{dev}$ is an eigenvalue of $\mathbf{L} \Rightarrow \lambda^{dev} = \lambda - \lambda^{sph}$.

Chapter 3

1. $\forall\mathbf{L}\in Lin(\mathcal{V}),\ (\mathbf{e}_i\otimes\mathbf{e}_j)\boxtimes(\mathbf{e}_k\otimes\mathbf{e}_l)\mathbf{L} = (\mathbf{e}_i\otimes\mathbf{e}_j)\mathbf{L}(\mathbf{e}_k\otimes\mathbf{e}_l)^\top = (\mathbf{e}_i\otimes\mathbf{e}_j)\mathbf{L}(\mathbf{e}_l\otimes\mathbf{e}_k) = (\mathbf{e}_i\otimes\mathbf{e}_j)((\mathbf{L}\mathbf{e}_l)\otimes\mathbf{e}_k) = \mathbf{e}_j\cdot(\mathbf{L}\mathbf{e}_l)\mathbf{e}_i\otimes\mathbf{e}_k = L_{jl}\mathbf{e}_i\otimes\mathbf{e}_k$; moreover $(\mathbf{e}_i\otimes\mathbf{e}_k\otimes\mathbf{e}_j\otimes\mathbf{e}_l)\mathbf{L} = (\mathbf{e}_i\otimes\mathbf{e}_k)\otimes(\mathbf{e}_j\otimes\mathbf{e}_l)\mathbf{L} = ((\mathbf{e}_j\otimes\mathbf{e}_l)\cdot\mathbf{L})(\mathbf{e}_i\otimes\mathbf{e}_k) = ((\mathbf{e}_j\otimes\mathbf{e}_l)\cdot(L_{pq}\mathbf{e}_p\otimes\mathbf{e}_q))(\mathbf{e}_i\otimes\mathbf{e}_k) = L_{pq}\delta_{jp}\delta_{lq}(\mathbf{e}_i\otimes\mathbf{e}_k) = L_{jl}\mathbf{e}_i\otimes\mathbf{e}_k \Rightarrow (\mathbf{e}_i\otimes\mathbf{e}_j)\boxtimes(\mathbf{e}_k\otimes\mathbf{e}_l) = \mathbf{e}_i\otimes\mathbf{e}_k\otimes\mathbf{e}_j\otimes\mathbf{e}_l$.
2. $\forall\mathbf{L},\mathbf{M}\in Lin(\mathcal{V}),\ \mathbf{L}\cdot(\mathbb{A}\mathbb{B})\mathbf{M} = \mathbb{A}^\top\mathbf{L}\cdot\mathbb{B}\mathbf{M} = \mathbb{B}^\top\mathbb{A}^\top\mathbf{L}\cdot\mathbf{M} \Rightarrow (\mathbb{A}\mathbb{B})^\top = \mathbb{B}^\top\mathbb{A}^\top$.
3. $\forall\mathbf{C}\in Lin(\mathcal{V}),\ (\mathbf{A}\otimes\mathbf{B}\mathbb{L})\mathbf{C} = (\mathbf{A}\otimes\mathbf{B})\mathbb{L}\mathbf{C} = \mathbf{B}\cdot\mathbb{L}\mathbf{C}\mathbf{A} = \mathbb{L}^\top\mathbf{B}\cdot\mathbf{C}\mathbf{A} = (\mathbf{A}\otimes\mathbb{L}^\top\mathbf{B})\mathbf{C}$.
4. $\forall\mathbf{L}\in Lin(\mathcal{V}),\ ((\mathbf{A}\boxtimes\mathbf{B})(\mathbf{C}\boxtimes\mathbf{D}))\mathbf{L} = \mathbf{A}\boxtimes\mathbf{B}\mathbf{C}\mathbf{L}\mathbf{D}^\top = \mathbf{A}\mathbf{C}\mathbf{L}\mathbf{D}^\top\mathbf{B}^\top = (\mathbf{A}\mathbf{C})\boxtimes(\mathbf{D}^\top\mathbf{B}^\top)^\top\mathbf{L} = (\mathbf{A}\mathbf{C})\boxtimes(\mathbf{B}\mathbf{D})\mathbf{L}$.
5. Let $\mathbb{A} = A_{ijkl}\mathbf{e}_i\otimes\mathbf{e}_j\otimes\mathbf{e}_k\otimes\mathbf{e}_l = A_{ijkl}(\mathbf{e}_i\otimes\mathbf{e}_k)\boxtimes(\mathbf{e}_j\otimes\mathbf{e}_l)$ and $\mathbb{B} = B_{pqrs}\mathbf{e}_p\otimes\mathbf{e}_q\otimes\mathbf{e}_r\otimes\mathbf{e}_s = B_{pqrs}(\mathbf{e}_p\otimes\mathbf{e}_r)\boxtimes(\mathbf{e}_q\otimes\mathbf{e}_s)$.
Then, $\mathbb{A}\mathbb{B} = A_{ijkl}B_{pqrs}((\mathbf{e}_i\otimes\mathbf{e}_k)\boxtimes(\mathbf{e}_j\otimes\mathbf{e}_l))((\mathbf{e}_p\otimes\mathbf{e}_r)\boxtimes(\mathbf{e}_q\otimes\mathbf{e}_s)) = A_{ijkl}B_{pqrs}((\mathbf{e}_i\otimes\mathbf{e}_k)(\mathbf{e}_p\otimes\mathbf{e}_r))\boxtimes((\mathbf{e}_j\otimes\mathbf{e}_l)(\mathbf{e}_q\otimes\mathbf{e}_s)) = A_{ijkl}B_{pqrs}\delta_{kp}\delta lq(\mathbf{e}_i\otimes\mathbf{e}_r)\boxtimes(\mathbf{e}_j\otimes\mathbf{e}_s) = A_{ijkl}B_{klrs}(\mathbf{e}_i\otimes\mathbf{e}_j\otimes\mathbf{e}_r\otimes\mathbf{e}_s)$.

6. $\forall \mathbf{L} \in Lin(\mathcal{V})$, $(\mathbf{A} \otimes \mathbf{B})(\mathbf{C} \boxtimes \mathbf{D})\mathbf{L} = (\mathbf{A} \otimes \mathbf{B})\mathbf{C}\mathbf{L}\mathbf{D}^\top = \mathbf{B} \cdot (\mathbf{C}\mathbf{L}\mathbf{D}^\top)\mathbf{A} = \mathbf{B}\mathbf{D} \cdot (\mathbf{C}\mathbf{L})\mathbf{A} = \mathbf{C}^\top\mathbf{B}\mathbf{D} \cdot \mathbf{L}\,\mathbf{A} = (\mathbf{C}^\top \boxtimes \mathbf{D}^\top)\mathbf{B} \cdot \mathbf{L}\,\mathbf{A} = \mathbf{A} \otimes ((\mathbf{C}^\top \boxtimes \mathbf{D}^\top)\mathbf{B})\mathbf{L}$.
7. $\forall \mathbf{L} \in Lin(\mathcal{V})$, $(\mathbf{A} \boxtimes \mathbf{B})(\mathbf{C} \otimes \mathbf{D})\mathbf{L} = \mathbf{D} \cdot \mathbf{L}(\mathbf{A} \boxtimes \mathbf{B})\mathbf{C} = (\mathbf{D} \cdot \mathbf{L})\mathbf{A}\mathbf{C}\mathbf{B}^\top = \mathbf{A}\mathbf{C}\mathbf{B}^\top(\mathbf{D} \cdot \mathbf{L}) = ((\mathbf{A} \boxtimes \mathbf{B})\mathbf{C})(\mathbf{D} \cdot \mathbf{L}) = (((\mathbf{A} \boxtimes \mathbf{B})\mathbf{C}) \otimes \mathbf{D})\mathbf{L}$.
8. $(\mathbf{P} \otimes \mathbf{P})_{ijhk} = P_{ij}P_{hk} = (\mathbf{p} \otimes \mathbf{p})_{ij}(\mathbf{p} \otimes \mathbf{p})_{hk} = p_i p_j p_h p_k$ $= p_i p_h p_j p_k = (\mathbf{p} \otimes \mathbf{p})_{ih}(\mathbf{p} \otimes \mathbf{p})_{jk} = P_{ih}P_{jk} = (\mathbf{P} \boxtimes \mathbf{P})_{ijhk}$.
9. (i) $\mathbb{I}\mathbb{A} = (\mathbf{I} \boxtimes \mathbf{I})\mathbb{A} = (\mathbf{I} \boxtimes \mathbf{I})A_{ijkl}(\mathbf{e}_i \otimes \mathbf{e}_j) \otimes (\mathbf{e}_k \otimes \mathbf{e}_l) = A_{ijkl}((\mathbf{I} \boxtimes \mathbf{I})(\mathbf{e}_i \otimes \mathbf{e}_j)) \otimes (\mathbf{e}_k \otimes \mathbf{e}_l) = A_{ijkl}(\mathbf{I}(\mathbf{e}_i \otimes \mathbf{e}_j)\mathbf{I}^\top) \otimes (\mathbf{e}_k \otimes \mathbf{e}_l) = A_{ijkl}(\mathbf{e}_i \otimes \mathbf{e}_j \otimes \mathbf{e}_k \otimes \mathbf{e}_l) = \mathbb{A}$.
   (ii) $\mathbb{A}\mathbb{I} = A_{ijkl}(\mathbf{e}_i \otimes \mathbf{e}_j) \otimes (\mathbf{e}_k \otimes \mathbf{e}_l)(\mathbf{I} \boxtimes \mathbf{I}) = A_{ijkl}(\mathbf{e}_i \otimes \mathbf{e}_j) \otimes ((\mathbf{I} \boxtimes \mathbf{I})\mathbf{e}_k \otimes \mathbf{e}_l) = A_{ijkl}(\mathbf{e}_i \otimes \mathbf{e}_j) \otimes (\mathbf{I}(\mathbf{e}_k \otimes \mathbf{e}_l))\mathbf{I}^\top) = A_{ijkl}(\mathbf{e}_i \otimes \mathbf{e}_j \otimes \mathbf{e}_k \otimes \mathbf{e}_l) = \mathbb{A}$.
10. $(\mathbf{A} \otimes \mathbf{B}) \cdot (\mathbf{C} \otimes \mathbf{D}) = \mathrm{tr}_4((\mathbf{A} \otimes \mathbf{B})^\top(\mathbf{C} \otimes \mathbf{D})) = \mathrm{tr}_4((\mathbf{B} \otimes \mathbf{A})(\mathbf{C} \otimes \mathbf{D}))$ $= \mathrm{tr}_4((\mathbf{B} \otimes \mathbf{A})\mathbf{C}) \otimes \mathbf{D} = \mathrm{tr}_4(\mathbf{A} \cdot \mathbf{C}\,\mathbf{B} \otimes \mathbf{D}) = \mathbf{A} \cdot \mathbf{C}\,\mathrm{tr}_4(\mathbf{B} \otimes \mathbf{D}) = \mathbf{A} \cdot \mathbf{C}\,\mathbf{B} \cdot \mathbf{D}$.
11. $\frac{\mathbf{I}}{|\mathbf{I}|} \otimes \frac{\mathbf{I}}{|\mathbf{I}|} = \frac{\mathbf{I}}{\sqrt{3}} \otimes \frac{\mathbf{I}}{\sqrt{3}} = \frac{1}{3}\mathbf{I} \otimes \mathbf{I} = \mathbb{S}^{sph}$.
12. $\forall \mathbf{L} \in Lin(\mathcal{V})$, $\mathbf{L} = \mathbf{L}^{sph} + \mathbf{L}^{dev}$ and $\mathbf{L}^{sph} = \frac{1}{3}\mathrm{tr}\mathbf{L}\,\mathbf{I} \rightarrow$, just one number is sufficient to determine $\mathbf{L}^{sph} \Rightarrow \dim(Sph(\mathcal{V})) = 1$. Then, $\mathbf{L}^{dev} = \mathbf{L} - \mathbf{L}^{sph}$ is determined by five numbers: $\dim(Dev(\mathcal{V})) = \dim(Lin(\mathcal{V}) - Sph(\mathcal{V})) = 6 - 1 = 5$.
13. (i) $\mathbb{S}^{sph}\mathbb{S}^{sph} = \left(\frac{1}{3}\mathbf{I} \otimes \mathbf{I}\right)\left(\frac{1}{3}\mathbf{I} \otimes \mathbf{I}\right) = \frac{1}{9}(\mathbf{I} \otimes \mathbf{I}) = \frac{1}{9}\mathbf{I} \cdot \mathbf{I}\,\mathbf{I} \otimes \mathbf{I} = \frac{1}{3}\mathbf{I} \otimes \mathbf{I} = \mathbb{S}^{sph}$.
    (ii) $\mathbb{D}^{dev}\mathbb{D}^{dev} = (\mathbb{I}^s - \mathbb{S}^{sph})(\mathbb{I}^s - \mathbb{S}^{sph}) = \mathbb{I}^s - 2\mathbb{S}^{sph} + \mathbb{S}^{sph}\mathbb{S}^{sph} = \mathbb{I}^s - \mathbb{S}^{sph} = \mathbb{D}^{dev}$.
    (iii) $\mathbb{S}^{sph}\mathbb{D}^{dev} = \mathbb{S}^{sph}(\mathbb{I}^s - \mathbb{S}^{sph}) = \mathbb{S}^{sph} - \mathbb{S}^{sph}\mathbb{S}^{sph} = \mathbb{S}^{sph} - \mathbb{S}^{sph} = \mathbb{O}$.
14. (i) $S^{sym}_{ijkl} = (\mathbf{e}_i \otimes \mathbf{e}_j) \cdot \mathbb{S}^{sym}(\mathbf{e}_k \otimes \mathbf{e}_l) = (\mathbf{e}_i \otimes \mathbf{e}_j) \cdot \frac{\mathbf{e}_k \otimes \mathbf{e}_l + \mathbf{e}_l \otimes \mathbf{e}_k}{2}$ $= \frac{1}{2}(\delta_{ik}\delta_{jl} + \delta_{il}\delta_{jk})$, $W^{skw}_{ijkl} = (\mathbf{e}_i \otimes \mathbf{e}_j) \cdot \mathbb{W}^{skw}(\mathbf{e}_k \otimes \mathbf{e}_l)$ $= (\mathbf{e}_i \otimes \mathbf{e}_j) \cdot \frac{\mathbf{e}_k \otimes \mathbf{e}_l - \mathbf{e}_l \otimes \mathbf{e}_k}{2} = \frac{1}{2}(\delta_{ik}\delta_{jl} - \delta_{il}\delta_{jk})$, $\rightarrow S^{sym}_{ijkl} + W^{skw}_{ijkl} = \delta_{ik}\delta_{jl} = I_{ijkl} \Rightarrow \mathbb{S}^{sym} + \mathbb{W}^{dev} = \mathbb{I}$.
    (ii) $S^{sym}_{ijkl} - W^{skw}_{ijkl} = \delta_{il}\delta_{jk} = T^{trp}_{ijkl} \Rightarrow \mathbb{S}^{sym} - \mathbb{W}^{dev} = \mathbb{T}^{trp}$.
15. (i) $\mathbb{S}^{sph} \cdot \mathbb{S}^{sph} = \left(\frac{1}{3}\mathbf{I} \otimes \mathbf{I}\right) \cdot \left(\frac{1}{3}\mathbf{I} \otimes \mathbf{I}\right) = \frac{1}{9}\mathrm{tr}_4((\mathbf{I} \otimes \mathbf{I})^\top(\mathbf{I} \otimes \mathbf{I}))$ $= \frac{1}{9}\mathrm{tr}_4((\mathbf{I} \otimes \mathbf{I})(\mathbf{I} \otimes \mathbf{I})) = \frac{1}{9}\mathrm{tr}_4((\mathbf{I} \otimes \mathbf{I})\mathbf{I}) \otimes \mathbf{I} = \frac{1}{9}\mathrm{tr}_4(\mathbf{I} \cdot \mathbf{I}\,\mathbf{I} \otimes \mathbf{I})$ $= \frac{1}{3}\mathrm{tr}_4(\mathbf{I}^\top \otimes \mathbf{I}) = \frac{1}{3}\mathbf{I} \cdot \mathbf{I} = 1$.
    (ii) $\mathbb{D}^{dev} \cdot \mathbb{D}^{dev} = (\mathbb{I}^s - \mathbb{S}^{sph}) \cdot (\mathbb{I}^s - \mathbb{S}^{sph}) = \mathbb{I}^s \cdot \mathbb{I}^s - 2\mathbb{I}^s \cdot \mathbb{S}^{sph} + \mathbb{S}^{sph} \cdot \mathbb{S}^{sph}$ $= \frac{1}{4}\left(\delta_{ih}\delta_{jk} + \delta_{ik}\delta_{jh}\right)\left(\delta_{ih}\delta_{jk} + \delta_{ik}\delta_{jh}\right) - 2\frac{\delta_{ih}\delta_{jk} + \delta_{ik}\delta_{jh}}{2}\frac{1}{3}\delta_{ij}\delta_{hk} + 1$ $= \frac{1}{4}\left(\delta_{ih}\delta_{ih}\,\delta_{jk}\delta_{jk} + 2\delta_{ik}\delta_{jh}\delta_{ih}\delta_{jk} + \delta_{ik}\delta_{ik}\,\delta_{jh}\delta_{jh}\right) - \frac{1}{3}\left(\delta_{ij}\delta_{hk}\delta_{ih}\delta_{jk} + \delta_{ij}\delta_{hk}\delta_{ik}\delta_{jh}\right) + 1$. Now, one should check that $\delta_{ih}\delta_{ih} = \delta_{jk}\delta_{jk}$ $= \delta_{ik}\delta_{ik} = \delta_{jh}\delta_{jh} = \delta_{ik}\delta_{jh}\delta_{ih}\delta_{jk} = \delta_{ij}\delta_{hk}\delta_{ih}\delta_{jk} = \delta_{ij}\delta_{hk}\delta_{ik}\delta_{jh}$ $= 3 \Rightarrow \mathbb{D}^{dev} \cdot \mathbb{D}^{dev} = 5$.

(iii) $\mathbb{S}^{sph} \cdot \mathbb{D}^{dev} = \mathbb{S}^{sph} \cdot (\mathbb{I}^s - \mathbb{S}^{sph}) = \mathbb{I}^s \cdot \mathbb{S}^{sph} - \mathbb{S}^{sph} \cdot \mathbb{S}^{sph} = \frac{\delta_{ih}\delta_{jk}+\delta_{ik}\delta_{jh}}{2}\frac{1}{3}\delta_{ij}\delta_{hk} - 1 = \frac{1}{6}(\delta_{ih}\delta_{jk}\delta_{ij}\delta_{hk} + \delta_{ik}\delta_{jh}\delta_{ij}\delta_{hk}) - 1 = \frac{1}{6}(3+3) - 1 = 0.$

16. $\mathbf{S}_R = \mathbf{I} - 2\mathbf{n}\otimes\mathbf{n} \rightarrow \mathbb{S}_R = (\mathbf{I} - 2\mathbf{n}\otimes\mathbf{n}) \boxtimes (\mathbf{I} - 2\mathbf{n}\otimes\mathbf{n})$
$= \mathbf{I}\boxtimes\mathbf{I} - 2(\mathbf{n}\otimes\mathbf{n}\boxtimes\mathbf{I} + \mathbf{I}\boxtimes\mathbf{n}\otimes\mathbf{n} + 4\mathbf{n}\otimes\mathbf{n}\boxtimes\mathbf{n}\otimes\mathbf{n})$
$= \mathbb{I} - 2(\mathbf{n}\otimes\mathbf{n}\boxtimes\mathbf{e}_i\otimes\mathbf{e}_i + \mathbf{e}_i\otimes\mathbf{e}_i\boxtimes\mathbf{n}\otimes\mathbf{n} + 4\mathbf{n}\otimes\mathbf{n}\otimes\mathbf{n}\otimes\mathbf{n})$
$= \mathbb{I} - 2(\mathbf{n}\otimes\mathbf{e}_i\otimes\mathbf{n}\otimes\mathbf{e}_i + \mathbf{e}_i\otimes\mathbf{n}\otimes\mathbf{e}_i\otimes\mathbf{n} + 4\mathbf{n}\otimes\mathbf{n}\otimes\mathbf{n}\otimes\mathbf{n}).$
By components, $(\mathbb{S}_R)_{ijkl} = (\mathbf{I} - 2\mathbf{n}\otimes\mathbf{n})_{ik}(\mathbf{I} - 2\mathbf{n}\otimes\mathbf{n})_{jl}$
$= (\delta_{ik} - 2n_in_k)(\delta_{jl} - 2n_jn_l) = \delta_{ik}\delta_{jl} - 2(\delta_{ik}n_jn_l + \delta_{jl}n_in_k) + 4n_in_jn_kn_l.$

17. (i) $R_0 = 0$: This is the case of the so-called R_0-orthotropy.
(ii) $R_1 = 0$: This is the case of the square symmetry (all the components depend upon 4θ).
(iii) $\Phi_0 - \Phi_1 = k\frac{\pi}{4},\ k \in \{0,1\}$: This is the case of the two ordinary orthotropies.
(iv) $R_0 = R_1 = 0$: This is the condition for isotropy. Nothing depends upon $\theta \Rightarrow$ all the directions are equivalent and thus, at the same time, the axes of elastic symmetry.

Chapter 4

1. (i) $(\mathbf{u}(t)+\mathbf{v}(t)) - (\mathbf{u}(t_0)+\mathbf{v}(t_0)) = (t-t_0)(\mathbf{u}+\mathbf{v})' + o(t-t_0)$, and also, $\mathbf{u}(t) - \mathbf{u}(t_0) + \mathbf{v}(t) - \mathbf{v}(t_0) = (t-t_0)\mathbf{u}' + (t-t_0)\mathbf{v}' + o(t-t_0) \Rightarrow (\mathbf{u}+\mathbf{v})' = \mathbf{u}' + \mathbf{v}'$.

   The proof that $(\mathbf{L}+\mathbf{M})' = \mathbf{L}'+\mathbf{M}'$ and that $(\mathbb{L}+\mathbb{M})' = \mathbb{L}'+\mathbb{M}'$ can be done in a similar way.

   (ii) We indicate, in short, $\alpha(t) = \alpha, \mathbf{v}(t) = \mathbf{v}, \alpha(t_0) = \alpha_0, \mathbf{v}(t_0) = \mathbf{v}_0, \alpha'(t_0) = \alpha_0', \mathbf{v}'(t_0) = \mathbf{v}_0' \rightarrow \alpha\mathbf{v} - \alpha_0\mathbf{v}_0 = (\alpha\mathbf{v})_0'(t-t_0) + o(t-t_0)$; moreover, $\alpha = \alpha_0 + \alpha_0'(t-t_0) + o(t-t_0), \mathbf{v} = \mathbf{v}_0 + \mathbf{v}_0'(t-t_0) + o(t-t_0) \Rightarrow \alpha\mathbf{v} - \alpha_0\mathbf{v}_0 = (\alpha_0 + \alpha_0'(t-t_0) + o(t-t_0))(\mathbf{v}_0 + \mathbf{v}_0'(t-t_0) + o(t-t_0)) - \alpha_0\mathbf{v}_0 = \alpha_0'\mathbf{v}_0(t-t_0) + \mathbf{v}_0 o(t-t_0) + \alpha_0\mathbf{v}_0'(t-t_0) + \alpha_0'\mathbf{v}_0'(t-t_0)^2 + \mathbf{v}_0'(t-t_0)o(t-t_0) + \alpha_0 o(t-t_0) + \alpha_0'(t-t_0)o(t-t_0) + o(t-t_0)^2 = (\alpha_0'\mathbf{v}_0 + \alpha_0\mathbf{v}_0')(t-t_0) + o(t-t_0)$, so by comparison, $(\alpha\mathbf{v})' = \alpha'\mathbf{v} + \alpha\mathbf{v}'$.

   By the same technique, one can easily prove the differentiation rule for all the product-like quantities: $(\mathbf{u}\cdot\mathbf{v})', (\mathbf{u}\times\mathbf{v})', (\mathbf{u}\otimes\mathbf{v})', (\alpha\mathbf{L})', (\mathbf{L}\mathbf{v})', (\mathbf{L}\mathbf{M})', (\mathbf{L}\cdot\mathbf{M})', (\mathbf{L}\otimes\mathbf{M})', (\mathbf{L}\boxtimes\mathbf{M})', (\alpha\mathbb{L})', (\mathbb{L}\mathbf{L})', (\mathbb{L}\mathbb{M})', (\mathbb{L}\cdot\mathbb{M})'$.

2. The proof is the same for all the cases; we just write that for $\mathbf{v}(t)$: Using the two properties shown in the previous exercise, we get

$\mathbf{v}(t) = v_i(t)\mathbf{e}_i \Rightarrow \mathbf{v}'(t) = (v_i(t)\mathbf{e}_i)' = v_i'(t)\mathbf{e}_i + v_i(t)\mathbf{e}_i' = v_i'(t)\mathbf{e}_i$ because $\mathbf{e}_i$ does not depend on t.

3. (i) $p(\theta) = (a\theta\cos\theta, a\theta\sin\theta) \rightarrow c = \frac{2+\theta^2}{a(1+\theta^2)^{\frac{3}{2}}}$.
   (ii) $\ell = \frac{a}{2}(\theta\sqrt{1+\theta^2} + \ln(\theta + \sqrt{1+\theta^2}))$.
   (iii) Let p_i, p_{i+1} be two consecutive intersection points of the spiral (i denotes the order of the intersection point) with a straight line passing through the origin and inclined at θ; their distances from the origin are $r_i = a(\theta+2\pi i), r_{i+1} = a(\theta+2\pi(i+1)) \Rightarrow |p_{i+1}-p_i| = r_{i+1} - r_i = 2\pi a$ that does not depend upon θ.
4. (i) $r = a\ e^{b\theta} \Rightarrow r = 0 \iff \theta \rightarrow +\infty$, for $b < 0, -\infty$ for $b > 0$.
   (ii) $p(\theta) = (a\ e^{b\theta}\cos\theta, a\ e^{b\theta}\sin\theta) \rightarrow c = \frac{1}{a\ e^{b\theta}\sqrt{1+b^2}}$.
   (iii) $\ell = \frac{a}{b}\sqrt{1+b^2}e^{b\theta}$.
   (iv) With the same meaning as in the previous exercise, $|p_{i+1} - p_i| = r_{i+1} - r_i = a(e^{2\pi b} - 1)e^{b(\theta+2\pi i)}$, i.e. the distance depends upon the order of the intersection: $\frac{r_{i+1}}{r_i} = e^{2\pi b}$, which is a geometric progression.
   (v) $\boldsymbol{\tau} = \frac{1}{\sqrt{1+b^2}}(b\cos\theta - \sin\theta, b\sin\theta + \cos\theta) \Rightarrow (p-o)\cdot\boldsymbol{\tau} = \frac{ab\ e^{b\theta}}{\sqrt{1+b^2}} \Rightarrow \cos\varphi = \frac{(p-o)\cdot\boldsymbol{\tau}}{|p-o||\boldsymbol{\tau}|} = \frac{b}{\sqrt{1+b^2}} \Rightarrow \varphi$, the angle between τ and $p-o$, is constant.
   (vi) $\boldsymbol{\nu} = \frac{1}{\sqrt{1+b^2}}(-b\sin\theta - \cos\theta, b\cos\theta - \sin\theta) \Rightarrow q = p + \frac{1}{c}\boldsymbol{\nu} = ab\ e^{b\theta}(-\sin\theta, \cos\theta)$ is a point of the evolute, whose polar equation is hence $r = ab\ e^{b\theta}$, which is still a logarithmic spiral.
5. (i) $c = \frac{1}{a\theta}$.
   (ii) $\ell = \frac{1}{2}a\theta^2$.
   (iii) $\boldsymbol{\nu} = (-\sin\theta, \cos\theta) \Rightarrow p + \rho\boldsymbol{\nu} = p + \frac{1}{c}\boldsymbol{\nu} = a(\cos\theta, \sin\theta)$, which is the parametric equation of a circle of center o and radius a.
6. (i) $\boldsymbol{\tau} = \frac{1}{\sqrt{a^2+b^2}}(-a\sin\omega\theta, a\cos\omega\theta, b) \Rightarrow \cos\varphi = \boldsymbol{\tau}\cdot\mathbf{e}_3 = \frac{b}{\sqrt{a^2+b^2}}$, which is independent of θ.
   (ii) $\ell = \omega\theta\sqrt{a^2+b^2}$.
   (iii) $c = \frac{a}{a^2+b^2}$.
   (iv) $\vartheta = -\frac{b}{a^2+b^2}$.
   (v) Let p_i, p_{i+1} be two points, the intersection of the same generatrix of the cylinder with the helix, i.e. for, say, $\theta + i\frac{2\pi}{\omega}$ and $\theta + (i+1)\frac{2\pi}{\omega} \Rightarrow d = (p_{i+1} - p_i)\cdot\mathbf{e}_3 = 2\pi b$.
   (vi) By definition, a curve is a helix $\iff \boldsymbol{\tau}\cdot\mathbf{e}_3 = const$; differentiating gives $\boldsymbol{\tau}'\cdot\mathbf{e}_3 + \boldsymbol{\tau}\cdot\mathbf{e}_3 = \boldsymbol{\tau}'\cdot\mathbf{e}_3 = 0 \Rightarrow$ by the first equation of Frenet–Serret, $c\boldsymbol{\nu}\cdot\mathbf{e}_3 = 0 \Rightarrow \boldsymbol{\nu}\cdot\mathbf{e}_3 = 0 \Rightarrow \boldsymbol{\beta}$ is tangent to the

cylinder and $\boldsymbol{\beta} \cdot \mathbf{e}_3 = const.$ Moreover, differentiating again, $\boldsymbol{\nu} \cdot \mathbf{e}_3 + \boldsymbol{\nu} \cdot \mathbf{e}_3' = \boldsymbol{\nu}' \cdot \mathbf{e}_3 = 0 \Rightarrow$ by the third equation of Frenet–Serret $(-c\boldsymbol{\tau} - \vartheta\boldsymbol{\beta}) \cdot \mathbf{e}_3 = 0 \Rightarrow -c\boldsymbol{\tau} \cdot \mathbf{e}_3 = \vartheta\boldsymbol{\beta} \cdot \mathbf{e}_3 \Rightarrow \frac{c}{\vartheta} = -\frac{\boldsymbol{\beta}\cdot\mathbf{e}_3}{\boldsymbol{\tau}\cdot\mathbf{e}_3} = const.$ Conversely, if $p(s)$ is a curve with $\frac{c}{\vartheta} = \alpha = const.$, through the first and second equations of Frenet–Serret, we get $\boldsymbol{\nu} = \frac{1}{c}\boldsymbol{\tau}' = \frac{1}{\vartheta}\boldsymbol{\beta}' \Rightarrow \boldsymbol{\tau}' = \frac{c}{\vartheta}\boldsymbol{\beta}' = \alpha\boldsymbol{\beta}' \Rightarrow (\boldsymbol{\tau} - \alpha\boldsymbol{\beta})' = 0 \Rightarrow \mathbf{v} = \boldsymbol{\tau} - \alpha\boldsymbol{\beta} = const. \Rightarrow \boldsymbol{\tau} \cdot \mathbf{v} = 1 \Rightarrow \boldsymbol{\tau}$ forms with $\mathbf{v}$ a constant angle, and because $\mathbf{v}$ is a constant vector, $\boldsymbol{\tau} \cdot \mathbf{e}_3 = const. \Rightarrow$ the curve is a helix.

(vii) $p' \times p'' = \omega^3(ab\sin\omega\theta, -ab\cos\omega\theta, a^2) \Rightarrow A = ab\omega^3, B = a^2\omega^3$.

7. (i) $p(\theta) = R(\theta - \sin\theta, 1 - \cos\theta)$.
   (ii) $\ell(\theta) = 4R\left(1 - \cos\frac{\theta}{2}\right) \Rightarrow \ell(2\pi) = 8R$.
   (iii) $c = \frac{1}{2R\sqrt{2(1-\cos\theta)}}$.
   (iv) $\boldsymbol{\nu} = \frac{1}{\sqrt{2(1-\cos\theta)}}(\sin\theta, \cos\theta - 1) \Rightarrow q(\theta) = p(\theta) + \frac{1}{c}\boldsymbol{\nu} = R(\theta + \sin\theta, \cos\theta - 1)$; this curve is the evolute of the cycloid, and it can also be obtained as $q(\theta) = p(\theta + \pi) - (\pi, 2R)$, i.e. it is the same cycloid $p(\theta)$ translated by $-(\pi, 2R)$.

8. (i) $c = \frac{1}{\cosh^2 t}$.
   (ii) $\boldsymbol{\nu} = \left(-\frac{\sinh t}{\cosh t}, \frac{1}{\cosh t}\right) \Rightarrow q(t) = p(t) + \frac{1}{c}\boldsymbol{\nu} = (t - \sinh t\cosh t, 2\cosh t)$ is the evolute.
   (iii) $s = \int |p'(t)|dt = \sinh t$, $\boldsymbol{\tau} = \left(\frac{1}{\cosh t}, \frac{\sinh t}{\cosh t}\right) \Rightarrow b(t) = p(t) + (a - s)\boldsymbol{\tau} = \left(t + \frac{a-\sinh t}{\cosh t}, \cosh t + \sinh t\frac{a-\sinh t}{\cosh t}\right)$ is the equation of the involutes.

9. (i) $\boldsymbol{\tau} = (\cos t, \sin t)$; tangent to the tractrix at p: $z = p + w\boldsymbol{\tau} = \left((1+w)\cos t + \ln\left(\tan\frac{t}{2}\right), (1+w)\sin t\right)$. Intersection of g with x_1 axis for $w = -1 \rightarrow g = \left(\ln\left(\tan\frac{t}{2}\right), 0\right) \Rightarrow |p - g| = 1\ \forall t$.
   (ii) $\ell = \ln\frac{\sin t_2}{\sin t_1}$.
   (iii) $c = \tan t$.
   (iv) $\boldsymbol{\nu} = (-\sin t, \cos t) \Rightarrow q = p + \frac{1}{c}\boldsymbol{\nu} = \left(\ln\left(\tan\frac{t}{2}\right), \frac{1}{\sin t}\right)$. Setting $\sigma = \ln\left(\tan\frac{t}{2}\right) \Rightarrow \tan\frac{t}{2} = e^\sigma, \frac{1}{\sin t} = \cosh\sigma \Rightarrow q = (\sigma, \cosh\sigma)$, which is the equation of a catenary.

10. (i) $p(\theta) = (\cos\theta, \sin\theta, \sin\theta) \Rightarrow p'(= -\sin\theta, \cos\theta, \cos\theta), p'' = (-\cos\theta, -\sin\theta, -\sin\theta) \Rightarrow c = \frac{\sqrt{2}}{(1+\cos^2\theta)^{\frac{3}{2}}} \Rightarrow c_{max} = \sqrt{2}$.
    (ii) $p''' = (\sin\theta, -\cos\theta, -\cos\theta) \Rightarrow p' \times p'' \cdot p''' = \cos\theta - \cos\theta = 0 \Rightarrow \vartheta = 0 \Rightarrow$ the curve is planar.

11. (i) $\mathbf{v} = \dot{p}, \boldsymbol{\tau} = \frac{\dot{p}}{|\dot{p}|} \Rightarrow \dot{p} = |\dot{p}|\boldsymbol{\tau} \Rightarrow \mathbf{v} = v\boldsymbol{\tau}, v = |\dot{p}|$ is the scalar velocity.
    (ii) $\mathbf{a} = \ddot{p} = \dot{\mathbf{v}} = (v\boldsymbol{\tau})^\cdot = \dot{v}\boldsymbol{\tau} + v\dot{\boldsymbol{\tau}}; \dot{\boldsymbol{\tau}} = |\dot{\boldsymbol{\tau}}|\boldsymbol{\nu}; c = \frac{|\dot{\boldsymbol{\tau}}|}{|\dot{p}|} \Rightarrow |\dot{\boldsymbol{\tau}}| = |\dot{p}|c = \frac{v}{\rho} \Rightarrow \mathbf{a} = \dot{v}\boldsymbol{\tau} + \frac{v^2}{\rho}\boldsymbol{\nu}$; $\dot{v}$ is the *tangential acceleration* and $\frac{v^2}{\rho}$ the *centripetal acceleration*.

(iii) $f = m\mathbf{a} = m\dot{v}\boldsymbol{\tau} + \frac{mv^2}{\rho}\boldsymbol{\nu} = f_\tau\boldsymbol{\tau} + f_\nu\boldsymbol{\nu}$; $f_\tau = m\dot{v}$ is the *tangential force*, which is responsible for the change in the scalar velocity; $f_\nu = \frac{mv^2}{\rho}$ is the *centripetal force*, which is responsible for the path change.

Chapter 5

1. (i) By Eq. $(5.1)_3$, $\text{grad}(\mathbf{v}\cdot\mathbf{w}) = (\text{grad}\mathbf{w})^\top\mathbf{v} + (\text{grad}\mathbf{v})^\top\mathbf{w} = (\text{grad}\mathbf{w})^\top\mathbf{v} + (\text{grad}\mathbf{w})\mathbf{v} - (\text{grad}\mathbf{w})\mathbf{v} + (\text{grad}\mathbf{v})^\top\mathbf{w} + (\text{grad}\mathbf{v})\mathbf{w} - (\text{grad}\mathbf{v})\mathbf{w} = (\text{grad}\mathbf{w})\mathbf{v} + (\text{grad}\mathbf{v})\mathbf{w} + ((\text{grad}\mathbf{w})^\top - \text{grad}\mathbf{w})\mathbf{v} + ((\text{grad}\mathbf{v})^\top - \text{grad}\mathbf{v})\mathbf{w} = (\text{grad}\mathbf{w})\mathbf{v} + (\text{grad}\mathbf{v})\mathbf{w} - (\text{curl}\mathbf{w})\times\mathbf{v} - (\text{curl}\mathbf{v})\times\mathbf{w} = (\text{grad}\mathbf{w})\mathbf{v} + (\text{grad}\mathbf{v})\mathbf{w} + \mathbf{v}\times(\text{curl}\mathbf{w}) + \mathbf{w}\times(\text{curl}\mathbf{v})$.

   (ii) By Eqs. $(5.1)_{2,3}$ and Exercise 3(iii), Chapter 2, $\text{grad}(\mathbf{u}\cdot\mathbf{v}\ \mathbf{w}) = \mathbf{u}\cdot\mathbf{v}\ \text{grad}\mathbf{w} + \mathbf{w}\otimes\text{grad}(\mathbf{u}\cdot\mathbf{v}) = \mathbf{u}\cdot\mathbf{v}\ \text{grad}\mathbf{w} + \mathbf{w}\otimes((\text{grad}\mathbf{u})^\top\mathbf{v} + (\text{grad}\mathbf{v})^\top\mathbf{u}) = \mathbf{u}\cdot\mathbf{v}\ \text{grad}\mathbf{w} + \mathbf{w}\otimes(\text{grad}\mathbf{u})^\top\mathbf{v} + \mathbf{w}\otimes(\text{grad}\mathbf{v})^\top\mathbf{u} = \mathbf{u}\cdot\mathbf{v}\ \text{grad}\mathbf{w} + (\mathbf{w}\otimes\mathbf{v})\text{grad}\mathbf{u} + (\mathbf{w}\otimes\mathbf{u})\text{grad}\mathbf{v}$.

   (iii) By Theorem 30(i), $\text{div}((\text{grad}\mathbf{v})\mathbf{v} - (\text{div}\mathbf{v})\mathbf{v}) + (\text{div}\mathbf{v})^2 = \text{div}((\text{grad}\mathbf{v})\mathbf{v}) - \text{div}((\text{div}\mathbf{v})\mathbf{v}) + (\text{div}\mathbf{v})^2 = (\text{grad}\mathbf{v})^\top\cdot\text{grad}\mathbf{v} + \mathbf{v}\cdot\text{div}(\text{grad}\mathbf{v})^\top - (\text{div}\mathbf{v})^2 - \mathbf{v}\cdot\text{grad}(\text{div}\mathbf{v}) + (\text{div}\mathbf{v})^2 = (\text{grad}\mathbf{v})^\top\cdot\text{grad}\mathbf{v} + \mathbf{v}\cdot\text{div}(\text{grad}\mathbf{v})^\top - \mathbf{v}\cdot\text{div}(\text{grad}\mathbf{v})^\top = (\text{grad}\mathbf{v})^\top\cdot\text{grad}\mathbf{v}$.

2. (i) By Theorem 30(i), $\text{div}(\text{grad}\mathbf{v})^\top = \text{div}(v_{j,i}\mathbf{e}_i\otimes\mathbf{e}_j) = v_{j,i}\text{div}(\mathbf{e}_i\otimes\mathbf{e}_j) + \mathbf{e}_i\otimes\mathbf{e}_j\text{grad}v_{j,i} = \mathbf{e}_i\otimes\mathbf{e}_j v_{j,ik}\mathbf{e}_k = v_{j,ik}\delta_{jk}\mathbf{e}_i = v_{j,ij}\mathbf{e}_i = v_{j,ji}\mathbf{e}_i = (\text{div}\mathbf{v})_{,i}\mathbf{e}_i = \text{grad}(\text{div}\mathbf{v})$.

   (ii) By (iii) of the previous exercise and Theorem 30(i), $\text{div}((\text{grad}\mathbf{v})\mathbf{v} - (\text{div}\mathbf{v})\mathbf{v}) = \text{grad}\mathbf{v}\cdot(\text{grad}\mathbf{v})^\top - (\text{div}\mathbf{v})^2$ and also $\text{div}((\text{grad}\mathbf{v})\mathbf{v} - (\text{div}\mathbf{v})\mathbf{v}) = \text{div}(\text{grad}\mathbf{v})\mathbf{v}) - \text{div}((\text{div}\mathbf{v})\mathbf{v}) = \text{div}((\text{grad}\mathbf{v})\mathbf{v}) - (\text{div}\mathbf{v})^2 - \mathbf{v}\cdot\text{grad}(\text{div}\mathbf{v})$, so comparing the two results, $\text{div}((\text{grad}\mathbf{v})\mathbf{v}) = \text{grad}\mathbf{v}\cdot(\text{grad}\mathbf{v})^\top + \mathbf{v}\cdot\text{grad}(\text{div}\mathbf{v})$.

   (iii) By Theorem 30(i) and (iv), $\text{div}(\varphi\mathbf{L}\mathbf{v}) = \varphi\text{div}(\mathbf{L}\mathbf{v}) + \mathbf{L}\mathbf{v}\cdot\text{grad}\varphi = \mathbf{L}\mathbf{v}\cdot\text{grad}\varphi + \varphi\mathbf{L}^\top\cdot\text{grad}\mathbf{v} + \varphi\mathbf{v}\cdot\text{div}\mathbf{L}^\top$.

3. (i) By Theorem 28(ii) and Eq. (2.33), $\forall\mathbf{a} = const. \in \mathcal{V}$, $(\text{curl}(\varphi\mathbf{v}))\times\mathbf{a} = (\text{grad}(\varphi\mathbf{v}) - (\text{grad}(\varphi\mathbf{v})^\top)\mathbf{a} = (\varphi\text{grad}\mathbf{v} + \mathbf{v}\otimes\text{grad}\varphi - (\varphi\text{grad}\mathbf{v} + \mathbf{v}\otimes\text{grad}\varphi)^\top)\mathbf{a} = (\varphi\text{grad}\mathbf{v} + \mathbf{v}\otimes\text{grad}\varphi - \varphi(\text{grad}\mathbf{v})^\top - \text{grad}\varphi\otimes\mathbf{v})\mathbf{a} = \varphi(\text{curl}\mathbf{v})\times\mathbf{a} + \mathbf{a}\cdot\text{grad}\varphi\ \mathbf{v} - \mathbf{a}\cdot\mathbf{v}\ \text{grad}\varphi = \varphi(\text{curl}\mathbf{v})\times\mathbf{a} + \mathbf{a}\times(\mathbf{v}\times\text{grad}\varphi) = \varphi(\text{curl}\mathbf{v})\times\mathbf{a} - (\mathbf{v}\times\text{grad}\varphi)\times\mathbf{a} = (\varphi\text{curl}\mathbf{v} - \mathbf{v}\times\text{grad}\varphi)\times\mathbf{a} \Rightarrow \text{curl}(\varphi\mathbf{v}) = \varphi\text{curl}\mathbf{v} + \text{grad}\varphi\times\mathbf{v}$.

(ii) Using Ricci's alternator for the cross product, $\mathbf{v}\times\mathbf{w}=\epsilon_{pqr}v_q w_r\mathbf{e}_p$ and $\text{curl}(\mathbf{v}\times\mathbf{w})=\epsilon_{ijk}(\mathbf{v}\times\mathbf{w})_{k,j}\mathbf{e}_i=\epsilon_{ijk}\epsilon_{kqr}(v_q w_r)_{,j}\mathbf{e}_i=\epsilon_{ijk}\epsilon_{kqr}(v_{q,j}w_r+v_q w_{r,j})\mathbf{e}_i=\epsilon_{kij}\epsilon_{kqr}(v_{q,j}w_r+v_q w_{r,j})\mathbf{e}_i$. Then, because $\epsilon_{kij}\epsilon_{kqr}=\delta_{iq}\delta_{jr}-\delta_{ir}\delta_{jq}$, we get $\text{curl}(\mathbf{v}\times\mathbf{w})=(\delta_{iq}\delta_{jr}-\delta_{ir}\delta_{jq})(v_{q,j}w_r+v_q w_{r,j})\mathbf{e}_i=\delta_{iq}\delta_{jr}(v_{q,j}w_r+v_q w_{r,j})\mathbf{e}_i-\delta_{ir}\delta_{jq}(v_{q,j}w_r+v_q w_{r,j})\mathbf{e}_i=(v_{i,j}w_j+v_i w_{j,j})\mathbf{e}_i-(v_{j,j}w_i+v_j w_{i,j})\mathbf{e}_i=\text{grad}\mathbf{v}\,\mathbf{w}-\text{grad}\mathbf{w}\,\mathbf{v}+\mathbf{v}\text{div}\mathbf{w}-\mathbf{w}\text{div}\mathbf{v}$.

4. (i) Through Theorem 30(iv), we get $\int_{\partial\Omega}\mathbf{v}\cdot\mathbf{L}\mathbf{n}\,dA=\int_{\partial\Omega}\mathbf{L}^\top\mathbf{v}\cdot\mathbf{n}\,dA=\int_\Omega\text{div}(\mathbf{L}^\top\mathbf{v})dV=\int_\Omega\mathbf{L}\cdot\text{grad}\mathbf{v}+\mathbf{v}\cdot\text{div}\mathbf{L}\,dV$.
   (ii) By Theorems 29 and 30(i), $\forall\mathbf{a}=const.\in\mathcal{V}$, $\int_{\partial\Omega}(\mathbf{L}\mathbf{n})\otimes\mathbf{v}\,\mathbf{a}\,dA=\int_{\partial\Omega}\mathbf{a}\cdot\mathbf{v}\;\mathbf{L}\mathbf{n}\;dA=\int_\Omega\text{div}(\mathbf{a}\cdot\mathbf{v}\;\mathbf{L})dV=\int_\Omega\mathbf{a}\cdot\mathbf{v}\;\text{div}\mathbf{L}+\mathbf{L}\text{grad}(\mathbf{a}\cdot\mathbf{v})dV=\int_\Omega\mathbf{a}\cdot\mathbf{v}\;\text{div}\mathbf{L}+\mathbf{L}(\text{grad}\mathbf{a})^\top\mathbf{v}+\mathbf{L}(\text{grad}\mathbf{v})^\top\mathbf{a}\,dV=\int_\Omega((\text{div}\mathbf{L})\otimes\mathbf{v}+\mathbf{L}(\text{grad}\mathbf{v})^\top)\mathbf{a}\,dV$.
   (iii) By Theorem 30(ii), $\int_{\partial\Omega}(\mathbf{w}\cdot\mathbf{n})\mathbf{v}\,dA=\int_{\partial\Omega}(\mathbf{v}\otimes\mathbf{w})\mathbf{n}\,dA=\int_\Omega\text{div}(\mathbf{v}\otimes\mathbf{w})dV=\int_\Omega\mathbf{v}\;\text{div}\mathbf{w}+(\text{grad}\mathbf{v})\mathbf{w}\;dV$.

5. (i) Take $\mathbf{u}=\alpha\mathbf{n}$, $\mathbf{n}\in\mathcal{S}\rightarrow\varphi(p+\alpha\mathbf{n})=\varphi(p)+\alpha\;\text{grad}\varphi\cdot\mathbf{n}+o(\alpha)\Rightarrow$ $\frac{d\varphi}{d\mathbf{n}}:=\lim_{\alpha\to0}\frac{\varphi(p+\alpha\mathbf{n})-\varphi(\mathbf{p})}{\alpha}=\text{grad}\varphi\cdot\mathbf{n}$.
   (ii) In a similar way, $\mathbf{v}(p+\alpha\mathbf{n})=\mathbf{v}(p)+\alpha\;\text{grad}\mathbf{v}\;\mathbf{n}+o(\alpha)\Rightarrow$ $\frac{d\mathbf{v}}{d\mathbf{n}}:=\lim_{\alpha\to0}\frac{\mathbf{v}(p+\alpha\mathbf{n})-\mathbf{v}(\mathbf{p})}{\alpha}=\text{grad}\mathbf{v}\;\mathbf{n}$.

6. (i) Applying the first proof of the previous exercise to $\mathbf{n}=\mathbf{e}_i$, $i=1,2,3$, we get immediately $\frac{df}{d\mathbf{e}_i}:=f_{,i}=\text{grad}f\cdot\mathbf{e}_i:=(\text{grad}f)_{,i}\Rightarrow\text{grad}f=f_{,i}\mathbf{e}_i$.
   (ii) Applying the second proof of the previous exercise to $\mathbf{n}=\mathbf{e}_j$, $j=1,2,3$, we obtain $\frac{d\mathbf{v}}{d\mathbf{e}_j}:=\mathbf{v}_{,j}=v_{i,j}\mathbf{e}_i=\text{grad}\mathbf{v}\;\mathbf{e}_j=(\text{grad}\mathbf{v})_{ik}\mathbf{e}_i\otimes\mathbf{e}_k\;\mathbf{e}_j=\delta_{kj}(\text{grad}\mathbf{v})_{ik}\mathbf{e}_i=(\text{grad}\mathbf{v})_{ij}\mathbf{e}_i$
   $\Rightarrow(\text{grad}\mathbf{v})_{ij}=v_{i,j}\Rightarrow\text{grad}\mathbf{v}=v_{i,j}\mathbf{e}_i\otimes\mathbf{e}_j$.
   (iii) $\text{div}\mathbf{v}:=\text{tr}(\text{grad}\mathbf{v})=\text{tr}(v_{i,j}\mathbf{e}_i\otimes\mathbf{e}_j)=v_{i,j}\text{tr}(\mathbf{e}_i\otimes\mathbf{e}_j)=\delta_{ij}v_{i,j}=v_{i,i}$.
   (iv) $\forall\mathbf{u}=const.\in\mathcal{V}$, $(\text{div}\mathbf{L})\cdot\mathbf{u}:=\text{div}(\mathbf{L}^\top\mathbf{u})\Rightarrow(\text{div}\mathbf{L})_i u_i=(\mathbf{L}^\top\mathbf{u})_{j,j}=(L_{ij}u_i)_{,j}=L_{ij,j}u_i\Rightarrow(\text{div}\mathbf{L})_i=L_{ij,j}\Rightarrow\text{div}\mathbf{L}=L_{ij,j}\mathbf{e}_i$.
   (v) $\Delta f:=\text{div}(\text{grad}f)=\text{div}(f_{,i}\mathbf{e}_i)=f_{,ii}$.
   (vi) $\Delta\mathbf{v}:=\text{div}(\text{grad}\mathbf{v})=\text{div}(v_{i,j}\mathbf{e}_i\otimes\mathbf{e}_j)=v_{i,jj}\mathbf{e}_i$.
   (vii) For the sake of brevity, let $\mathbf{w}=\text{curl}\mathbf{v}$; $\forall\mathbf{u}\in\mathcal{V}$,

$$(\text{curl}\mathbf{v})\times\mathbf{u}:=(\text{grad}\mathbf{v}-\text{grad}\mathbf{v}^\top)\mathbf{u}\Rightarrow\begin{bmatrix}0&-w_3&w_2\\w_3&0&-w_1\\-w_2&w_1&0\end{bmatrix}\begin{Bmatrix}u_1\\u_2\\u_3\end{Bmatrix}$$

$$=\left(\begin{bmatrix}v_{1,1}&v_{1,2}&v_{1,3}\\v_{2,1}&v_{2,2}&v_{2,3}\\v_{3,1}&v_{3,2}&v_{3,3}\end{bmatrix}-\begin{bmatrix}v_{1,1}&v_{2,1}&v_{3,1}\\v_{1,2}&v_{2,2}&v_{3,2}\\v_{1,3}&v_{2,3}&v_{3,3}\end{bmatrix}\right)\begin{Bmatrix}u_1\\u_2\\u_3\end{Bmatrix}$$

$$= \begin{bmatrix} 0 & v_{1,2}-v_{2,1} & v_{1,3}-v_{3,1} \\ v_{2,1}-v_{1,2} & 0 & v_{2,3}-v_{3,2} \\ v_{3,1}-v_{1,3} & v_{3,2}-v_{2,3} & 0 \end{bmatrix} \begin{Bmatrix} u_1 \\ u_2 \\ u_3 \end{Bmatrix} \Rightarrow \text{curl}\mathbf{v}$$

$$= \begin{Bmatrix} v_{3,2}-v_{2,3} \\ v_{1,3}-v_{3,1} \\ v_{2,1}-v_{1,2} \end{Bmatrix}.$$

7. (i) $\mathbf{v}(p) = \mathbf{v}(p_0) + \boldsymbol{\omega} \times (p - p_0) \Rightarrow \exists \mathbf{W}_\omega \in Skw(\mathcal{V}) |\ \mathbf{v}(p) = \mathbf{v}(p_0) + \mathbf{W}_\omega(p - p_0)$, with $\mathbf{W}_\omega$ the axial tensor of $\boldsymbol{\omega}$. Moreover, by the definition of gradient, $\mathbf{v}(p) = \mathbf{v}(p_0) + (\text{grad}\mathbf{v})(p - p_0) \Rightarrow \mathbf{W}_\omega = \text{grad}\mathbf{v} \Rightarrow \text{grad}\mathbf{v} = \frac{\text{grad}\mathbf{v}+\text{grad}\mathbf{v}^\top}{2} + \frac{\text{grad}\mathbf{v}-\text{grad}\mathbf{v}^\top}{2} = \mathbf{W}_\omega = -\mathbf{W}_\omega^\top \iff \frac{\text{grad}\mathbf{v}+\text{grad}\mathbf{v}^\top}{2} = \mathbf{O} \Rightarrow \text{grad}\mathbf{v} = \frac{\text{grad}\mathbf{v}-\text{grad}\mathbf{v}^\top}{2} \Rightarrow \mathbf{v}(p) = \mathbf{v}(p_0) + \frac{\text{grad}\mathbf{v}-\text{grad}\mathbf{v}^\top}{2}(p - p_0)$ so, by the definition of curl, $\mathbf{v}(p) = \mathbf{v}(p_0) + \frac{1}{2}\text{curl}\mathbf{v} \times (p - p_0)$, and comparing the two results, we get $\boldsymbol{\omega} = \frac{1}{2}\text{curl}$.
   (ii) By the definition of divergence and Eq. (2.8), $\text{div}\mathbf{v} = \text{tr}(\text{grad}\mathbf{v}) = \text{tr}\mathbf{W}_\omega = 0$.
8. (i) $\text{div}\mathbf{u} = 3\alpha \to$ nowhere isochoric.
   (ii) $\text{div}\mathbf{u} = 0 \to$ globally isochoric.
   (iii) $\text{div}\mathbf{u} = \gamma(x_1 + x_2 + x_3) = 0$: isochoric on the points of the plane $x_1 + x_2 + x_3 = 0$.
   (iv) $\text{div}\mathbf{u} = \delta(\cos x_1 + \sin \mathbf{x}_2 + \cos x_3) = 0$: isochoric on the points of the surface $\cos x_1 + \sin \mathbf{x}_2 + \cos x_3 = 0$.
9. Using Eq. (5.14), we get:

   (i) $v_\theta = v_z = 0,\ v_\rho = \frac{\alpha}{\rho} \Rightarrow \text{div}\mathbf{v} = -\frac{\alpha}{\rho^2} + \frac{\alpha}{\rho^2} = 0.$
   (ii) $v_\rho = v_z = 0,\ v_\theta = \frac{\alpha}{\rho} \Rightarrow \text{div}\mathbf{v} = 0.$
   (iii) $v_\rho = \frac{\alpha\cos\theta}{\rho^2}, v_\theta = \frac{\alpha\sin\theta}{\rho^2}, v_z = 0 \Rightarrow \text{div}\mathbf{u} = -\frac{2\alpha\cos\theta}{\rho^3} + \frac{2\alpha\cos\theta}{\rho^3} = 0.$

10. Using Eq. (5.15), we get:

   (i) $v_{\rho,\theta} = v_{\rho,z} = v_{\theta,\rho} = v_{\theta,z} = v_{z,\rho} = v_{z,\theta} = 0 \Rightarrow \text{curl}\mathbf{v} = \mathbf{o}.$
   (ii) $v_{\rho,\theta} = v_{\rho,z} = v_{\theta,z} = v_{z,\rho} = v_{z,\theta} = 0, v_{\theta,\rho} = -\frac{\alpha}{\rho^2} \Rightarrow \text{curl}\mathbf{v} = \left(0, 0, \frac{\alpha}{\rho^2} - \frac{\alpha}{\rho^2}\right) = \mathbf{o}.$
   (iii) $v_{\rho,\theta} = -\frac{\alpha\cos\theta}{\rho^2}, v_{\rho,z} = 0, v_{\theta,\rho} = -\frac{2\alpha\sin\theta}{\rho^3}, v_{\theta,z} = 0, v_{z,\rho} = v_{z,\theta} = 0 \Rightarrow \text{curl}\mathbf{v} = \left(0, 0, \frac{\alpha\cos\theta}{\rho^3} - 2\frac{\alpha\cos\theta}{\rho^3} + \frac{\alpha\cos\theta}{\rho^3}\right) = \mathbf{o}.$

Chapter 6

1. Setting $\rho = z^1, \theta = z^2, z = z^3$, by Eq. (6.3), we get $\mathbf{g} = \begin{bmatrix} 1 & 0 & 0 \\ 0 & \rho^2 & 0 \\ 0 & 0 & 1 \end{bmatrix} \Rightarrow$ $ds = \sqrt{g_{hk}dz^h dz^k} = \sqrt{d\rho^2 + \rho^2 d\theta^2 + dz^2}$.
2. Setting $r = z^1, \theta = z^2, \varphi = z^3$, proceeding in a similar way, we get $\mathbf{g} = \begin{bmatrix} 1 & 0 & 0 \\ 0 & r^2\sin^2\varphi & 0 \\ 0 & 0 & r^2 \end{bmatrix} \Rightarrow ds = \sqrt{g_{hk}dz^h dz^k}$ $= \sqrt{dr^2 + r^2\sin^2\varphi d\theta^2 + r^2 d\varphi^2}$.
3. For cylindrical coordinates, cf. Exercise 1, $ds = \sqrt{d\rho^2 + \rho^2 d\theta^2 + dz^2}$, and for a curve on a circular cylinder, $\rho = R \Rightarrow d\rho = 0 \Rightarrow ds = \sqrt{R^2 d\theta^2 + dz^2}$; if the equation of the helix is $p(\theta) = R\cos\theta\mathbf{e}_1 + R\sin\theta\mathbf{e}_2 + b\theta\mathbf{e}_3$, then $\frac{dz}{d\theta} = b \Rightarrow dz = b\ d\theta \Rightarrow$ $ds = \sqrt{R^2 + b^2}d\theta \Rightarrow \ell = \int_\theta^{\theta+2\pi}\sqrt{R^2 + b^2}d\theta = 2\pi\sqrt{R^2 + b^2}$.
4. (i) $r = \frac{2R}{\pi}\theta \Rightarrow dr = \frac{2R}{\pi}d\theta;\ ds = \sqrt{dr^2 + r^2 d\theta^2} = \frac{2R}{\pi}\sqrt{1+\theta^2}d\theta \Rightarrow$ $\ell = \int_0^{\frac{\pi}{2}} ds = \frac{2R}{\pi}\int_0^{\frac{\pi}{2}}\sqrt{1+\theta^2}d\theta = \frac{R}{\pi}[\theta\sqrt{1+\theta^2} + \text{arcsinh}\theta]_0^{\frac{\pi}{2}} =$ $\left(\frac{1}{4}\sqrt{4+\pi^2} + \frac{1}{\pi}\text{arcsinh}\frac{\pi}{2}\right)R \sim 1.324R$.
   (ii) $\frac{\pi}{2}R = 2\pi R_0 \Rightarrow R_0 = \frac{R}{4} \Rightarrow h = \sqrt{R^2 - R_0^2} = \frac{15}{4}R;\ \rho(z) = \frac{R_0}{h}z,\ z = \frac{h}{2\pi}\theta \Rightarrow \rho(\theta) = \frac{R_0}{2\pi}\theta \Rightarrow \rho(\theta) = \frac{R}{8\pi}\theta,\ z(\theta) = \frac{\sqrt{15}}{8\pi}R\theta \Rightarrow d\rho = \frac{R}{8\pi}d\theta,\ dz = \frac{\sqrt{15}}{8\pi}R\ d\theta$; equation of the conical helix: $p(\theta) = \rho(z)\cos\theta\mathbf{e}_1 + \rho(z)\sin\theta\mathbf{e}_2 + \sqrt{15}\theta\mathbf{e}_3 = \frac{R}{8\pi}(\theta\cos\theta\mathbf{e}_1 + \theta\sin\theta\mathbf{e}_2 + \sqrt{15}\theta\mathbf{e}_3) \Rightarrow ds = \sqrt{d\rho^2 + \rho^2 d\theta^2 + dz^2} = \frac{R}{8\pi}\sqrt{d\theta^2 + \theta^2 d\theta^2 + 15 d\theta^2} = \frac{R}{8\pi}\sqrt{16+\theta^2}d\theta \Rightarrow \ell = \int_0^{2\pi} ds = \frac{R}{8\pi}\int_0^{2\pi}\sqrt{16+\theta^2}d\theta =$ $\frac{R}{8\pi}\left[\frac{1}{2}\theta\sqrt{16+\theta^2} + 8\text{arcsinh}\frac{\theta}{4}\right]_0^{2\pi} \sim 1.324R$.
5. (i) Referring to Fig. 6.5 and by Eq. (6.3), $x_1 = z^1\cos\alpha_1 + z^2\cos\alpha_2$, $x_2 = z^1\sin\alpha_1 + z^2\sin\alpha_2 \Rightarrow \mathbf{g} = \begin{bmatrix} 1 & \cos(\alpha_2 - \alpha_1) \\ \cos(\alpha_2 - \alpha_1) & 1 \end{bmatrix}$.
   (ii) By Eq. (6.5), $\mathbf{g}_1 = \cos\alpha_1\mathbf{e}_1 + \sin\alpha_1\mathbf{e}_2,\ \mathbf{g}_2 = \cos\alpha_2\mathbf{e}_1 + \sin\alpha_2\mathbf{e}_2$.
   (iii) $z^1 = h(x_1\sin\alpha_2 - x_2\cos\alpha_2),\ z^2 = h(-x_1\sin\alpha_1 + x_2\cos\alpha_1),\ h = \frac{1}{\sin(\alpha_2-\alpha_1)} \Rightarrow$ by Eq. (6.14), $\mathbf{g}^1 = \frac{\sin\alpha_2}{\sin(\alpha_2-\alpha_1)}\mathbf{e}_1 - \frac{\cos\alpha_2}{\sin(\alpha_2-\alpha_1)}\mathbf{e}_2,\ \mathbf{g}^2 = -\frac{\sin\alpha_1}{\sin(\alpha_2-\alpha_1)}\mathbf{e}_1 + \frac{\cos\alpha_1}{\sin(\alpha_2-\alpha_1)}\mathbf{e}_2$.
   (iv) $\mathbf{g}_1 \cdot \mathbf{g}^1 = \frac{\cos\alpha_1\sin\alpha_2 - \sin\alpha_1\cos\alpha_2}{\sin(\alpha_2-\alpha_1)} = 1$,
   $\mathbf{g}_2 \cdot \mathbf{g}^2 = \frac{-\sin\alpha_1\cos\alpha_2 + \sin\alpha_2\cos\alpha_1}{\sin(\alpha_2-\alpha_1)} = 1$,

$$\mathbf{g}_1 \cdot \mathbf{g}^2 = \frac{-\cos\alpha_1 \sin\alpha_1 + \sin\alpha_1 \cos\alpha_1}{\sin(\alpha_2-\alpha_1)} = 0,$$
$$\mathbf{g}_2 \cdot \mathbf{g}^1 = \frac{\cos\alpha_2 \sin\alpha_2 - \sin\alpha_2 \cos\alpha_2}{\sin(\alpha_2-\alpha_1)} = 0.$$

(v) $|\mathbf{g}_1| = |\mathbf{g}_2| = 1,\ |\mathbf{g}^1| = |\mathbf{g}^2| = \frac{1}{|\sin(\alpha_2-\alpha_1)|}$.

(vi)

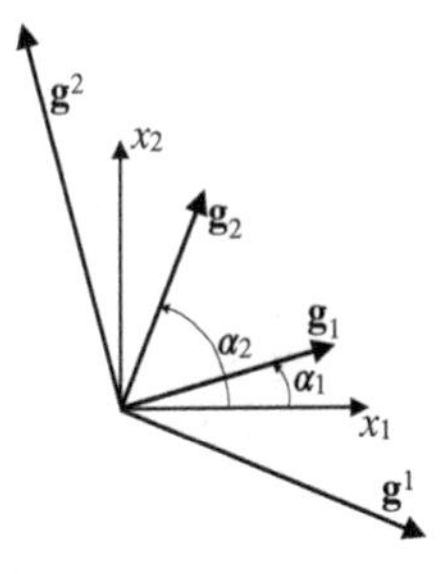

6. Referring to Exercise 2 and by Eq. (6.5), we get $\mathbf{g}_1 = \cos\theta\sin\varphi\mathbf{e}_1 + \sin\theta\sin\varphi\mathbf{e}_2 + \cos\varphi\mathbf{e}_3$, $\mathbf{g}_2 = -r\sin\theta\sin\varphi\mathbf{e}_1 + r\cos\theta\sin\varphi\mathbf{e}_2$, $\mathbf{g}_3 = r\cos\theta\cos\varphi\mathbf{e}_1 + r\sin\theta\cos\varphi\mathbf{e}_2 - r\sin\varphi\mathbf{e}_3$.
7. (i) If $z^1 = const. \Rightarrow x_1 = a\cos z^2,\ x_2 = b\sin z^2$, with $a = c\ \cosh z^1 = const.$, $b = c\ \sinh z^1 = const. \Rightarrow \frac{x_1^2}{a^2} + \frac{x_2^2}{b^2} = 1$: family of ellipses all with the same focuses $x_e = \pm\sqrt{a^2 - b^2} = \pm c$.
   (ii) If $z^2 = const. \Rightarrow x_1 = A\cosh z^1,\ x_2 = B\sinh z^1$, with $A = c\ \cos z^2 = const.$, $B = c\ \sin z^2 = const. \Rightarrow \frac{x_1^2}{A^2} - \frac{x_2^2}{B^2} = 1$: family of hyperbolae all with the same focuses $x_h = \pm\sqrt{A^2 + B^2} = \pm c \Rightarrow x_e = x_h$.
   (iii) The axes of the ellipses are $2a = 2c\ \cosh z^1$ and $2b = 2c\ \sinh z^1$.
   (iv) A crack along the horizontal axis corresponds to $b \to 0$, which happens $\iff z^1 \to 0 \Rightarrow \cosh z^1 \to 1$ and $a \to c \Rightarrow$ length of the crack: $2c$.
   (v) Applying Eq. (6.3), we get $\mathbf{g} = \frac{c^2}{2}\big(\cosh 2z^1 - \cos 2z^2\big)\mathbf{I}$.
   (vi) By Eq. (6.5), $\mathbf{g}_1 = c\ \sinh z^1 \cos z^2 \mathbf{e}_1 + c\ \cosh z^1 \sin z^2 \mathbf{e}_2$, $\mathbf{g}_2 = -c\ \cosh z^1 \sin z^2 \mathbf{e}_1 + c\ \sinh z^1 \cos z^2 \mathbf{e}_2$. We note that $\mathbf{g}_1 \cdot \mathbf{g}_2 = 0$.
8. (i) By Eq. $(6.16)_1$, setting $z^1 = \rho, z^2 = \theta, z^3 = z$, we get

$$L^{11} = L^x_{11}\cos^2\theta + (L^x_{12} + L^x_{21})\sin\theta\cos\theta + L^x_{22}\sin^2\theta,$$
$$L^{12} = \tfrac{1}{\rho}\big((L^x_{22} - L^x_{11})\sin\theta\cos\theta + L^x_{12}\cos^2\theta - L^x_{21}\sin^2\theta\big),$$
$$L^{13} = L^x_{13}\cos\theta + L^x_{23}\sin\theta,$$
$$L^{21} = \tfrac{1}{\rho}\big((L^x_{22} - L^x_{11})\sin\theta\cos\theta - L^x_{12}\sin^2\theta + L^x_{21}\cos^2\theta\big),$$

$$L^{22} = \frac{1}{\rho^2}\left(L^x_{11}\sin^2\theta - (L^x_{12} + L^x_{21})\sin\theta\cos\theta + L^x_{22}\cos^2\theta\right),$$
$$L^{23} = -L^x_{13}\sin\theta + L^x_{23}\cos\theta,$$
$$L^{31} = L^x_{31}\cos\theta + L^x_{32}\sin\theta,$$
$$L^{32} = -L^x_{31}\sin\theta + L^x_{32}\cos\theta,$$
$$L^{33} = L^x_{33}.$$

(ii) The covariant components can alternatively be found by Eq. $(6.16)_2$ or, using the results of the previous point, by Eq. $(6.19)_2$; by this latter way, using the result of Ex. 1, we get easily $L_{11} = L^{11}$, $L_{12} = \rho^2 L^{12}$, $L_{13} = L^{13}$, $L_{21} = \rho^2 L^{21}, L_{22} = \rho^4 L^{22}$, $L_{23} = \rho^2 L^{23}$, $L_{31} = L^{31}$, $L_{32} = \rho^2 L^{32}$, $L_{33} = L^{33}$.

9. (i) First, the covariant components: By Eq. $(6.16)_2$, setting $z^1 = r, z^2 = \theta, z^3 = \varphi$, we get:

$$\begin{aligned}L_{11} = {}& L^x_{11}\cos^2\theta\sin^2\varphi + (L^x_{12} + L^x_{21})\sin\theta\cos\theta\sin^2\varphi \\ &+ (L^x_{13} + L^x_{31})\cos\theta\sin\varphi\cos\varphi + L^x_{22}\sin^2\theta\sin^2\varphi \\ &+(L^x_{23} + L^x_{32})\sin\theta\sin\varphi\cos\varphi + L^x_{33}\cos^2\varphi,\end{aligned}$$

$$\begin{aligned}L_{12} = {}& -r\cos\theta\sin\theta\sin^2\varphi L^x_{11} + r\sin^2\varphi(L^x_{12}\cos^2\theta - L^x_{21}\sin^2\theta) \\ &+ r\sin\theta\cos\theta\sin^2\varphi L^x_{22} + r\sin\varphi\cos\varphi(L^x_{32}\cos\theta - L^x_{31}\sin\theta),\end{aligned}$$

$$\begin{aligned}L_{13} = {}& r\cos^2\theta\sin\varphi\cos\varphi L^x_{11} + r\sin\theta\cos\theta\sin\varphi\cos\varphi(L^x_{12} + L^x_{21}) \\ &+ r\sin^2\theta\cos\varphi\sin\varphi L^x_{22} - r\sin^2\varphi\sin\theta L^x_{23} \\ &+ r\cos\theta(L^x_{31}\cos^2\varphi - L^x_{13}\sin^2\varphi) \\ &- r\cos\varphi\sin\varphi L^x_{33},\end{aligned}$$

$$\begin{aligned}L_{21} = {}& -r\sin\theta\cos\theta\sin^2\varphi L^x_{11} + r\sin^2\varphi(L^x_{21}\cos^2\theta - L^x_{12}\sin^2\theta) \\ &+ r\sin\theta\cos\theta\sin^2\varphi L^x_{22} + r\sin\varphi\cos\varphi(L^x_{23}\cos\theta - L^x_{13}\sin\theta),\end{aligned}$$

$$\begin{aligned}L_{22} = {}& r^2\sin^2\theta\sin^2\varphi L^x_{11} - r^2\sin\theta\cos\theta\sin^2\varphi(L^x_{12} + L^x_{21}) \\ &+ r^2\cos^2\theta\sin^2\varphi L^x_{22},\end{aligned}$$

$$\begin{aligned}L_{23} = {}& -r^2\sin\theta\cos\theta\sin\varphi\cos\varphi(L^x_{22} - L^x_{11}) + r^2\sin\varphi\cos\varphi \\ &(L^x_{21}\cos^2\theta - L^x_{12}\sin^2\theta) \\ &+ r^2\sin^2\varphi(L^x_{13}\sin\theta - L^x_{23}\cos\theta),\end{aligned}$$

$$\begin{aligned}L_{31} = {}& r\sin\varphi\cos\varphi(L^x_{11}\cos^2\theta + L^x_{22}\sin^2\theta) \\ &+ r\sin\theta\cos\theta\sin\varphi\cos\varphi(L^x_{12} + L^x_{21}) \\ &+ r\cos\theta(L^x_{13}\cos^2\varphi - L^x_{31}\sin^2\varphi) \\ &+ r\sin\theta(L^x_{23}\cos^2\varphi - L^x_{32}\sin^2\varphi) - r\sin\varphi\cos\varphi L^x_{33},\end{aligned}$$

$$\begin{aligned}L_{32} = {}& r^2\sin\theta\cos\theta\sin\varphi\cos\varphi(L^x_{22} - L^x_{11}) \\ &+ r^2\sin\varphi\cos\varphi(L^x_{12}\cos^2\theta - L^x_{21}\sin^2\theta) \\ &+ r^2\sin^2\varphi(L^x_{31}\sin\theta - L^x_{32}\cos\theta),\end{aligned}$$

$$L_{33} = r^2\cos^2\varphi(L^x_{11}\cos^2\theta + L^x_{22}\sin^2\theta) + r^2\sin\theta\cos\theta\cos^2\varphi(L^x_{12} + L^x_{21}) - r^2\cos\theta\sin\varphi\cos\varphi(L^x_{13} + L^x_{31}) - r^2\sin\theta\sin\varphi\cos\varphi(L^x_{23} + L^x_{32}) + r^2\sin^2\varphi L^x_{33}.$$

(ii) For the contravariant components, we use Eq. $(6.19)_1$, after having calculated $\mathbf{g}^{cont}$; this can be done either using Eq. (6.11) or simply observing that $\mathbf{g}^{cov}$ is diagonal (see Exercise 2) and that $g^{pq} = \frac{1}{g_{pq}} \Rightarrow \mathbf{g}^{cont} = \begin{bmatrix} 1 & 0 & 0 \\ 0 & \frac{1}{r^2\sin^2\varphi} & 0 \\ 0 & 0 & \frac{1}{r^2} \end{bmatrix} \Rightarrow$

$L^{11} = L_{11}$, $L^{12} = \frac{L_{12}}{r^2\sin^2\varphi}$, $L^{13} = \frac{L_{13}}{r^2}$, $L^{21} = \frac{L_{21}}{r^2\sin^2\varphi}$, $L^{22} = \frac{L_{22}}{r^4\sin^4\varphi}$, $L^{23} = \frac{L_{23}}{r^4\sin^2\varphi}$, $L^{31} = \frac{L_{31}}{r^2}$, $L^{32} = \frac{L_{32}}{r^4\sin^2\varphi}$, $L^{33} = \frac{L_{33}}{r^4}$.

10. (i) $\mathrm{tr}\mathbf{L} = L^x_{hh}$, Eq. (2.7).
(ii) By Eq. $(2.7)_1 \Rightarrow L^x_{hh} = \frac{\partial x_h}{\partial z^i}\frac{\partial x_k}{\partial z^j}L^{ij}\delta_{hk} = g_{ij}L^{ij}$.
(iii) By Eq. $(2.7)_2 \Rightarrow L^x_{hh} = \frac{\partial z^i}{\partial x_h}\frac{\partial z^j}{\partial x_k}L_{ij}\delta_{hk} = g^{ij}L_{ij}$.
(iv) By Eq. $(2.7)_3 \Rightarrow L^x_{hh} = \frac{\partial x_h}{\partial z^i}\frac{\partial z^j}{\partial x_k}L^i{}_j\delta_{hk} = \delta_i{}^j L^i{}_j = L^i{}_i$.
(v) By Eq. $(2.7)_4 \Rightarrow L^x_{hh} = \frac{\partial z^i}{\partial x_h}\frac{\partial x_k}{\partial z^j}L_i{}^j\delta_{hk} = \delta^i{}_j L_i{}^j = L_j{}^j$.

11. $\frac{1}{2}g^{hm}\left(\frac{\partial g_{mk}}{\partial z^l} + \frac{\partial g_{ml}}{\partial z^k} - \frac{\partial g_{kl}}{\partial z^m}\right) = \frac{1}{2}\frac{\partial z^h}{\partial x_p}\frac{\partial z^m}{\partial x_p}\left(\frac{\partial}{\partial z^l}\frac{\partial x_p}{\partial z^m}\frac{\partial x_p}{\partial z^k} + \frac{\partial}{\partial z^k}\frac{\partial x_p}{\partial z^m}\frac{\partial x_p}{\partial z^l} - \frac{\partial}{\partial z^m}\frac{\partial x_p}{\partial z^k}\frac{\partial x_p}{\partial z^l}\right) = \frac{1}{2}\frac{\partial z^h}{\partial x_p}\frac{\partial z^m}{\partial x_p}\left(\frac{\partial^2 x_p}{\partial z^l\partial z^m}\frac{\partial x_p}{\partial z^k} + \frac{\partial x_p}{\partial z^m}\frac{\partial^2 x_p}{\partial z^l\partial z^k} + \frac{\partial^2 x_p}{\partial z^k\partial z^m}\frac{\partial x_p}{\partial z^l} + \frac{\partial x_p}{\partial z^m}\frac{\partial^2 x_p}{\partial z^k\partial z^l} - \frac{\partial^2 x_p}{\partial z^m\partial z^k}\frac{\partial x_p}{\partial z^l} - \frac{\partial^2 x_p}{\partial z^m\partial z^l}\frac{\partial x_p}{\partial z^k}\right) = \frac{\partial z^h}{\partial x_p}\frac{\partial z^m}{\partial x_p}\frac{\partial x_p}{\partial z^m}\frac{\partial^2 x_p}{\partial z^k\partial z^l} = \frac{\partial z^h}{\partial x_p}\frac{\partial^2 x_p}{\partial z^k\partial z^l} = \Gamma^h_{kl}$.

12. First, we remark that $g^{im}g_{ik} = \frac{\partial z^i}{\partial x_p}\frac{\partial z^m}{\partial x_p}\frac{\partial x_q}{\partial z^i}\frac{\partial x_q}{\partial z^k} = \delta_{pq}\frac{\partial z^m}{\partial x_p}\frac{\partial x_q}{\partial z^k} = \frac{\partial z^m}{\partial z^k} = \delta_{mk}$, and similarly, $g^{im}g_{ij} = \delta_{mj}$. Then, $\Gamma^i_{jh}g_{ik} + \Gamma^i_{kh}g_{ji}$
$= \frac{1}{2}\left(g^{im}\left(\frac{\partial g_{mj}}{\partial z^h} + \frac{\partial g_{mh}}{\partial z^j} - \frac{\partial g_{jh}}{\partial z^m}\right)g_{ik} + g^{im}\left(\frac{\partial g_{mk}}{\partial z^h} + \frac{\partial g_{mh}}{\partial z^k} - \frac{\partial g_{kh}}{\partial z^m}\right)g_{ji}\right)$
$= \frac{1}{2}\left(g^{im}g_{ik}\left(\frac{\partial g_{mj}}{\partial z^h} + \frac{\partial g_{mh}}{\partial z^j} - \frac{\partial g_{jh}}{\partial z^m}\right) + g^{im}g_{ij}\left(\frac{\partial g_{mk}}{\partial z^h} + \frac{\partial g_{mh}}{\partial z^k} - \frac{\partial g_{kh}}{\partial z^m}\right)\right)$
$= \frac{1}{2}\left(\delta_{mk}\left(\frac{\partial g_{mj}}{\partial z^h} + \frac{\partial g_{mh}}{\partial z^j} - \frac{\partial g_{jh}}{\partial z^m}\right) + \delta_{mj}\left(\frac{\partial g_{mk}}{\partial z^h} + \frac{\partial g_{mh}}{\partial z^k} - \frac{\partial g_{kh}}{\partial z^m}\right)\right)$
$= \frac{1}{2}\left(\frac{\partial g_{kj}}{\partial z^h} + \frac{\partial g_{kh}}{\partial z^j} - \frac{\partial g_{jh}}{\partial z^k} + \frac{\partial g_{jk}}{\partial z^h} + \frac{\partial g_{jh}}{\partial z^k} - \frac{\partial g_{kh}}{\partial z^j}\right) = \frac{\partial g_{jk}}{\partial z^h}$.

13. (i) $g_{\rho\rho} = g_{zz} = 1, g_{\theta\theta} = \rho^2$, and the other components are null $\Rightarrow g^{\rho\rho} = g^{zz} = 1, g^{\theta\theta} = \frac{1}{\rho^2} \Rightarrow \Gamma^\rho_{\theta\theta} = \frac{1}{2}g^{\rho m}\left(\frac{\partial g_{m\theta}}{\partial\theta} + \frac{\partial g_{m\theta}}{\partial\theta} - \frac{\partial g_{\theta\theta}}{\partial z^m}\right) = -\frac{1}{2}g^{\rho\rho}\frac{\partial g_{\theta\theta}}{\partial\rho} = -\rho, \Gamma^\theta_{\rho\theta} = \frac{1}{2}g^{\theta m}\left(\frac{\partial g_{m\rho}}{\partial\theta} + \frac{\partial g_{m\theta}}{\partial\rho} - \frac{\partial g_{\rho\theta}}{\partial z^m}\right) = \frac{1}{2}g^{\theta\theta}\frac{\partial g_{\theta\theta}}{\partial\rho} = \frac{1}{\rho}$, and the other Γ^k_{ij} are null.

(ii) $g_{rr} = 1, g_{\varphi\varphi} = r^2 \sin^2 \varphi, g_{\theta\theta} = r^2 \Rightarrow g^{rr} = 1, g^{\varphi\varphi} = \frac{1}{r^2 \sin^2 \varphi}, g^{\theta\theta} = \frac{1}{r^2}$, and the other components are null $\Rightarrow$
$\Gamma^{\varphi}_{\varphi r} = \frac{1}{2} g^{\varphi m} \left(\frac{\partial g_{m\varphi}}{\partial r} + \frac{\partial g_{mr}}{\partial \varphi} - \frac{\partial g_{\varphi r}}{\partial z^m} \right) = \frac{1}{2} g^{\varphi\varphi} \frac{\partial g_{\varphi\varphi}}{\partial r} = \frac{1}{r}$,
$\Gamma^{\theta}_{\theta r} = \frac{1}{2} g^{\theta m} \left(\frac{\partial g_{\theta m}}{\partial r} + \frac{\partial g_{rm}}{\partial \theta} - \frac{\partial g_{\theta r}}{\partial z^m} \right) = \frac{1}{2} g^{\theta\theta} \frac{\partial g_{\theta\theta}}{\partial r} = \frac{1}{r}$,
$\Gamma^{r}_{\varphi\varphi} = \frac{1}{2} g^{rm} \left(\frac{\partial g_{\varphi m}}{\partial \varphi} + \frac{\partial g_{\varphi m}}{\partial \varphi} - \frac{\partial g_{\varphi\varphi}}{\partial z^m} \right) = -\frac{1}{2} g^{rr} \frac{\partial g_{\varphi\varphi}}{\partial r} = -r$,
$\Gamma^{r}_{\theta\theta} = \frac{1}{2} g^{rm} \left(\frac{\partial g_{\theta m}}{\partial \theta} + \frac{\partial g_{\theta m}}{\partial \theta} - \frac{\partial g_{\theta\theta}}{\partial z^m} \right) = -\frac{1}{2} g^{rr} \frac{\partial g_{\theta\theta}}{\partial r} = -r \sin^2 \varphi$,
$\Gamma^{\theta}_{\theta\varphi} = \frac{1}{2} g^{\theta m} \left(\frac{\partial g_{\theta m}}{\partial \varphi} + \frac{\partial g_{\varphi m}}{\partial \theta} - \frac{\partial g_{\theta\varphi}}{\partial z^m} \right) = \frac{1}{2} g^{\theta\theta} \frac{\partial g_{\theta\theta}}{\partial \varphi} = \cot \varphi$,
$\Gamma^{\varphi}_{\theta\theta} = \frac{1}{2} g^{\varphi m} \left(\frac{\partial g_{\theta m}}{\partial \theta} + \frac{\partial g_{\theta m}}{\partial \theta} - \frac{\partial g_{\theta\theta}}{\partial z^m} \right) = -\frac{1}{2} g^{\varphi\varphi} \frac{\partial g_{\theta\theta}}{\partial \varphi} = -\sin\varphi \cos\varphi$,
$\Gamma^{\varphi}_{r\varphi} = \Gamma^{\varphi}_{\varphi r}, \Gamma^{\theta}_{r\theta} = \Gamma^{\theta}_{\theta r}, \Gamma^{\theta}_{\varphi\theta} = \Gamma^{\theta}_{\theta\varphi}$, and the other Γ^k_{ij} are null.

(iii) $g_{11} = g_{22} = \frac{c^2}{2}(\cosh 2z^1 - \cos 2z^2) \Rightarrow g^{11} = g^{22} = \frac{2}{c^2(\cosh 2z^1 - \cos 2z^2)}$, and the other components are null
$\Rightarrow \Gamma^1_{11} = \frac{1}{2} g^{1m} \left(\frac{\partial g_{1m}}{\partial z^1} + \frac{\partial g_{1m}}{\partial z^1} - \frac{\partial g_{11}}{\partial z^m} \right) = \frac{1}{2} g^{11} \frac{\partial g_{11}}{\partial z^1}$
$= \frac{\sinh 2z^1}{\cosh 2z^1 - \cos 2z^2}$,
$\Gamma^1_{12} = \frac{1}{2} g^{1m} \left(\frac{\partial g_{1m}}{\partial z^2} + \frac{\partial g_{2m}}{\partial z^1} - \frac{\partial g_{12}}{\partial z^m} \right) = \frac{1}{2} g^{11} \frac{\partial g_{11}}{\partial z^2}$
$= \frac{\sin 2z^2}{\cosh 2z^1 - \cos 2z^2}$,
and $\Gamma^2_{12} = \Gamma^2_{21} = \Gamma^1_{11} = -\Gamma^1_{22}, \Gamma^2_{22} = \Gamma^1_{12} = \Gamma^1_{21} = -\Gamma^2_{11}$.

14. Applying Eq. (6.27), we get:

(i) $\Delta f = \frac{\partial}{\partial \rho} \left(g^{\rho k} \frac{\partial f}{\partial z^k} \right) + \Gamma^{\rho}_{\rho j} g^{jk} \frac{\partial f}{\partial z^k} + \frac{\partial}{\partial \theta} \left(g^{\theta k} \frac{\partial f}{\partial z^k} \right) + \Gamma^{\theta}_{\theta j} g^{jk} \frac{\partial f}{\partial z^k} + \frac{\partial}{\partial z} \left(g^{zk} \frac{\partial f}{\partial z^k} \right) + \Gamma^{z}_{zj} g^{jk} \frac{\partial f}{\partial z^k} = \frac{\partial}{\partial \rho} g^{\rho\rho} \frac{\partial f}{\partial \rho} + \Gamma^{\rho}_{\rho\theta} g^{\theta\theta} \frac{\partial f}{\partial \theta} + \frac{\partial}{\partial \theta} g^{\theta\theta} \frac{\partial f}{\partial \theta} + \Gamma^{\theta}_{\theta\theta} g^{\theta\theta} \frac{\partial f}{\partial \theta} + \Gamma^{\theta}_{\theta\rho} g^{\rho\rho} \frac{\partial f}{\partial \rho} + \frac{\partial}{\partial z} g^{zz} \frac{\partial f}{\partial z} = \frac{\partial^2 f}{\partial \rho^2} + \frac{1}{\rho^2} \frac{\partial^2 f}{\partial \theta^2} + \frac{1}{\rho} \frac{\partial f}{\partial \rho} + \frac{\partial^2 f}{\partial z^2}$
$= \frac{1}{\rho}(\rho\ f_{,\rho})_{,\rho} + \frac{1}{\rho^2} f_{,\theta\theta} + f_{,zz}$, which is the same already found in Section 5.6.

(ii) $\Delta f = \frac{\partial}{\partial r} \left(g^{rk} \frac{\partial f}{\partial z^k} \right) + \Gamma^{r}_{rj} g^{jk} \frac{\partial f}{\partial z^k} + \frac{\partial}{\partial \theta} \left(g^{\theta k} \frac{\partial f}{\partial z^k} \right) + \Gamma^{\theta}_{\theta j} g^{jk} \frac{\partial f}{\partial z^k} + \frac{\partial}{\partial \varphi} \left(g^{\varphi k} \frac{\partial f}{\partial z^k} \right) + \Gamma^{\varphi}_{\varphi j} g^{jk} \frac{\partial f}{\partial z^k} = \frac{\partial}{\partial r} g^{rr} \frac{\partial f}{\partial r} + \Gamma^{\theta}_{\theta r} g^{rk} \frac{\partial f}{\partial z^k} + \frac{\partial}{\partial \theta} g^{\theta\theta} \frac{\partial f}{\partial \theta} + \Gamma^{\theta}_{\theta\varphi} g^{\varphi k} \frac{\partial f}{\partial z^k} + \frac{\partial}{\partial \varphi} g^{\varphi\varphi} \frac{\partial f}{\partial \varphi} + \Gamma^{\varphi}_{\varphi r} g^{rk} \frac{\partial f}{\partial z^k} = \frac{\partial^2 f}{\partial r^2} + \frac{1}{r^2 \sin^2 \varphi} \frac{\partial^2 f}{\partial \theta^2} + \frac{2}{r} \frac{\partial f}{\partial r} + \cot\varphi \frac{1}{r^2} \frac{\partial f}{\partial \varphi} + \frac{1}{r^2} \frac{\partial^2 f}{\partial \varphi^2} = \frac{1}{r^2}(r^2 f_{,r})_{,r} + \frac{1}{r^2 \sin\varphi} \left(\frac{f_{,\theta\theta}}{\sin\varphi} + (f_{,\varphi} \sin\varphi)_{,\varphi} \right)$, which is to be compared to the one given in Section 5.7.

15. (i) By Eqs. (6.11), (6.25), and (6.29), $g^{np}_{;h} = \frac{\partial g^{np}}{\partial z^h} + \Gamma^n_{hr} g^{rp} + \Gamma^p_{hr} g^{nr}$, $g^{np} = \frac{\partial z^n}{\partial x_k} \frac{\partial z^p}{\partial x_k}, \Gamma^n_{hr} = \frac{\partial z^n}{\partial x_m} \frac{\partial^2 x_m}{\partial z^h \partial z^r}$, $\Gamma^p_{hr} = \frac{\partial z^p}{\partial x_t} \frac{\partial^2 x_t}{\partial z^h \partial z^r}$
$\Rightarrow g^{np}_{;h} = \frac{\partial}{\partial x_k} \frac{\partial z^n}{\partial x_h} \frac{\partial z^p}{\partial x_k} + \frac{\partial z^n}{\partial x_k} \frac{\partial}{\partial x_k} \frac{\partial z^p}{\partial z^h} + \frac{\partial z^n}{\partial x_m} \frac{\partial}{\partial z^r} \frac{\partial x_m}{\partial z^h} \frac{\partial z^r}{\partial x_q} \frac{\partial z^p}{\partial x_q}$
$+ \frac{\partial z^p}{\partial x_t} \frac{\partial}{\partial z^r} \frac{\partial x_t}{\partial z^h} \frac{\partial z^n}{\partial x_s} \frac{\partial z^r}{\partial x_s} = \frac{\partial z^n}{\partial x_m} \frac{\partial}{\partial z^h} \frac{\partial x_m}{\partial x_q} \frac{\partial z^p}{\partial x_q} + \frac{\partial z^p}{\partial x_t} \frac{\partial}{\partial z^h} \frac{\partial x_t}{\partial x_s} \frac{\partial z^n}{\partial x_s} = 0$

because, e.g., $\frac{\partial z^n}{\partial z^h}=\delta_{nh}$, $\frac{\partial x_m}{\partial x_q}=\delta_{mq}$ etc., so their derivatives are null.

(ii) By eqs. (6.3), (6.25), and (6.30), $g_{np;h}=\frac{\partial^2 x_k}{\partial z^h\partial z^n}\frac{\partial x_k}{\partial z^p}+\frac{\partial x_k}{\partial z^n}\frac{\partial^2 x_k}{\partial z^h\partial z^p}-\frac{\partial z^r}{\partial x_m}\frac{\partial^2 x_m}{\partial z^p\partial z^h}\frac{\partial x_q}{\partial z^n}\frac{\partial x_q}{\partial z^r}-\frac{\partial z^r}{\partial x_t}\frac{\partial^2 x_t}{\partial z^n\partial z^h}\frac{\partial x_s}{\partial z^p}\frac{\partial x_s}{\partial z^r}=\frac{\partial^2 x_k}{\partial z^h\partial z^n}\frac{\partial x_k}{\partial z^p}+\frac{\partial x_k}{\partial z^n}\frac{\partial^2 x_k}{\partial z^h\partial z^p}-\delta_{qm}\frac{\partial^2 x_m}{\partial z^p\partial z^h}\frac{\partial x_q}{\partial z^n}-\delta_{st}\frac{\partial^2 x_t}{\partial z^n\partial z^h}\frac{\partial x_s}{\partial z^p}=\frac{\partial^2 x_k}{\partial z^h\partial z^n}\frac{\partial x_k}{\partial z^p}+\frac{\partial x_k}{\partial z^n}\frac{\partial^2 x_k}{\partial z^h\partial z^p}-\frac{\partial^2 x_q}{\partial z^p\partial z^h}\frac{\partial x_q}{\partial z^n}-\frac{\partial^2 x_s}{\partial z^n\partial z^h}\frac{\partial x_s}{\partial z^p}=0.$

Chapter 7

1. It is sufficient to pose $x_1=u, x_2=v, x_3=f(u,v)\Rightarrow p(u,v)$ defines a surface because as $f(u,v)$ is smooth, $p(u,v)$ is also smooth, and because the Jacobian is $[J]=\begin{bmatrix}1&0\\0&1\\f_{,u}&f_{,v}\end{bmatrix}$, then $\text{rank}[J]=2$.

2. Catenoid: $\mathbf{f}(u,v):\begin{cases}x_1=\cosh u\cos v,\\x_2=\cosh u\sin v,\\x_3=u;\end{cases}$ Meridians: $v=const.$; if, for example, $v=0\Rightarrow\begin{cases}x_1=\cosh u,\\x_2=0,\\x_3=u,\end{cases}$ is a catenary in the plane (x_1,x_3).

$\mathbf{f}_{,u}=\begin{Bmatrix}\sinh u\cos v\\\sinh u\sin v\\1\end{Bmatrix}, \mathbf{f}_{,v}=\begin{Bmatrix}-\cosh u\sin v\\\cosh u\cos v\\0\end{Bmatrix}\Rightarrow\mathbf{g}=\begin{bmatrix}\cosh^2 u&0\\0&\cosh^2 u\end{bmatrix}.$

$\mathbf{f}_{,u}\times\mathbf{f}_{,v}=\begin{Bmatrix}-\cosh u\cos v\\-\cosh u\sin v\\\sinh u\cosh u\end{Bmatrix}, |\mathbf{f}_{,u}\times\mathbf{f}_{,v}|=\cosh^2 u\Rightarrow\mathbf{N}=\frac{1}{\cosh u}\begin{Bmatrix}-\cos v\\-\sin v\\\sinh u\end{Bmatrix};$

$\mathbf{f}_{,uv}=\mathbf{f}_{,vu}=\begin{Bmatrix}-\sinh u\sin v\\\sinh u\cos v\\0\end{Bmatrix}, \mathbf{f}_{,uu}=\begin{Bmatrix}\cosh u\cos v\\\cosh u\sin v\\0\end{Bmatrix},$

$\mathbf{f}_{,vv}=\begin{Bmatrix}-\cosh u\cos v\\-\cosh u\sin v\\0\end{Bmatrix}\Rightarrow\mathbf{B}=\begin{bmatrix}-1&0\\0&1\end{bmatrix}\Rightarrow K=-\frac{1}{\cosh^4 u}.$

3. Pseudo-sphere: $\mathbf{f}(u,v) \ : \left\{ \begin{array}{l} x_1 = \sin u \cos v, \\ x_2 = \sin u \sin v, \\ x_3 = \cos u + \ln\left(\tan \frac{u}{2}\right). \end{array} \right.$ Meridians: $v =$ *const.*; if, for example, $v = 0 \Rightarrow \left\{ \begin{array}{l} x_1 = \sin u, \\ x_2 = 0, \\ x_3 = \cos u + \ln\left(\tan \frac{u}{2}\right), \end{array} \right.$ is a tractrix in the plane (x_1, x_3).
$\mathbf{f}_{,u} = \left\{ \begin{array}{c} \cos u \cos v \\ \cos u \sin v \\ -\sin u + \frac{1}{\sin u} \end{array} \right\}, \mathbf{f}_{,v} = \left\{ \begin{array}{c} -\sin u \sin v \\ \sin u \cos v \\ 0 \end{array} \right\} \Rightarrow \mathbf{g} = \left[\begin{array}{cc} \frac{\cos^2 u}{\sin^2 u} & 0 \\ 0 & \sin^2 u \end{array} \right].$
$\mathbf{f}_{,u} \times \mathbf{f}_{,v} = \left\{ \begin{array}{c} -\cos^2 u \cos v \\ -\cos^2 u \sin v \\ \sin u \cos u \end{array} \right\}, |\mathbf{f}_{,u} \times \mathbf{f}_{,v}| = |\cos u| \Rightarrow \mathbf{N} = \frac{1}{|\cos u|} \left\{ \begin{array}{c} -\cos^2 u \cos v \\ -\cos^2 u \sin v \\ \sin u \cos u \end{array} \right\};$
$\mathbf{f}_{,uu} = \left\{ \begin{array}{c} -\sin u \cos v \\ -\sin u \sin v \\ -\cos u - \frac{\cos u}{\sin^2 u} \end{array} \right\}, \mathbf{f}_{,uv} = \mathbf{f}_{,vu} = \left\{ \begin{array}{c} -\cos u \sin v \\ \cos u \cos v \\ 0 \end{array} \right\},$
$\mathbf{f}_{,vv} = \left\{ \begin{array}{c} -\sin u \cos v \\ -\sin u \sin v \\ 0 \end{array} \right\} \Rightarrow \mathbf{B} = \left[\begin{array}{cc} -\frac{\cos^2 u}{\sin u |\cos u|} & 0 \\ 0 & \frac{\cos^2 u \sin u}{|\cos u|} \end{array} \right] \Rightarrow$
$K = -1.$
4. Cone: $\mathbf{f}(u,v) = v\boldsymbol{\gamma}(u) \Rightarrow \mathbf{f}_{,u} = v\boldsymbol{\gamma}', \mathbf{f}_{,v} = \boldsymbol{\gamma} \rightarrow \mathbf{N} \neq \mathbf{o} \iff v \neq 0$ and $\boldsymbol{\gamma} \neq \alpha\boldsymbol{\gamma}'$, i.e. everywhere except at the apex of the cone and on straight lines tangent to $\boldsymbol{\gamma}$.
5. The most general equation of the hyperbolic hyperboloid is
$\mathbf{f}(u,v) : \left\{ \begin{array}{l} x_1 = a(\cos u - v \sin u), \\ x_2 = b(\sin u + v \cos u), \\ x_3 = c\, v, \end{array} \right. \ a, b, c \in \mathbb{R} \Rightarrow \mathbf{f}(u,v) = \boldsymbol{\gamma}(u) + v\boldsymbol{\lambda}(u),$
with $\boldsymbol{\gamma}(u) = a \cos u \mathbf{e}_1 + b \sin u \mathbf{e}_2, \boldsymbol{\lambda}(u) = -a \sin u \mathbf{e}_1 + b \cos u \mathbf{e}_2 + c\mathbf{e}_3 \Rightarrow$
$\mathbf{f}(u,v)$ is a ruled surface. Fixing $u = u_0 \Rightarrow \left\{ \begin{array}{l} x_1 = a(\cos u_0 - v \sin u_0), \\ x_2 = b(\sin u_0 + v \cos u_0), \\ x_3 = c\, v, \end{array} \right.$
equation of a bundle of straight lines belonging to $\mathbf{f}(u,v)$ as well as

$\begin{cases} x_1 = a(\cos u_0 - v \sin u_0), \\ x_2 = b(\sin u_0 + v \cos u_0), \\ x_3 = -c\, v. \end{cases}$ The angle formed by the two straight lines of the two sets is $\theta = \arccos \frac{a^2 \sin^2 u_0 + b^2 \cos^2 u_0 - c^2}{a^2 \sin^2 u_0 + b^2 \cos^2 u_0 + c^2}$.

6. $x_3 = x_1 x_2$; setting $u = x_1, v = x_2 \Rightarrow \mathbf{f}(u,v) : \begin{cases} x_1 = u, \\ x_2 = v, \\ x_3 = u\, v, \end{cases}$ is of the type $\mathbf{f}(u,v) = \boldsymbol{\gamma}(u) + v\boldsymbol{\lambda}(v)$, with $\boldsymbol{\gamma}(u) = u\mathbf{e}_1, \boldsymbol{\lambda}(u) = \mathbf{e}_2 + u\mathbf{e}_3$.
The straight lines $\begin{cases} x_1 = u_0, \\ x_2 = v, \\ x_3 = u_0\ v, \end{cases}$ and $\begin{cases} x_1 = u, \\ x_2 = v_0, \\ x_3 = u\ v_0, \end{cases}$ belong of course to $\mathbf{f}(u,v)$; they form the angle $\theta = \arccos \frac{u_0 v_0}{\sqrt{(1+u_0^2)(1+v_0^2)}}; \theta = \frac{\pi}{2} \iff 0 = v_0 = 0$, i.e. at $(0,0,0)$.
7. (i) $\boldsymbol{\gamma}(u) = (\cos u, \sin u, -1), \boldsymbol{\lambda}(u) = (\cos u, \sin u, 1) \Rightarrow \mathbf{f}(u,v) = (\cos u, \sin u, 2v-1)$, which is of the form $\mathbf{f}(u,v) = \boldsymbol{\gamma}_1(u) + v\boldsymbol{\lambda}_1(u)$, with $\boldsymbol{\gamma}_1(u) = (\cos u, \sin u, -1), \boldsymbol{\lambda}_1(u) = (0,0,2) = const. \Rightarrow \mathbf{f}(u,v)$ being a cylinder whose Cartesian equation is $x_1^2 + x_2^2 = 1$.
(ii) $\boldsymbol{\gamma}(u) = (\sin u, -\cos u, -1), \boldsymbol{\lambda}(u) = (-\sin u, \cos u, 1) \Rightarrow$ $\mathbf{f}(u,v) = (2v-1)(-\sin u, \cos u, 1)$, which is of the form $\mathbf{f}(u,v) = \boldsymbol{\gamma}_2(u) + v\boldsymbol{\lambda}_2(u)$, with $\boldsymbol{\gamma}_2(u) = (\sin u, -\cos u, -1), \boldsymbol{\lambda}_1(u) = (-2\sin u, 2\cos u, 2) = -2\boldsymbol{\gamma}_2(u) \Rightarrow \mathbf{f}(u,v)$ being a cone whose Cartesian equation is $x_1^2 + x_2^2 = x_3^2$.
(iii) $\begin{cases} x_1 = (1-v)\cos(u-\alpha) + v\cos(u+\alpha), \\ x_2 = (1-v)\sin(u-\alpha) + v\sin(u+\alpha), x_3 = -(1-v) + v, \end{cases} \Rightarrow$
$\begin{cases} x_1 = \cos u \cos\alpha - \sin u \sin\alpha(2v-1), \\ x_2 = \sin u \cos\alpha + \cos u \sin\alpha(2v-1), x_3 = 2v-1. \end{cases}$
Change in parameter $w = \frac{\sin\alpha}{\cos\alpha}(2v-1) \Rightarrow$
$\begin{cases} x_1 = \cos\alpha(\cos u - w\sin u), \\ x_2 = \cos\alpha(\sin u + w\cos u), \\ x_3 = w\frac{\cos\alpha}{\sin\alpha}, \end{cases} \Rightarrow \begin{cases} x_1 = a(\cos u - w\sin u), \\ x_2 = a(\sin u + w\cos u), \\ x_3 = cw, \end{cases}$ $a = \cos\alpha, c = \frac{\cos\alpha}{\sin\alpha}$, which is the parametric equation of a hyperbolic hyperboloid with Cartesian equation $\frac{x_1^2}{a^2} + \frac{x_2^2}{a^2} - \frac{x_3^2}{c^2} = 1 \to \frac{x_1^2 + x_2^2}{\cos^2\alpha} - \frac{x_3^2}{\cot^2\alpha} = 1$.
8. (i) A sphere of radius $R : x_1^2 + x_2^2 + x_3^2 = R^2 \Rightarrow$ using the spherical coordinates $\theta = u, \varphi = v$ for expressing the x_is, we get the parametric equation $\mathbf{f}(u,v) : \begin{cases} x_1 = R\cos\theta\sin\varphi, \\ x_2 = R\sin\theta\sin\varphi, \\ x_3 = R\cos\varphi, \end{cases} \Rightarrow \mathbf{f}_{,u} =$

$R(-\sin u\sin v, \cos u\sin v, 0)$,

$\mathbf{f}_{,v} = R(\cos u\cos v, \sin u\cos v, -\sin v) \Rightarrow \mathbf{g} = \begin{bmatrix} R^2\sin^2 v & 0 \\ 0 & R^2 \end{bmatrix}$.

(ii) If $\mathbf{w} = a\mathbf{f}_{,u} + b\mathbf{f}_{,v} \in T_p\Sigma, I(\mathbf{w}) = \mathbf{w}\cdot\mathbf{g}\mathbf{w} = R^2(a^2\sin^2 v + b^2)$.

(iii) $A = \int_{\theta_1}^{\theta_2}\int_0^{\pi}\sqrt{\det\mathbf{g}}du\,dv = \int_{\theta_1}^{\theta_2}\int_0^{\pi}\sqrt{R^4\sin^2 v}du\,dv = 2R^2(\theta_2-\theta_1)$.

(iv) Parallel: setting $u = t, v = \frac{\pi}{4}, \gamma(t) : \begin{cases} x_1 = \frac{R}{\sqrt{2}}\cos t, \\ x_2 = \frac{R}{\sqrt{2}}\sin t, \\ x_3 = \frac{R}{\sqrt{2}}, \end{cases} \Rightarrow$

$\gamma'(t) : \begin{cases} x_1 = -\frac{R}{\sqrt{2}}\sin t, \\ x_2 = \frac{R}{\sqrt{2}}\cos t, \\ x_3 = 0, \end{cases} \Rightarrow \gamma'(t) = \frac{du}{dt}\mathbf{f}_{,u} + \frac{dv}{dt}\mathbf{f}_{,v} = \mathbf{f}_{,u} \Rightarrow$ in the natural basis of $T_p\Sigma, \mathbf{w} = (1,0)$ is the tangent vector to the parallel $\gamma(t) \Rightarrow I(\mathbf{w}) = \mathbf{w}\cdot\mathbf{g}\mathbf{w} = R^2\sin^2 v = \frac{R^2}{2} \Rightarrow \ell = \int_{\theta_1}^{\theta_2}\sqrt{I(\mathbf{w})}dt = \frac{R}{\sqrt{2}}(\theta_2 - \theta_1)$.

9. (i) $\mathbf{x}_1^2 + x_2^2 + x_3^2 = \frac{\cos^2 v}{\cosh^2 u} + \frac{\sin^2 v}{\cosh^2 u} + \frac{\sinh^2 u}{\cosh^2 u} = \frac{1}{\cosh^2 u} + \frac{\sinh^2 u}{\cosh^2 u} = \frac{\cosh^2 u}{\cosh^2 u} = 1 \rightarrow$ Cartesian equation of a sphere of centre $(0,0,0)$ and radius $R = 1$.

(ii) Straight line in Ω : $\begin{cases} u = u_0 + a\,t, \\ v = v_0 + b\,t; \end{cases} \Rightarrow$ curve on Σ : $\gamma(t)$:

$\begin{cases} x_1 = \frac{\cos(v_0+bt)}{\cosh(u_0+at)}, \\ x_2 = \frac{\sin(v_0+bt)}{\cosh(u_0+at)}, \\ x_3 = \frac{\sinh(u_0+at)}{\cosh(u_0+at)}, \end{cases}$ or also, $\gamma(t) = \mathbf{f}(u(t), v(t)) \Rightarrow \gamma'(t) = \frac{du}{dt}\mathbf{f}_{,u} + \frac{dv}{dt}\mathbf{f}_{,v} = a\mathbf{f}_{,u} + b\mathbf{f}_{,v}$.

Meridians: setting $v = const. = \hat{v} \Rightarrow \boldsymbol{\mu}(u) = \mathbf{f}(u,\hat{v})$; in fact, $\frac{x_2}{x_1} = \tan\hat{v} = const. \rightarrow$ equation of a vertical plane. Tangent to the meridian $\boldsymbol{\mu}(u) : \boldsymbol{\mu}'(u) = \mathbf{f}_{,u} \Rightarrow$ in the natural basis $\{\mathbf{f}_{,u}, \mathbf{f}_{,v}\}, \gamma'(t) = (a,b), \boldsymbol{\mu}'(u) = (1,0); \cos\theta = \frac{I(\gamma',\boldsymbol{\mu}')}{\sqrt{I(\gamma')I(\boldsymbol{\mu}')}}$.

$\mathbf{f}_u = \left(-\cosh v\frac{\sinh u}{\cosh^2 u}, -\sin v\frac{\sinh u}{\cosh^2 u}, \frac{1}{\cosh^2 u}\right), \mathbf{f}_{,v} = \left(-\frac{\sin v}{\cosh u}, \frac{\cos v}{\cosh u}, 0\right)$

$\Rightarrow \mathbf{g} = \frac{1}{\cosh^2 u}\mathbf{I} \Rightarrow I(\gamma',\boldsymbol{\mu}') = \gamma'\cdot\mathbf{g}\boldsymbol{\mu}' = \frac{a}{\cosh^2 u}, I(\gamma') = \gamma'\cdot\mathbf{g}\gamma' = \frac{a^2+b^2}{\cosh^2 u}, I(\boldsymbol{\mu}') = \boldsymbol{\mu}'\cdot\mathbf{g}\boldsymbol{\mu}' = \frac{1}{\cosh^2 u} \Rightarrow \cos\theta = \frac{a}{\sqrt{a^2+b^2}} = const. \Rightarrow$ $\gamma(t)$ is a loxodromic line on the sphere.

10. (i) Catenoid $\mathbf{f}(u,v) : \begin{cases} x_1 = \varphi(u)\cos v, \\ x_2 = \varphi(u)\sin v, \\ x_3 = \psi(u). \end{cases}$ with $\varphi(u) = \cosh u, \psi(u) = u \Rightarrow \varphi'(u) = \sinh u, \varphi''(u) = \cosh u, \psi'(u) = 1, \psi''(u) = 0 \Rightarrow$

$\mathbf{f}_{,u} = (\sinh u \cos v, \sinh u \sin v, 1), \mathbf{f}_{,v} =$
$(-\cosh u \sin v, \cosh u \cos v, 0)$.

(ii) $\mathbf{g} = \cosh^2 u\mathbf{I}$.

(iii) $\mathbf{f}_{,u} \times \mathbf{f}_{,v} = (-\cos v, -\sin v, -\sinh u), |\mathbf{f}_{,u} \times \mathbf{f}_{,v}| = \cosh u \Rightarrow$
$\mathbf{N} = \left(-\frac{\cos v}{\cosh u}, -\frac{\sin v}{\cosh u}, -\frac{\sinh u}{\cosh u}\right)$.
$\mathbf{f}_{,uu} = (\cosh u \cos v, \cosh u \sin v, 0), \mathbf{f}_{,uv} = \mathbf{f}_{,vu} = (-\sinh u \sin v,$
$\sinh u \cos v, 0), \mathbf{f}_{,vv} = -\mathbf{f}_{,uu} \Rightarrow \mathbf{B} = \begin{bmatrix} -1 & 0 \\ 0 & 1 \end{bmatrix}$.

(iv) $\mathbf{g}^{-1} = \frac{1}{\cosh^2 u}\mathbf{I} \Rightarrow \mathbf{X} = \mathbf{g}^{-1}\mathbf{B} = \frac{1}{\cosh^2 u}\mathbf{B}$.

(v) Let $\mathbf{w} = (a, b) \in T_p\Sigma \Rightarrow I(\mathbf{w}) = \mathbf{w} \cdot \mathbf{g}\mathbf{w} = \cosh^2 u(a^2 + b^2)$.

(vi) $II(\mathbf{w}) = \mathbf{w} \cdot \mathbf{B}\mathbf{w} = b^2 - a^2$.

11. (i) Helicoid $\mathbf{f}(u, v) : \begin{cases} x_1 = v \cos u, \\ x_2 = v \sin u, \\ x_3 = u, \end{cases} \Rightarrow \mathbf{f}_{,u} = (-v \sin u, v \cos u, 1)$,
$\mathbf{f}_{,v} = (\cos u, \sin u, 0)$.

(ii) $\mathbf{g} = \begin{bmatrix} 1 + v^2 & 0 \\ 0 & 1 \end{bmatrix}$.

(iii) $\mathbf{f}_{,u} \times \mathbf{f}_{,v} = (-\sin u, \cos u, -v), |\mathbf{f}_{,u} \times \mathbf{f}_{,v}| = \sqrt{1 + v^2} \Rightarrow \mathbf{N} = \frac{1}{\sqrt{1+v^2}}(-\sin u, \cos u, -v)$.
$\mathbf{f}_{uu} = (-v \cos u, -v \sin u, 0), \mathbf{f}_{,uv} = \mathbf{f}_{,vu} = (-\sin u, \cos u, 0), \mathbf{f}_{,vv} =$
$(0, 0, 0) \Rightarrow \mathbf{B} = \frac{1}{\sqrt{1+v^2}} \begin{bmatrix} 0 & 1 \\ 1 & 0 \end{bmatrix}$.

(iv) $\mathbf{g}^{-1} = \begin{bmatrix} \frac{1}{1+v^2} & 0 \\ 0 & 1 \end{bmatrix}. \Rightarrow \mathbf{X} = \mathbf{g}^{-1}\mathbf{B} = \begin{bmatrix} 0 & \frac{1}{(1+v^2)^{\frac{3}{2}}} \\ \frac{1}{(1+v^2)^{\frac{1}{2}}} & 0 \end{bmatrix}$.

(v) Let $\mathbf{w} = (a, b) \in T_p\Sigma \Rightarrow I(\mathbf{w}) = \mathbf{w} \cdot \mathbf{g}\mathbf{w} = (1 + v^2)a^2 + b^2$.

(vi) $II(\mathbf{w}) = \mathbf{w} \cdot \mathbf{B}\mathbf{w} = \frac{2ab}{\sqrt{1+v^2}}$.

12. (i) Catenoid (see Exercise 2): $K = -\frac{1}{\cosh^4 u} < 0 \ \forall u \Rightarrow$ hyperbolic points.

(ii) Helicoid (see Exercise 11): $K = \frac{\det \mathbf{B}}{\det \mathbf{g}} = -\frac{1}{(1+v^2)^2} < 0 \ \forall v \Rightarrow$ hyperbolic points.

13. The parametric equation of a circular cylinder of radius $R \to \mathbf{f}(u, v)$:
$\begin{cases} x_1 = R \cos v, \\ x_2 = R \sin v, \\ x_3 = u, \end{cases}$ with $u = z, v = \theta$ of a system of cylindrical coordinates.

Referring to Eq. (7.7), $\varphi(u) = R, \psi(u) = u \Rightarrow$ Eq. (7.32) is $\begin{cases} u'' = 0, \\ v'' = 0, \end{cases} \Rightarrow$
$\begin{cases} u(t) = \alpha t + \alpha_1, \\ v(t) = \beta t + \beta_1, \end{cases}$ with $\alpha, \alpha_1, \beta, \beta_1 = const.$ If $\alpha_1 = \beta_1 = 0$, we get the geodesic $\gamma(t)$ passing through $(R, 0, 0)$ for $t = 0 \Rightarrow \gamma(t)$:
$\begin{cases} x_1 = R\cos v(t), \\ x_2 = R\sin v(t), \\ x_3 = u(t), \end{cases} \Rightarrow \begin{cases} x_1 = R\cos(\beta t), \\ x_2 = R\sin(\beta t), \\ x_3 = \alpha t, \end{cases} \Rightarrow$ equation of a helix if $\alpha, \beta \neq 0$, of a circle (cross section) if $\alpha = 0, \beta \neq 0$, and of a straight line on the cylinder (generatrix) if $\alpha \neq 0, \beta = 0$.

Index

A

adjugate, 35
analytical mechanics, 184
angular velocity, 80
anisotropic elasticity, 71
anomaly, 116, 121
antisymmetry of the cross product, 33
antisymmetry projector, 65
applied vector, 7
arc length, 83
Archimedes' spiral, 98
asymptotic line, 162
axial tensor, 31
axial tensor field, 109
axial vector, 32

B

basis, 5, 77, 119
basis of eigenvectors, 52
Bertrand's theorem, 99
binormal vector, 85
Bonnet's theorem, 95

C

calculus of variations, 183
canonic equations of a curve, 96
canonical decomposition, 16, 58
Cartesian components, 5, 117
Cartesian components of a tensor, 17
Cartesian coordinates, 130
catenary, 100
catenoid, 145
Cauchy–Poisson symmetries, 61
center of the moment, 7
central axis, 10
change in parameterization, 143
change of basis, 6, 41
change of parameter, 78
Christoffel symbols, 137, 173
Codazzi conditions, 178
colatitude, 44, 121
commutation theorem, 30
commutativity property, 43
compatibility conditions, 179
complementary projector, 54, 87
completely symmetric, 61
compliance tensor, 70
components, 118, 127
condition of parallelism, 33
cones, 146
conical curves of Dupin, 162
conical helix, 92
conjugated directions, 162
conjugation product, 59
continuity, 75
contravariant components, 128

coordinate lines, 142
couple, 9
covariant, 127
covariant components, 128
covariant coordinates, 135
covariant derivative, 137, 169
cross product, 32
cubic parabola, 97
curl, 107
curl of a curl, 110
curl of an axial vector, 109
curl theorem, 113
curvature of a curve, 86
curvature vector, 151
curve of vectors, 75
curves of minimal distance, 171
curves of points, 75
curvilinear abscissa, 83
curvilinear coordinates, 125–126
cusp, 97
cycloid, 100
cylinders, 146
cylindrical basis, 117

D

deformation, 103
derivative, 76
derivative of a point, 76
determinant, 22–23
developable surfaces, 158
deviatoric part, 21
deviatoric projector, 63
diffeomorphism, 143
differentiable function, 80
differentiable vector field, 109
differential geometry, 184
differential operators, vi, 121
dilatation factor of the areas, 149
directional derivative, 103
director cosines, 6, 67
distance, 4, 21
divergence, 104
divergence lemma, 111
divergence of a tensor field, 104
divergence of products, 107
divergence or Gauss theorem, 112
doubly ruled, 146
dual basis, 132
dyad, 16
dyadic tensors, 59

E

elasticity tensor, 71
elliptic point, 156
energy decomposition, 71
Enneper's surface, 166
envelope, 93–94
equiangular property, 98
Euclidean norm, 21, 75
Euler's angles, 45
Euler's rotation representation theorem, 39
Euler–Lagrange equations, 172
evolute, 93

F

first fundamental form, 147
flux theorem, 113
fourth-rank tensor, 57
Frenet–Serret formula, 88, 91
Frenet–Serret local basis, 85

G

Gauss condition, 179
Gauss equations, 164
Gauss theorem, 112
Gauss' basis, 163
Gauss–Codazzi compatibility conditions, 174
Gauss–Weingarten equations, 165
Gaussian curvature, 153
general parameterization, 144
generator line, 146
geodesic curvature, 170
geodesics, vii, 169
global rotation, 47
gradient, 104
gradient of the vector field, 137
Green's formula, 114

H

harmonic, 105
harmonic fields, 111
helicoids, 146
Hessian, 104
Hooke's law, 69
hyperbolic hyperboloid, 145
hyperbolic paraboloid, 179
hyperbolic point, 156

I

identity tensor, 15, 135
immersion, 141
incompressible flows, 124
infinitesimal theory of strain, 124
integral of a curve, 80
intrinsic local characteristic, 84
intrinsic property, 86
invariant under an orthogonal transformation, 68
inverse of a tensor, 22
invertibility theorem, 26
invertible tensor, 26
involute, 93–94
irrotational, 124
isochoric, 124
isomorphism, 30
isotropic, 68

K

Kelvin formalism, 69
kinematics of rigid bodies, 123
knots line, 45
Kronecker's delta, 5

L

Lamé's parameters, 174
Laplace's equation, 27
laplacian, 105, 110, 120
laplacian of products, 110
left minor symmetries, 61
left polar decomposition, 53
lemma of divergence, 112
lemma of Ricci, 140
length of the curve, 81
line of curvature, 161
linear application, 15
linear combination, 29
linear forms representation theorem, 13
linearly independent vectors, 26
local basis, 127
logarithmic spiral, 98
longitude, 44
loxodrome, 145

M

Möbius strip, 146
major symmetries, 60
map of Gauss, 149
material symmetries, 71
matrix, 17
matrix form, 118
matrix of Cartan, 89
mean curvature, 153
meridian, 144
metric, 75
metric tensor, 127
minimal surfaces, vii, 166
Minkowski's inequality, 82
minor right symmetries, 61
minor symmetries, 61
mirror symmetries, 37
mixed components, 133
mixed coordinates, 135
mixed product, 34
moment of the couple, 9
monkey's saddle, 157
multiplicity, 30
mutually orthogonal subspaces, 22

N

natural basis, 142
natural parameter, 83
Navier–Stokes equations, 107
norm of a vector, 4
normal basis, 28
normal curvature, 151

normal plane, 85, 152
normal section, 152
normal to the surface, 141
normal variation, 167–168
null form, 23
null Gaussian curvature, 158
null tensor, 15
nutation, 45

O

operator nabla, 115
orientation of a basis, 35
orientation of the space, 37
origin, 77
orthogonal conjugator, 66
orthogonal projector, 54, 65
orthogonal tensors, 36
orthogonality conditions, 133
orthogonality property, 33, 88
orthonormal, 38
orthonormal basis, 40
osculating circle, 92
osculating plane, 85
osculating sphere, 91

P

parabola, 97
parabolic point, 156
parallel, 144
parallel vector field, 169
parameter, 76
parametric point equation, 78
parametric vector equation, 78
partial derivative, 104
partition, 82
permutation, 25
planar point, 153
polar decomposition, 51
polar decomposition theorem, 53
polar formalism, 71
position vector, 78
positive definiteness, 3
potential theorem, 114
precession, 45
principal curvatures, 153
principal directions, 153
principal invariants, 25
principal normal, 94
principal normal vector, 84
profile curve, 146
proper rotations, 37
proper space, 27
pseudo-scalar, 51
pseudo-sphere, 145
pseudo-vector, 51

Q

quadratic form, 28–29

R

radius of curvature, 92
raising of the indices, 131
re-parameterization, 82
rectangular coordinates, 125
rectifying plane, 85
reflexion tensor, 50
regular curve, 81, 83, 170
regular point, 141
regular region, 113
resultant, 7
resultant moment, 9
rhumb line, 145
Ricci's alternator, 25
right polar decomposition, 53
rigid displacement, 96
rotation, 36
rotation axis, 39
rotation of the space, 42
rotation tensor, 36
rotation's amplitude, 39
ruled surface of the normals, 162
ruled surface of the tangents, 159
ruled surfaces, 146

S

scalar, 3
scalar field, 103
scalar product, 20
scalar product of fourth-rank tensors, 62

Schwarz's inequality, 4
Schwarz's surfaces, 166
scroll, 146
second fundamental form, 150
second gradient, 104
second principal invariant, 55
second-order derivatives, 178
second-rank tensor, 15
semigeodesic coordinates, 170
similitude of the triangles, 176
skew tensor, 19
skew trilinear form, 24
small rotations, 49
spatial derivatives of fields, 135
spectral theorem, 28
spectrum, 27
sphere, 145
spherical coordinates, 120
spherical part, 21
spherical projector, 63
spin tensor, 80
square root theorem, 51
Stokes theorem, 114
streamlines, 105
support, 141
surface in $\mathcal{E}$, 141
surface of revolution, 143
symmetric tensor, 18
symmetry, 3
symmetry projector, 65

T

tangent plane, 142
tangent vector, 84, 142
tangent vector space, 142
tensor algebra and analysis, 183
tensor field, 103
tensor invariant, 20
tensor product, 16, 17
tensor scalar product, 21
tensor theory, vi
tensor-invariant, 23
tensors, 15
theorem of Binet, 24
theorem of Rodrigues, 154, 175
theorema egregium, 165
theoretical mechanics, 183
torsion, 90
torsion of the curve, 89
trace, 19
trace for fourth-rank tensors, 62
tractrix, 101
translation, 1
transport of moment, 8
transpose, 18
transpose of a fourth-rank tensor, 60
transposition projector, 65
transpositions, 25

U

umbilical point, 153
uniqueness, 11, 52
uniqueness of the inverse, 26

V

vector, 1–2
vector field, 103
vector space, 2
vorticity equation, 110

W

Weingarten operator, 150

CPSIA information can be obtained
at www.ICGtesting.com
Printed in the USA
JSHW012352210323
39022JS00001B/2